SENSING TOOLS AND TECHNIQUES FOR COVID-19

SENSING TOOLS AND TECHNIQUES FOR COVID-19

Developments and Challenges in Analysis and Detection of Coronavirus

Edited by

Chaudhery Mustansar Hussain

Adjunct Professor, Academic Advisor and Director of Chemistry and EVSc Labs, Department of Chemistry and Environmental Sciences, New Jersey Institute of Technology (NJIT), Newark, New Jersey, USA

Sudheesh K. Shukla

Department of Biomedical Engineering, School of Biological Engineering and Life Science, Shobhit Institute of Engineering & Technology (Deemed-to-be University), Modipuram, Meerut, India; Department of Chemical Sciences, University of Johannesburg, Doornfontein Campus, Johannesburg, 2028, South Africa

ELSEVIER

Elsevier

Radarweg 29, PO Box 211, 1000 AE Amsterdam, Netherlands
The Boulevard, Langford Lane, Kidlington, Oxford OX5 1GB, United Kingdom
50 Hampshire Street, 5th Floor, Cambridge, MA 02139, United States

Notices
Knowledge and best practice in this field are constantly changing. As new research and experience broaden our understanding, changes in research methods, professional practices, or medical treatment may become necessary.

Practitioners and researchers must always rely on their own experience and knowledge in evaluating and using any information, methods, compounds, or experiments described herein. In using such information or methods they should be mindful of their own safety and the safety of others, including parties for whom they have a professional responsibility.

To the fullest extent of the law, neither the Publisher nor the authors, contributors, or editors, assume any liability for any injury and/or damage to persons or property as a matter of products liability, negligence or otherwise, or from any use or operation of any methods, products, instructions, or ideas contained in the material herein.

ISBN: 978-0-323-90280-9

For Information on all Elsevier publications visit our website at
https://www.elsevier.com/books-and-journals

Publisher: Candice Janco
Acquisitions Editor: Charlotte Rowley
Editorial Project Manager: Lena Sparks
Production Project Manager: Kumar Anbazhagan
Cover Designer: Vicky Pearson Esser

Typeset by Aptara, New Delhi, India

Contents

Contributors

Isabel Barea-Navarro Soil Erosion and Degradation Research Group, Department of Geography, Valencia University, Valencia, Spain

Manisha Bhardwaj Department of Biochemistry, Ch. Charan Singh University, Meerut, India

Prama Bhattacherjee Department of Chemistry, Indian Institute of Technology (ISM), Dhanbad, Jharkhand, India

Anupama Chaudhary School of Biomedical Engineering, Shobhit Institute of Engineering & Technology (Deemed to-be-University), Modipuram, Meerut, India

Meenakshi Choudhary The Swiss Institute for Dryland Environmental and Energy Research, The Jacob Blaustein Institutes for Desert Research, Ben-Gurion University of the Negev, Midreshet Ben-Gurion, Israel

Ruchika Chauhan Department of Pharmaceutical Chemistry, College of Health Sciences, University of KwaZulu-Natal, Durban, South Africa

Subrata K. Das Department of Biomedical Engineering, Shobhit Institute of Engineering & Technology (Deemed-to-be University), Modipuram, Meerut, India

Trupti R. Das CIPET: Institute of Petrochemicals Technology (IPT) – Bhubaneswar, Patia, Bhubaneswar, India

Ivan Franch-Pardo Universidad Nacional Autónoma de México, Escuela Nacional de Estudios Superiores, Morelia, Michoacan, Mexico

Mbuso Faya Discipline of Pharmaceutical Sciences, College of Health Sciences, University of KwaZulu-Natal, Durban, South Africa

Atal Gill Department of Pharmaceutical Chemistry, College of Health Sciences, University of KwaZulu-Natal, Durban, South Africa

Anchit Gugnani Department of Physiotherapy, Nims University, jaipur, Rajasthan, India

Mohd. Rahil Hasan Department of Biotechnology, Jamia Hamdard University, New Delhi, India

Blake N. Johnson Department of Industrial and Systems Engineering, Virginia Tech, Blacksburg, VA, USA; Department of Materials Science and Engineering, Virginia Tech, Blacksburg, VA, USA; Department of Chemical Engineering, Virginia Tech, Blacksburg, VA, USA; Macromolecules Innovation Institute, Virginia Tech, Blacksburg, VA, USA

Alpana Joshi School of Agriculture Technology and Agri-informatics, Shobhit Institute of Engineering & Technology (Deemed-to-be University), Modipuram, Meerut, India

Priyanka Kakkar Department of Chemistry, Ch. Charan Singh University, Meerut, India

Rajshekhar Karpoormath Department of Pharmaceutical Chemistry, College of Health Sciences, University of KwaZulu-Natal, Durban, South Africa

Sevinc Kurbanoglu Ankara University, Faculty of Pharmacy, Department of Analytical Chemistry, Yenimahalle, Ankara, Turkey

Cansu İlke Kuru, PhD Ege University, Faculty of Science, Department of Biochemistry, Izmir, Turkey; Buca Municipality Kızılçullu Science and Art Center, Izmir, Turkey

Yang Liu Department of Industrial and Systems Engineering, Virginia Tech, Blacksburg, VA, USA; Macromolecules Innovation Institute, Virginia Tech, Blacksburg, VA, USA

Amit Mall CIPET: Institute of Petrochemicals Technology (IPT) - Bhubaneswar, Patia, Bhubaneswar, India

Charu Mehandiratta Human Consciousness and Yogic Science, Department of Yoga and Naturopathy, Apex University, Jaipur, Rajasthan, India

Vishalkumar I. Naik Department of Chemistry, Gokhale Centenary College, Ankola, India

Jagriti Narang Department of Biotechnology, Jamia Hamdard University, New Delhi, India

Zondi Nate Discipline of Pharmaceutical Sciences, College of Health Sciences, University of KwaZulu-Natal, Durban, South Africa

Calvin A. Omolo Discipline of Pharmaceutical Sciences, College of Health Sciences, University of KwaZulu-Natal, Durban, South Africa; United States International University-Africa, School of Pharmacy and Health Sciences, Department of Pharmaceutics, Nairobi, Kenya

Sweety Pal School of Biomedical Engineering, Shobhit Institute of Engineering & Technology (Deemed to-be-University), Meerut, India

Santanu Patra Department of Environmental Engineering, Khon Kaen University, Khon Kaen, Thailand

Frieder W. Scheller University of Potsdam, Institute of Biochemistry and Biology, Potsdam, Germany

Shiva Sharma School of Biomedical Engineering, Shobhit Institute of Engineering & Technology (Deemed to-be-University), Modipuram, Meerut, India

Sudheesh K. Shukla Department of Biomedical Engineering, School of Biological Engineering and Life Science, Shobhit Institute of Engineering & Technology (Deemed-to-be University), Modipuram, Meerut, India; Department of Chemical Sciences, University of Johannesburg, Doornfontein Campus, Johannesburg, 2028, South Africa

Pooja Singh Department of Biotechnology, SRGI Institute of Management and Engineering, Lucknow, Uttar Pradesh, India

Emily Sturdivant Universidad Nacional Autónoma de México, Centro de Investigaciones en Geografía Ambiental, Morelia, Michoacan, Mexico

Shariq Suleman Department of Biotechnology, Jamia Hamdard University, New Delhi, India

Ranjita D. Tandel Department of Chemistry, Gokhale Centenary College, Ankola, India

Nagappa L. Teradal Department of Chemistry J. S. S. Arts, Science and Commerce College, Gokak, India

Nitika Thakur Department of Biotechnology, Faculty of Applied Sciences and Biotechnology, Shoolini University of Biotechnology and Management Sciences, Solan (HP), India

Vanshika AK Dev Department of Biotechnology, Faculty of Applied Sciences and Biotechnology, Shoolini University of Biotechnology and Management Sciences, Solan (HP), India

Fulden Ulucan-Karnak, PhD Ege University, Faculty of Science, Department of Biochemistry, Izmir, Turkey

Brij M Upreti Faculty of Agriculture and Veterinary Science, Jayoti Vidyapeeth Women's University, Jaipur, Rajasthan, India

Aysu Yarman University of Potsdam, Institute of Biochemistry and Biology, Potsdam, Germany; Molecular Biotechnology, Faculty of Science, Turkish-German University, Sahinkaya Cad. 86, 34820 Beykoz, Istanbul, Turkey

Zeynep Yilmaz-Sercinoglu, PhDMarmara University, Faculty of Engineering, Department of Bioengineering, Istanbul, Turkey

SECTION 1

Introduction (COVID-19-Detection and Solutions)

CHAPTER

1

Present state of MIP-based sensors for SARS-CoV-2

Aysu Yarman[a,b], *Sevinc Kurbanoglu*[c], *and Frieder W. Scheller*[a]

[a]University of Potsdam, Institute of Biochemistry and Biology, Potsdam, Germany, [b]Molecular Biotechnology, Faculty of Science, Turkish-German University, Sahinkaya Cad. 86, 34820 Beykoz, Istanbul, Turkey, [c]Ankara University, Faculty of Pharmacy, Department of Analytical Chemistry, Yenimahalle, Ankara, Turkey

1.1 Introduction

The outbreak of a COVID-19 was first reported in December in Wuhan, China, and since then, it has spread worldwide. WHO officially declared the pandemic on 12th March 2020. Therefore, its fast and accurate diagnosis of the virus is essential to prevent its spread. Diagnosis of COVID-19 uses three different approaches: i) viral gene detection, ii) detection of human antibodies, and iii) viral antigen assays, which have different fields of applications [1–3]. Up to now, detection of the viral genome by real-time reverse transcription-polymerase chain reaction (RT-PCR) is the most reliable procedure. Furthermore, it offers the detection of the uprising incidence by mutants of the virus. Antibody indication is the key in tracing the effect of vaccination and the immune response during infection. On the other hand, the indication of segments of the virus, so-called antigens, is the basis of the decentralized test (Point-of-Care) by dipsticks and lateral flow devices. Most immunoassays indicate the spike protein (S protein) or the nucleocapside protein (NP protein). The S protein is the binding place of the virus to the host receptor, the angiotensin-converting enzyme (ACE2), for entering the host cell (Fig. 1.1). This interaction has an affinity in the lower nanomolar range, and it was predicted from docking simulations

Sensing Tools and Techniques for COVID-19: Developments and Challenges in Analysis and Detection of Coronavirus
DOI: https://doi.org/10.1016/B978-0-323-90280-9.00007-9

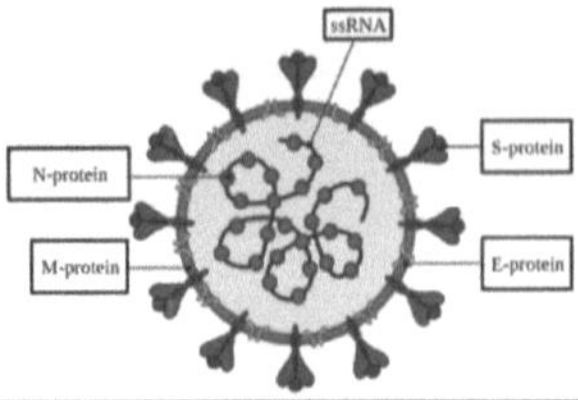

Protein	S-Protein	E-Protein	N-Protein	M-Protein
Subunits	S1 and S2	-	NTD and CTD	-
Mass	~150 kDa	~8–12 kDa	~15.4 kDa (NTD) and ~28.7 kDa (CTD)	~25–30 kDa
Function	Attachment, fusion and infection of a host cell.	Formation of viral envelope. Association and maturation of the virus.	Virion formation.	Shaping of the viral envelope.

FIGURE 1.1 Severe acute respiratory syndrome coronavirus 2 (SARS-CoV-2) structural proteins location, structure, mass, and function. Reprinted with permission [1].

that the peptide chain starting with Leu455, Phe486, to Tyr505 is involved in the binding of the receptor binding domain (RBD) of the S protein to ACE2 [4].

1.2 Molecularly imprinted polymers

The most important drivers in the field of molecular imprinting are Wulff and Mosbach ([58]; [59]). Nevertheless, the concept dates back to the 1930s, when Polyakov demonstrated specific adsorption properties of silica gel that recognized its target methyl orange ([60]).

MIPs are prepared by the copolymerization of functional monomer(s) and the target analyte (so-called template) in the presence or absence of cross-linkers, as demonstrated in Fig. 1.2 [5]. In the first step, a prepolymerization complex is formed. Template removal with an appropriate solvent leads to the formation of cavities, which mirror size, shape, and sometimes the functionality of the template molecules with a "molecular memory." These cavities in the polymer network mimic the active sites of biomolecules like enzymes and antibodies. In nature, molecular recognition occurs via a whole spectrum of 20 amino acids, whereas MIPs are prepared by one up to six functional monomers.

Two main approaches have been used depending on the interaction between the template and functional monomers to form a prepolymerization complex. In the covalent approach, reversible covalent bonds are formed between the template molecule and functional monomer, which must be cleaved and reestablished during the removal and rebinding steps. The interaction is highly specific, and homogenous binding sites are obtained; nevertheless, it has been applied for only a limited number of analytes. By contrast, the noncovalent approach relies on weaker noncovalent

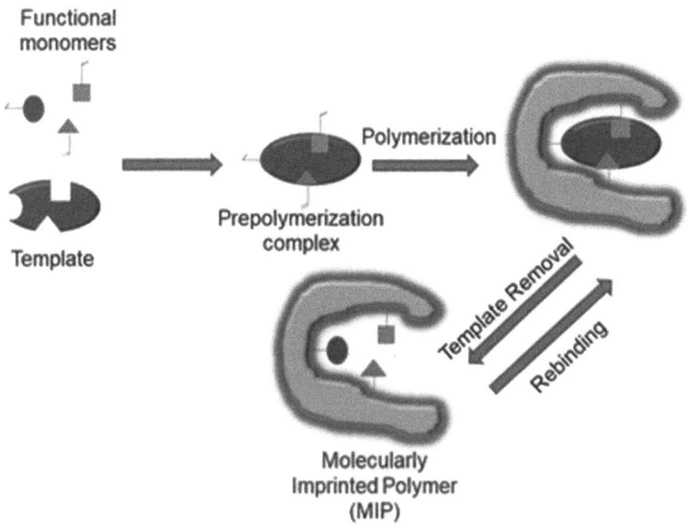

FIGURE 1.2 Schematic representation of steps of MIP preparation: Formation of prepolymerization complex, polymerization, and template removal. Reprinted with permission [5].

interactions such as hydrogen bonds, van der Waals forces, ionic bonds, and hydrophobic interactions. Since, in nature, most of the molecular recognition is based on noncovalent interactions, this approach is called the biochemist's approach, whereas the covalent approach is known as the chemist's approach. It serves a broader spectrum of analytes in comparison to the covalent approach. However, the yield of the binding sites is lower ([61]; [62]).

Various methods have been described for the MIP preparation (Table 1.1; [6]). Among them, bulk polymerization finds wide application. It results in monolithic structures, which must be subsequently grounded and sieved. Moreover, it is time-consuming, and slow binding kinetics are obtained. Alternatively, different MIP preparation methods, like suspension, precipitation, emulsion polymerization, surface imprinting techniques (microcontact-printing and electropolymerization), have been applied [7].

Due to structural complexity and fragility, the most successful results have been achieved for low-molecular weight templates compared to biomolecules such as proteins, nucleic acids, viruses, bacteria, and cells [8–21]. The common imprinting methods can disturb partially or completely their 3D structures. Furthermore, larger pores obtained after template removal can result in poor selectivity. To overcome these obstacles, only a small peptide portion or a fragment like a subunit has been described for larger molecules. This approach has been introduced for the

TABLE 1.1 General advantages and disadvantages of different types of polymerization types.

Polymerization type	General advantages and disadvantages
Bulk	√ Simple and universal type of polymerization. √ No need for sophisticated instrumentation. √ Obtaining spherical materials. √ Providing reproducible results √ Allowing a large-scale examination of products. √ Requiring lengthy procedures. √ Resulting in irregularity in size and shape. √ Low performance.
Precipitation	√ Providing uniform size and high yields of imprinted materials √ Creating homogeneous binding sites. √ One of the easiest and well-suited type with a high dilution factor. √ Requirement for a polymerization mixture in the presence of a much higher amount of porogen maker. √ The growing polymer chains are unable to occupy the entire volume.
Suspension	√ An organic-based medium is mixed with an excess of water and the amount of suspension stabilizer. √ Two phases are mixed by stirring to form a suspension of organic droplets in the aqueous phase. √ The imprinted materials are scarce because water might disrupt noncovalent interactions between the template molecule and the monomers.
Multistep swelling	√ Producing mono-disperse and outstanding materials with controlled diameter. √ The size of the imprinted materials might be controlled by changing the polymerization conditions. √ Requiring complex and long polymerization conditions. √ Requiring laborious procedure and aqueous emulsions.
Surface	√ Producing mono-disperse materials and thin imprinted layers. √ Creating more accessible binding sites. √ Allowing rapid binding and high desorption rates. √ Providing more effective ability to recognize the template molecules. √ Providing a large specific surface area for the particles, hence leading to excellent a nity and selectivity. √ Requiring a complicated system and time-consuming procedure.
In situ	√ Requiring a single-step preparation strategy. √ Being a cost-friendly fashion. √ Providing a, well-porous structure. √ Requiring a comprehensive and lengthy optimization procedure that needs to be optimized for every template molecules systems.

Reprinted with permission [6]

first time by Rachkov and Minoura and called epitope imprinting because of the similarity to the biological determinants ([22–28]a).

Two main approaches have been described for the preparation of MIP-based sensors. In the first approach, the MIP is prepared in-situ direct on the transducer, whereas, in the second approach, MIPs have to be integrated on the surface of the transducer using several methods like spin-coating, drop-coating, layer-by-layer assembly, or grafting.

1.3 MIP sensors for SARS-CoV-2

In addition to the broad spectrum of MIPs for small molecules and proteins, the recognition of virus particles has been described in the literature [13]. Table 1.2 summarizes the analytical parameters of MIP-sensors for pathogenic viruses. The majority of these MIPs were prepared using the virus particle as the template. Among them, a multifunctional MIP allows the simultaneous visual indication of Hepatitis A and Hepatitis B viruses by evaluating the different color signals of two quantum dots [29]. In another work, subtypes of Influenza could be detected by individual MIP-QCM sensors with high specificity down to 10^{-5}particles per mil [30].

On the other hand, MIP-sensors for the dengue virus and HIV, respectively, used the epitope approach for the preparation of MIPs [31]. A mixture of a 35 amino acid peptide of the immunodominant region of the HIV-1 related glycoprotein 41 (gp41) and dopamine was "self" polymerized on a QCM-chip. The MIP bound both the peptide and the parent protein gp41 in the lower nanomolar concentration range with a LOD of 2 ng/mL for the protein. The authors claim satisfactory results for the detection of gp41 in human urine [31].

Parisi et al. realized the idea to develop "monoclonal-type" plastic antibodies based on MIPs to recognize the RBD of SARS-CoV-2 to block the binding of the spike protein to the ACEII-receptor [32]. They developed noncovalent imprinting MIP-nanoparticles with a diameter of around 80 nm and an imprinting factor for the RBD of 7. Specificity of the MIP was demonstrated by the neglectable binding of the RBD of SARS-CoV-2. The final aim is the intravenous application by blocking the binding of the virus spike to the ACE II-receptor, as shown in Fig. 1.3 [33].

The broad spectrum of transducers applied in biosensors has also been used in MIP-based sensors. Among them, the electrochemical and optical readout is the most frequently applied in comparison with quartz crystal microbalance [34] piezoelectric [35,36], thermal [37], and micromechanical [38] transducer. Herein we focus on these two most frequently used transducers since the MIP-sensors developed since the outbreak of pandemic are based on these techniques. In following sections we concentrated on

TABLE 1.2 Molecular imprinting of some pathogenic viruses.

Viruses	Functional monomers	Cross-linkers	Templates	Method	Linear range	LOD/maximum detected concentration	Application	Ref.
Tobacco mosaic viruses	p,p′-diisocyanato-diphenylmethane containing 30% triisocyanate and bisphenol A	NS	tobacco mosaic virus	Quartz crystal microbalance	100 ng mL^{-1} –1 mg mL^{-1}	NS	Tobacco plant sap	Dickert et al. [39]
Dengue virus	Acrylamide and acrylic acid and N-benzylacrylamide	Ethylene glycol dimethacrylate	The dengue virus NS1 protein	Quartz crystal microbalance	NS	NS	NS	Tai, Lin, Wu, and Chen [40]
Dengue virus (flavivirus)	Acrylamide and acrylic acid and N-benzylacrylamide	NS	The dengue virus NS1 protein	Quartz crystal microbalance	NS	0.85 $ng.Hz^{-1}$	Serum	Tai, Lin, Wu, Huang, and Shu [41]
Picornaviruses	Polyurethane and bisphenol A	Phloroglucinol	Human rhinovirus serotype 2 (hrv2), Hrv14, Hrv1A	Quartz crystal microbalance	NS	5.4 × 10(10) virions	NS	Jenik et al. [42]
Hepatitis B virus	N-Methacryloyl-L-tyrosine methyl ester and hydroxyethyl methacrylate	Ethylene glycol dimethacrylate	HBV antibody	Surface plasmon resonance	NS	21.4 $mIU.mg^{-1}$	NS	Uzun, Say, Ünal, and Denizli [43]
Human immunodeficiency virus	Polydopamine	Polydopamine	HIV-1 gp41	Quartz crystal microbalance	5–200 $ng.mL^{-1}$	2 $ng.mL^{-1}$	Human urine	Lu et al. [31]

(*continued on next page*)

Influenza virus	Acrylamide and methylmethacrylate and N-vinylpyrrolidone and methacrylic acid	N,N′-(1,2-Dihydrox-yethylene) bisacrylamide	H1N3, H6N1, H5N1, H5N3 & H1N1	Quartz crystal microbalance	NS	5×10^5 particles.mL^{-1}	NS	Wangchareansak et al. [30]
Human immuno-deficiency virus	3-Aminobenzen-eboronic acid	3-Aminobenze-neboronic acid	HIV-1 antibody	Electrochemi-luminescence	1:20,000–1:50 anti-HIV-1 dilution ratios	1:60,000 anti-HIV-1 dilution ratios	Serum	Zhou et al. [44]
Japanese encephalitis virus	3-Aminopropyl triethoxysilane	Tetraethoxy-silicane	Freeze-dried JEV vaccine	Förster resonance energy transfer	2.5–45 nM	0.32 nM	Human serum	He et al. [45]
Japanese encephalitis virus	3-Aminopropyl triethoxysilane	Tetraethoxy-silicane	JEV vaccine	fluorescence resonance energy transfer	24–960 pM	9.6 pM	Human serum	Liang et al. [46]
Hepatitis A virus	Polydopamine	Polydopamine	HAV type 2	Resonance light scattering	0.04–6.0 nM	8.6 pM	Human serum	Yang, Gong, Chen, Chen, and Cai [47]
Swine fever virus	Acrylamide, methacrylic acid, methyl methacrylate and N-vinylpyrrolidone	Dihydroxy-ethylene-bisacrylamide	Swine fever (Hog cholera) live vaccine	Quartz crystal microbalance	4–21 μg.mL^{-1}	1.7 μg.mL^{-1}	NS	Klangprapan, Choke-arpornchai, Lieberzeit, and Choowong-komon [48]
Dengue virus	Acrylamide, methacrylic acid methyl methacrylate, and N-vinylpyrrolidone	N,N′-(1,2-dihydroxy-ethylene) bisacrylamide	Dengue virus stamps	Quartz crystal microbalance	10–10^3 pfu mL^{-1}	0.58 pfu mL^{-1}	NS	Navakul et al. [34]

Reproduced with permission [49]

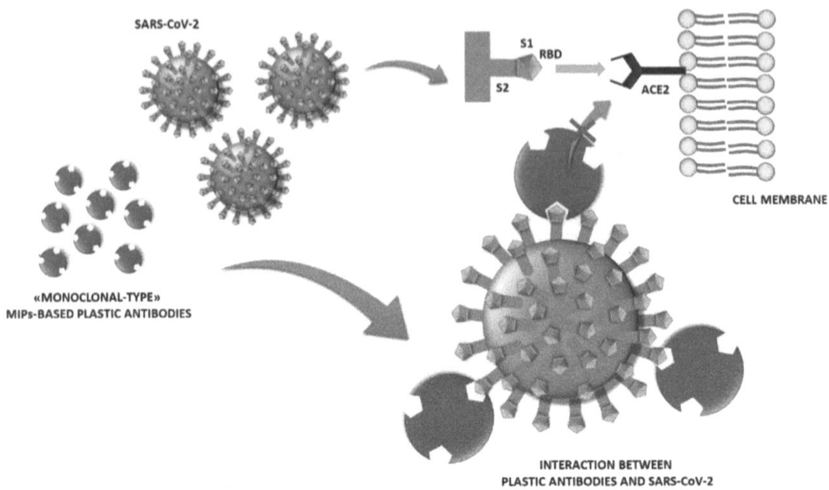

FIGURE 1.3 Schematic representation of the interaction between Molecularly Imprinted Polymers (MIP)-based "monoclonal-type" plastic antibodies and SARS-CoV-2 (severe acute respiratory syndrome coronavirus 2). Reprinted with permission [33].

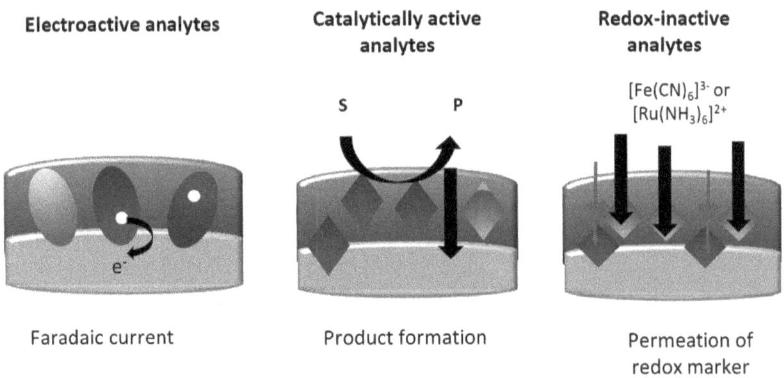

FIGURE 1.4 Main approaches used in an electrochemical readout of MIP-based sensors. Reprinted with permission [5].

very first examples and more information about the recent studies can be seen from Table 1.3.

1.3.1 Electrochemical MIP-sensor for SARS-CoV-2 nucleoprotein

The electrochemical readout exploits three main transduction principles, which we recently reviewed (Fig. 1.4). Most research is based on the "gating effect" of a redox marker like ferri-/ferrocyanide, ferrocene, or ruthenium complexes [50,5]. After template removal, cavities are formed in the polymer layer, through which the redox marker permeates and reaches

TABLE 1.3 MIP-based biomimetic sensors for SARS-CoV-2 detection.

Template	Monomer	Transducer	Detection Method	Linear range	LOD	Application	Ref.
SARS-CoV-2 whole virus	3-AP	CNT/WO_3-SPCE	EIS	7 to 320 pg/mL	57 pg/mL	Nasopharyngeal swabs	[63]
SARS-CoV-2 whole virus	NHMA MBAm (cross-linker)	SPE	EIS	3–7 log_{10} pfu/mL	4.9 log_{10} pfu/mL	Saliva samples	[64]
SARS-CoV-2 whole virus	AAM, MAA, MMA, and NVP; DHEBA (cross-linker)	GO integrated Ag-SPE	CV	0.01 fM to 100 fM	0.1 fM	Wastewater	[65]
SARS-CoV- 2 nucleoprotein	m-PD	4-ATP-modified Au-TFE	DPV	Up to 111 fM	15 fM (in lysis buffer)	Nasopharyngeal swab	[2]
SARS-CoV-2 nucleocapsid protein	Arginine	Au/Gr modified SPCE	DPV	10.0-200.0 fM	3 fM	Artificial nasal and saliva samples	[66]
SARS-CoV-2 spike protein	Pyrrole	Pt Electrode	CA	0 μg/mL to 25 μg/mL	NS	NS	[67]
SARS-CoV-2 RBD	o-PD	MP-Au-SPE	EIS	2.0 $pg.mL^{-1}$ -40 $pg.mL^{-1}$	0.7 $pg.mL^{-1}$	Saliva samples	[68]

(continued on next page)

TABLE 1.3 MIP-based biomimetic sensors for SARS-CoV-2 detection—cont'd

Template	Monomer	Transducer	Detection Method	Linear range	LOD	Application	Ref.
SARS-CoV-2 spike protein subunit S1	APBA	4-ATP-modified Au-TFME	SWV	0-200 fM	15 fM (in PBS) and 64 fM (patient's nasopharyngeal samples)	Nasopharyngeal swab	[69]
SARS-CoV-2 spike protein subunit S1	Aam, TBAm, and HEMA; BIS (cross-Linker)	POF-based SPR chip	SPR	NS	0.058 μM	Nasopharyngeal swabs in universal transport medium and physiological solution (0.9% NaCl)	[57]
SARS-CoV-2 spike protein RBD epitope (GFNCYFPLQ)	Scopoletin	Au- SPRi chips	SPR	NS		Artificial saliva samples	[70]

3-AP: 3-aminophenol; AAM: Acrylamide; APBA: 3-aminophenyl- boronic acid; Au-TFME: Thin-film Au metal electrodes, BIS: N,N′-methylene bisacrylamide; CA: Chronoamperometry; DHEBA: N,N′-(1,2-dihydroxy- ethylene) bisacrylamide; HEMA: 2-hydroxyethyl methacrylate; MAA: Methacrylic acid; MBAm: N,N′-methylenebisacrylamide; MMA: methyl methacrylate; m-PD: m-Phenylenediamine; NHMA: N-hydroxmethylacrylamide; NS: Not stated; NVP: N-vinylpyrrolidone; o-PD: o-Phenylenediamine; SPCE: screen-printed carbon electrode; TBAm: N-t-butylacrylamide the electrode. This procedure is mainly suitable for redox-inactive analytes (from small molecules like drugs, toxins, pesticides to biomacromolecules like proteins, nucleic acids, and viruses) or even for the majority of redox proteins where direct electron transfer is not possible due to steric hindrance of the redox-active center. In contrast, the binding of electroactive analytes direct generates a faradaic current [5]. Hence the measurement is based on the direct electron transfer between the template and electrode. The third measurement principle relies on the detection of the signal generated by catalytically active analytes ([51]b; [52,53]).

Among different electrochemical detection methods in MIP-based sensors, voltammetry is the most frequently used electrochemical readout method compared to potentiometric sensors like field-effect transistors or capacitors. In voltammetry, where the current is recorded as a function of potential, the effect of oxygen evolution and hydrogen generation should be considered. Moreover, false-positive results can be obtained in the presence of electroactive interfering substances, which can be suppressed by the application of more sensitive methods like differential pulse voltammetry (DPV) and square wave voltammetry (SWV) by the eliminating charging current [7,5].

The first biomimetic sensor described since the outbreak of Covid-19 utilized an electrochemical transducer, which targeted nucleoprotein (ncovNP) [2]. The sensor was prepared using electropolymerization. It can be prepared under mild conditions directly on the transducer, eliminating additional integration steps [54]. The use of cross-linkers is not needed. Further, the electrosynthesis is fast, and desired thickness can be achieved by controlling the charge passed through the electrode during polymerization. Controlling the thickness is one of the crucial parameters, especially for imprinting biomacromolecules, since the template molecules may be fully entrapped in the polymer matrix thus, hindering their removal and rebinding.

To obtain a more homogenous surface, the target analyte or their epitopes can be immobilized on the surface prior to electropolymerization. This method is called hierarchical imprinting and is more time-consuming as compared to random imprinting, which uses mixtures of the template and the functional monomer(s).

For the preparation of the ncovNP-MIP first, a self-assembled monolayer of 4-aminothiophenol (4-ATP) was formed on a gold thin film electrode (Au-TPE). Then it was incubated in DTSP solution for covalent immobilization of the template, ncovNP. As seen from the CVs, almost no current suppression was observed after incubation in 4-ATP, whereas significant signal suppression was obtained by subsequent immobilization of DTSP and ncovNP, respectively. According to molecular docking and quantum chemical calculations of the interactions between the ncovNP and the three candidate monomers m-phenylenediamine (mPD), dopamine, and EDOT, mPD was chosen as the optimal functional

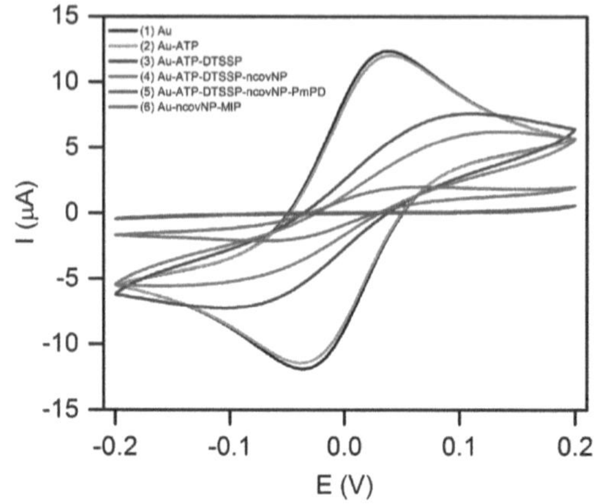

FIGURE 1.5 Cyclic voltammograms recorded in 1 M KCl solution containing a 4 mM redox probe K3[Fe(CN)6]/K4[Fe(CN)6] on bare Au (1), and after the subsequent fabrication steps of ncovNP-MIP: modification by 4-ATP (2), DTSSP (3), and ncovNP (4), electrodeposition of PmPD (5) and treatment in 2-ME and acetic acid (6). Reprinted with permission [2].

monomer. The film was electrodeposited by polarizing the Au-ATP-DTSP-ncovNP electrode at 0.6 V with an optimized charge density. As seen from Fig. 1.5, after electropolymerization, further signal suppression was obtained, which is due to the formation of the nonconducting film. Subsequent removal of the template molecules with an ethanolic solution of 2-mercaptoethanol and 10% acetic solution led to an increase in the redox marker's signal due to the formation of free cavities [2]. It was postulated that hydrogen bonds are formed between the accessible proton acceptor groups of the protein and the amino groups of mPD in the prepolymerization complex (Fig. 1.6; [2]).

Samples were prepared in lysis buffer (pH 7.2) containing 27.5 mM of Tris-HCl, 12.5 mM of EDTA, 1.5% (v/v) of TritonX-100, and 0.1% SDS diluted with MQ water to obtain the desired concentration. Rebinding of the template was measured in ferri/ferrocyanide solution by applying DPV (Fig. 1.7). The sensor showed a linear response up to 110 fM with a LOD and a LOQ of 15 fM and 50 fM, respectively, which are in the clinical range.

To illustrate the selectivity of the developed ncovNP-MIP sensor various proteins (of SARS-Cov-2 spike protein: S1, 75 kDa, pI: 6.0; Cluster of Differentiation 48: CD48, 22 kDa, pI:9.3; Hepatitis C virus surface viral antigen: E2 HCV, 47 kDa, pI: 8.2; bovine serum albumin: BSA, 66 kDa, pI: 4.7)with different size, molecular weight, and isoelectric point (pI) were studied. As seen from Fig. 1.7, the highest response was obtained for the target analyte ncovNP compared to the interfering proteins.

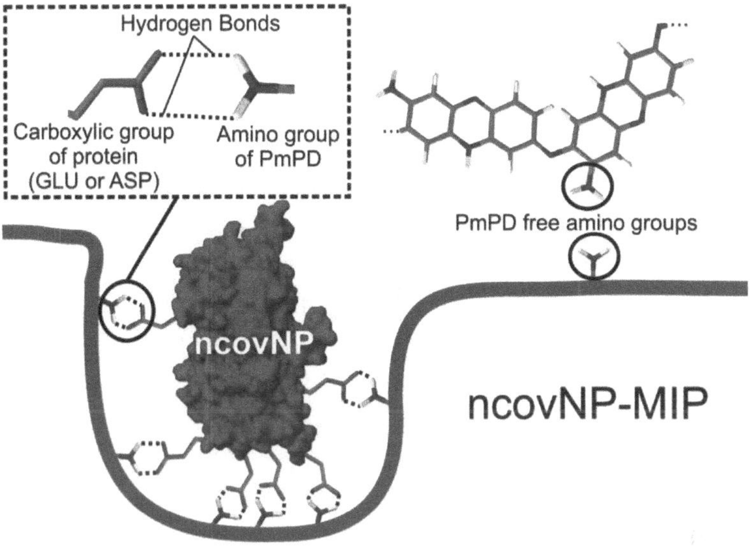

FIGURE 1.6 Arrangement of NH_2-groups in a molecular cavity of ncovNP-MIP to show a hypothetical possibility to achieve complementarity between the cavity and ncovNP. Reprinted with permission [7].

Moreover, the performance of the sensor was applied in nasopharyngeal swab specimens. The calibration graph was prepared first in spiked COVID-19 negative samples. The sensor had a pseudo linear response in the range of 0.22 fM and 333 fM. LOD and LOQ were calculated to be 27 fM and 90 fM, respectively. The sensor was further applied to COVID-19 positive patients. The performance of the MIP sensor was compared with RP-PCR, and a correlation was found. It was claimed that it could be applied in a portable sensing platform.

In a proof-of-concept study, we applied the epitope approach by utilizing a peptide from the RBD as the template for the electrosynthesis of MIPs for the S-protein [55]. After chemisorption of the heptapeptide from the receptor-binding domain of the spike protein of SARS-CoV-2 on the gold electrode, electrodeposition of the polymer, and template removal, both the target peptide and the RBD were bound in the lower nanomolar concentration range.

1.3.2 SPR-based MIP-sensor for SARS-CoV-2

Optical readout relies on the measurement of the optical changes like refractive index, fluorescent, and chemiluminescent upon the interaction of the (bio)recognition element with the target analyte.

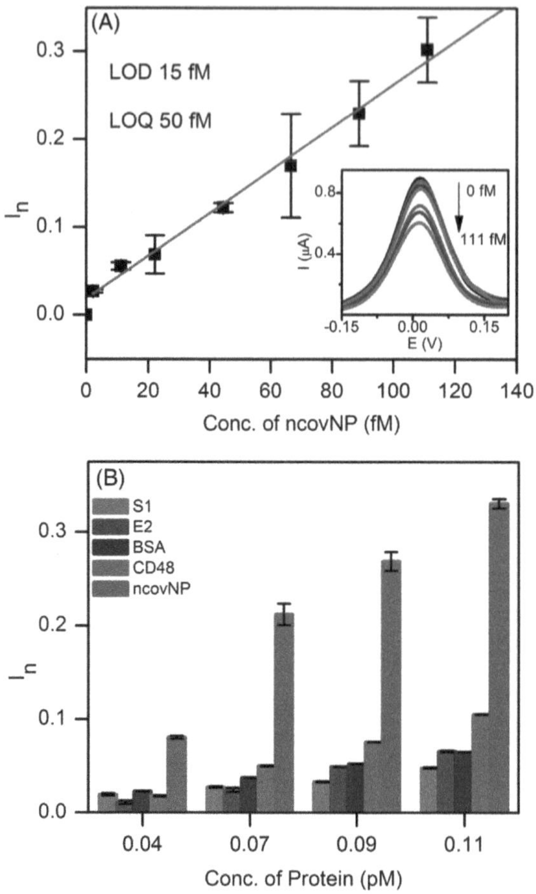

FIGURE 1.7 (A) Calibration plot of ncovNP sensor obtained at the low concentration range of ncovNP (2–111 fM) in LB. The inset shows typical DPV curves used to construct the calibration plot. (B) Selectivity test of ncovNP sensor showing its responses against the different proteins (S1, E2 HCV, BSA, CD48 and ncovNP) applied at concentrations (0.04, 0.07, 0.09, and 0.11 pM) in LB. Reprinted with permission [2].

Surface plasmon resonance (SPR) is an optical phenomenon, which is frequently used in optical MIP-sensors. It is based on the measurement of refractive index changes near the sensor surface ([9,56]; Fig. 1.8). It enables label-free and in-situ detection of the binding event, including the determination of the binding affinities and kinetics.

Cennamo et al. recently described the first prototype of a plastic optical fiber (POF)-based SPR sensor using a MIP for the specific recognition of the S1 subunit of SARS-CoV-2 spike protein [57]. To check and optimize the process, bovine serum albumin was first used as a template. The protocol

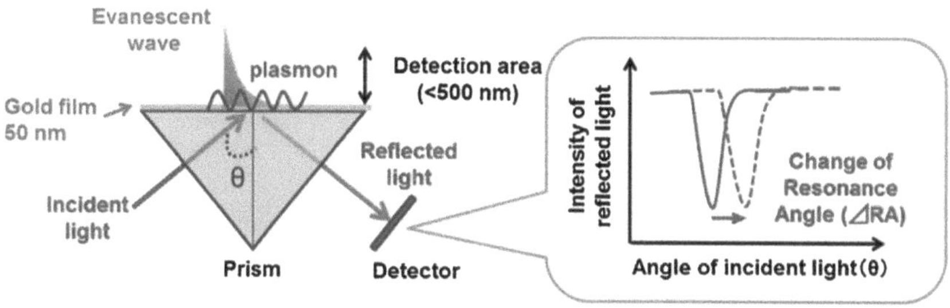

FIGURE 1.8 Schematic description of SPR sensor. Reprinted with permission [56].

was then transferred for the preparation of a MIP for the S1 subunit of SARS-CoV-2 spike protein. The MIP was prepared as follows: acrylamide, N-t-butyl acrylamide, 2-hydroxyethyl methacrylate were mixed with a molar ratio of 1:0.5:0.6, in 15 mM phosphate buffer (pH 7.4.) The final concentration of N, N′-methylene bisacrylamide in the monomeric mixture was 0.19 M. After the mixture had been ultrasonicated and degassed with nitrogen, the template was added. Finally, APS (0.08% w/v) and TEMED (0.06% w/v) were added. 50 μL of the final solution was dropped on the previously allythiol modified D-shaped POF covered gold SPR surface and polymerized. Polymerization was ended by washing the sensor with Milli-Q water. Template molecules were removed with trypsin and SDS. Fig. 1.9 illustrates the response of the bare electrode of the MIP before and after template removal. Fitting the data obtained upon the rebinding of the template to Hill equation, LOD and affinity constant were calculated to be 0.058 μM^{-1} and 2.318 μM^{-1}, respectively.

The authors further investigated the specificity of the developed sensor in a universal transport medium using MERS-CoV Spike S1 protein, which results in almost no detectable signal. Moreover, the sensor was applied to detect SARS-CoV-2 virions obtained by nasopharyngeal (NP) swabs in UTM and physiological solution (0.9% NaCl).

1.3.3 First commercial MIP-sensor for SARS-CoV-2

MIP diagnostics developed MIP–nanoparticles for the recognition of the spike-protein using the receptor-binding domain (RBD) as the template (https://www.mip-dx.com/covid19-nanomip). The RBD region of the spike glycoprotein is unique to SARS-CoV-2, and therefore the nanoMIP will not detect other human coronaviruses but should detect the current variants of SARS-CoV-2. The nanoMIPs are within the 40–80 nm range and allow for simple integration into sensor platforms because they can be readily functionalized with an amine group which enables attachment of a label of choice or on the sensor surface. The Nano-MIPs have been

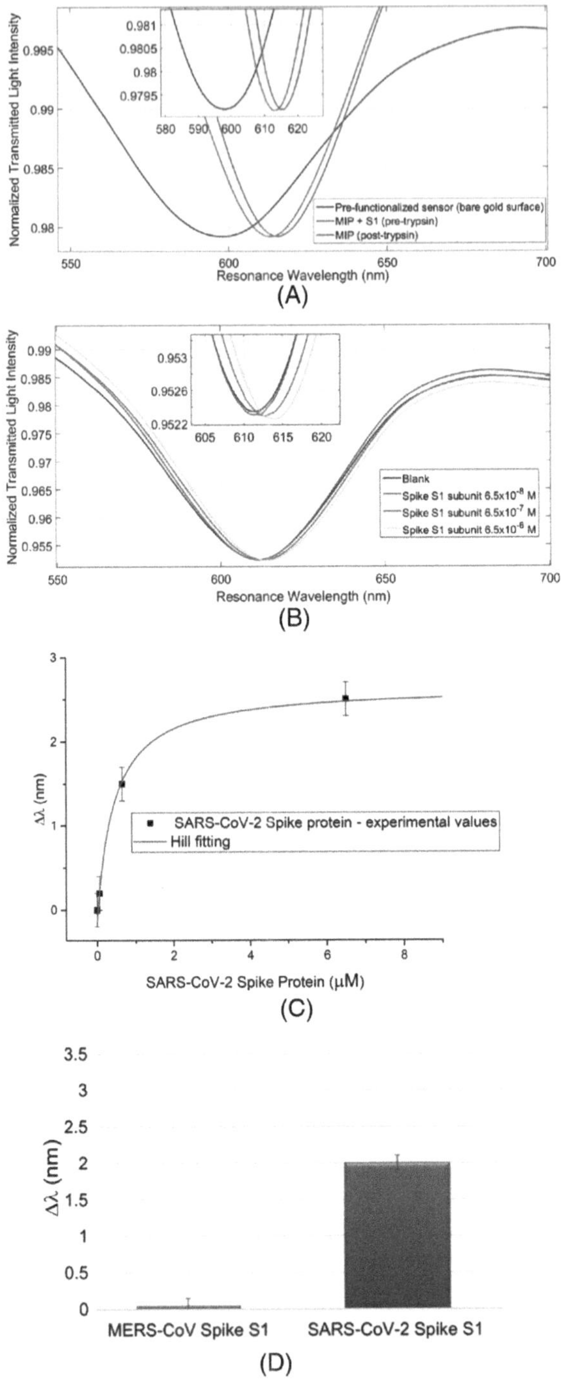

FIGURE 1.9 (A) Blue resonance peak: bare gold surface before the functionalization. Red resonance peak: red-shifted resonance due to the MIP layer before template extraction with

covalently coupled to an electrode surface and integrated into a thermal resistance sensor platform. Sensitivity has been demonstrated to be below 5 fg/mL for the RBD with an affinity constant (KD) ranging from 2 to 18 nM.

1.4 Conclusions

The general motivation to develop biomimetic recognition elements like aptamers and MIPs is the substitution of antibodies in diagnostics, therapy, and downstream processing. The recently published European Union Reference Laboratory for alternatives to animal testing: "Recommendation of Non-Animal-Derived Antibodies" underlines the importance of these biomimetic concepts. As reported in this chapter, a few MIP-based sensors for the determination of the viral load of SARS-COV-2 have been developed. However, it is questionable whether these sensors can displace or even partially substitute the "gold standard" of quantitative RT-PCR in SARS-CoV-2 diagnostics. On the other hand, the analytical performance of the presented MIPs should allow its application in lateral flow devices, which visually indicate the presence of the virus above a given threshold concentration.

Acknowledgments

This work was funded by the Deutsche Forschungsgemeinschaft (DFG, German Research Foundation) under Germany´s Excellence Strategy – EXC 2008/1 (UniSysCat) – 390540038 [Gefördert durch die Deutsche Forschungsgemeinschaft (DFG) im Rahmen der Exzellenzstrategie des Bundes und der Länder – EXC 2008/1 (UniSysCat) – 390540038].

References

[1] M. Drobysh, A. Ramanaviciene, R. Viter, A. Ramanavicius, Affinity Sensors for the Diagnosis of COVID-19, Micromachines 12 (2021) 1–19 390.

[2] A. Raziq, A. Kidakova, R. Boroznjak, J. Reut, A. Öpik, V. Syritski, Development of a portable MIP-based electrochemical sensor for detection of SARS-CoV-2 antigen, Biosensors and Bioelectronics 178 (2021) 1–7 113029.https://doi.org/10.1016/j.bios.2021.113029.

trypsin enzyme 4.2×10^{-8} M in buffer phosphate pH 7.4 and sodium dodecyl sulfate (SDS) 5%. Magenta resonance peak: Blueshift due to the freeing of sites from the template protein. (B) Response curves of SARS-Cov-2 Spike S1 subunit-MIP at different concentrations of protein. (C) SARS-CoV-2 Spike protein dose-response curve with the Hill fitting of the data. (D) Specificity test: sensor's responses for MERS-CoV Spike protein and SARS-CoV-2 Spike protein, both with a concentration of 1 μM in universal transport medium buffer. Reprinted with permission [57].

[3] M. Yüce, E. Filiztekin, K.G. Özkaya, COVID-19 diagnosis —A review of current methods, Biosensors and Bioelectronics 172 (2021) 1–15 112752. https://doi.org/10.1016/j.bios.2020.112752.
[4] Z. Liu, X. Xiao, X. Wei, J. Li, J. Yang, H. Tan, et al., Composition and divergence of coronavirus spike proteins and host ACE2 receptors predict potential intermediate hosts of SARS-CoV-2, Journal of Medical Virology 92 (6) (2020) 595–601. https://doi.org/10.1002/jmv.25726.
[5] A. Yarman, F.W. Scheller, How reliable is the electrochemical readout of MIP sensors? Sensors (Switzerland) 20 (9) (2020) 1–23 2677. https://doi.org/10.3390/s20092677 .
[6] Y. Saylan, Ö. Erdem, F. Inci, A. Denizli, Advances in biomimetic systems for molecular recognition and biosensing, Biomimetics 5 (2) (2020) 1–16 20. https://doi.org/10.3390/BIOMIMETICS5020020.
[7] A. Yarman, S. Kurbanoglu, I. Zebger, F.W. Scheller, Simple and robust: The claims of protein sensing by molecularly imprinted polymers, Sensors and Actuators B: Chemical 330 (2021) 1–12 129369. https://doi.org/10.1016/j.snb.2020.129369 .
[8] O.S. Ahmad, T.S. Bedwell, C. Esen, A. Garcia-Cruz, S.A. Piletsky, Molecularly imprinted polymers in electrochemical and optical sensors, Trends in Biotechnology 37 (3) (2019) 294–309. https://doi.org/10.1016/j.tibtech.2018.08.009.
[9] N. Bereli, S. Akgönüllü, S. Aslıyüce, D. Cimen, İ. Gokturk, D. Turkmen, et al., Molecular imprinting technology for biomimetic assemblies, Hacettepe Journal of Biology and Chemistry 48 (5) (2020) 575–601. https://doi.org/10.15671/hjbc.801427.
[10] M. Feroz, P. Vadgama, Molecular imprinted polymer modified electrochemical sensors for small drug analysis: Progress to practical application, Electroanalysis 32 (11) (2020) 2361–2386. https://doi.org/10.1002/elan.202060276.
[11] R.A. Hand, E. Piletska, T. Bassindale, G. Morgan, N. Turner, Application of molecularly imprinted polymers in the anti-doping field: sample purification and compound analysis, Analyst 145 (2020) 4716–4736. https://doi.org/10.1039/d0an00682c.
[12] O. Hayden, P.A. Lieberzeit, D. Blaas, F.L. Dickert, Artificial antibodies for bioanalyte detection - Sensing viruses and proteins, Advanced Functional Materials 16 (10) (2006) 1269–1278. https://doi.org/10.1002/adfm.200500626.
[13] G. Jamalipour Soufi, S. Iravani, R.S. Varma, Molecularly imprinted polymers for the detection of viruses: Challenges and opportunities, Analyst 146 (10) (2021) 3087–3100. https://doi.org/10.1039/d1an00149c.
[14] J. Kalecki, Z. Iskierko, M. Cieplak, P.S. Sharma, Oriented immobilization of protein templates: A new trend in surface imprinting, ACS Sensors 5 (12) (2020) 3710–3720. https://doi.org/10.1021/acssensors.0c01634.
[15] M.A.R. Khan, T.C. Moreira, J. F., Riu, F. Sales, G M., Plastic antibody for the electrochemical detection of bacterial surface proteins, Sensors and Actuators B: Chemical 233 (2016) 697–704. https://doi.org/10.1016/j.snb.2016.04.075.
[16] C. Malitesta, E. Mazzotta, R.A. Picca, A. Poma, I. Chianella, S.A. Piletsky, MIP sensors - The electrochemical approach, Analytical and Bioanalytical Chemistry 402 (5) (2012) 1827–1846. https://doi.org/10.1007/s00216-011-5405-5.
[17] G. Moro, F. Bottari, N. Sleegers, A. Florea, T. Cowen, L.M. Moretto, et al., Conductive imprinted polymers for the direct electrochemical detection of β-lactam antibiotics: The case of cefquinome, Sensors and Actuators B: Chemical 297 (2019) 1–9 126786. https://doi.org/10.1016/j.snb.2019.126786 .
[18] G. Ozcelikay, S. Kurbanoglu, A. Yarman, F.W. Scheller, S.A. Ozkan, Au-Pt nanoparticles based molecularly imprinted nanosensor for electrochemical detection of the lipopeptide antibiotic drug Daptomycin, Sensors and Actuators B: Chemical 320 (2020) 1–7 128285. https://doi.org/10.1016/j.snb.2020.128285 .
[19] J.G. Pacheco, P. Rebelo, M. Freitas, H.P.A. Nouws, C. Delerue-Matos, Breast cancer biomarker (HER2-ECD) detection using a molecularly imprinted electrochemical sensor, Sensors and Actuators B: Chemical 273 (2018) 1008–1014. https://doi.org/10.1016/j.snb.2018.06.113.

[20] F.W. Scheller, X. Zhang, A. Yarman, U. Wollenberger, R.E. Gyurcsányi, Molecularly imprinted polymer-based electrochemical sensors for biopolymers, Current Opinion in Electrochemistry 14 (2019) 53–59. https://doi.org/10.1016/j.coelec.2018.12.005.

[21] A. Yarman, F.W. Scheller, The first electrochemical MIP sensor for tamoxifen, Sensors (Switzerland) 14 (5) (2014) 7647–7654. https://doi.org/10.3390/s140507647.

[22] L. Cenci, R. Tatti, R. Tognato, E. Ambrosi, C. Piotto, A.M. Bossi, Synthesis and characterization of peptide-imprinted nanogels of controllable size and affinity, European Polymer Journal 109 (2018) 453–459. https://doi.org/10.1016/j.eurpolymj.2018.08.031.

[23] D. Dechtrirat, K.J. Jetzschmann, W.F.M. Stöcklein, F.W. Scheller, N. Gajovic-Eichelmann, Protein rebinding to a surface-confined imprint, Advanced Functional Materials 22 (24) (2012) 5231–5237. https://doi.org/10.1002/adfm.201201328.

[24] K.J. Jetzschmann, A. Yarman, L. Rustam, P. Kielb, V.B. Urlacher, A. Fischer, et al., Molecular LEGO by domain-imprinting of cytochrome P450 BM3, Colloids and Surfaces B: Biointerfaces 164 (2018) 240–246. https://doi.org/10.1016/j.colsurfb.2018.01.047.

[25] H. Nishino, C.S. Huang, K.J. Shea, Selective protein capture by epitope imprinting, Angewandte Chemie - International Edition 45 (15) (2006) 2393–2396. https://doi.org/10.1002/anie.200503760.

[26] L. Pasquardini, A.M. Bossi, Molecularly imprinted polymers by epitope imprinting: a journey from molecular interactions to the available bioinformatics resources to scout for epitope templates, Analytical and Bioanalytical Chemistry 413 (2021) 6101–6115. https://doi.org/10.1007/s00216-021-03409-1.

[27] A. Rachkov, N. Minoura, Recognition of oxytocin and oxytocin-related peptides in aqueous media using a molecularly imprinted polymer synthesized by the epitope approach, Journal of Chromatography A 889 (1–2) (2000) 111–118. https://doi.org/10.1016/S0021-9673(00)00568-9.

[28] K. Yang, S. Li, L. Liu, Y. Chen, W. Zhou, J. Pei, et al., Epitope imprinting technology: Progress, applications, and perspectives toward artificial antibodies, Advanced Materials 31 (50) (2019) 1–17. https://doi.org/10.1002/adma.201902048.

[29] L. Luo, F. Zhang, C. Chen, C. Cai, Visual simultaneous detection of hepatitis A and B viruses based on a multifunctional molecularly imprinted fluorescence sensor, Analytical Chemistry 91 (24) (2019) 15748–15756. https://doi.org/10.1021/acs.analchem.9b04001.

[30] T. Wangchareansak, A. Thitithanyanont, D. Chuakheaw, M.P. Gleeson, P.A. Lieberzeit, C. Sangma, Influenza A virus molecularly imprinted polymers and their application in virus sub-type classification, Journal of Materials Chemistry B 1 (16) (2013) 2190–2197. https://doi.org/10.1039/c3tb00027c.

[31] C.H. Lu, Y. Zhang, S.F. Tang, Z.Bin Fang, H.H. Yang, X. Chen, et al., Sensing HIV related protein using epitope imprinted hydrophilic polymer coated quartz crystal microbalance, Biosensors and Bioelectronics 31 (1) (2012) 439–444. https://doi.org/10.1016/j.bios.2011.11.008.

[32] O.I. Parisi, M. Dattilo, F. Patitucci, R. Malivindi, S. Delbue, P. Ferrante, et al., Design and development of plastic antibodies against SARS-CoV-2 RBD based on molecularly imprinted polymers that inhibit in vitro virus infection, Nanoscale 12 (2021) 16885–16899 2020.05.28.120709. https://doi.org/10.1101/2020.05.28.120709 .

[33] F. Puoci, Monoclonal-type" plastic antibodies for covid-19 treatment: What is the idea? Journal of Functional Biomaterials 11 (2) (2020) 9–12. https://doi.org/10.3390/jfb11020043.

[34] K. Navakul, C. Sangma, P. Yenchitsomanus, S. Chunta, P.A. Lieberzeit, Enhancing sensitivity of QCM for dengue type 1 virus detection using graphene-based polymer composites, Analytical and Bioanalytical Chemistry 413 (2021) 6191–6198. https://doi.org/10.1007/s00216-021-03410-8.

[35] E.A. Beliaeva, B. Pluhar, T.N. Ermolaeva, B. Mizaikoff, N.A. Karaseva, Synthesis and application of molecularly imprinted polymers for trypsin piezoelectric sensors, Sensors and Actuators B: Chemical 280 (2018) 272–279. https://doi.org/10.1016/j.snb.2018.10.022.

[36] S. Chunta, R. Suedee, W. Boonsriwong, P.A. Lieberzeit, Biomimetic sensors targeting oxidized-low-density lipoprotein with molecularly imprinted polymers, Analytica Chimica Acta 1116 (2020) 27–35. https://doi.org/10.1016/j.aca.2020.04.017.

[37] R.D. Crapnell, F. Canfarotta, J. Czulak, R. Johnson, K. Betlem, F. Mecozzi, et al., Thermal detection of cardiac biomarkers heart-fatty acid binding protein and ST2 using a molecularly imprinted nanoparticle-based multiplex sensor platform, ACS Sensors 4 (10) (2019) 2838–2845. https://doi.org/10.1021/acssensors.9b01666.

[38] K. El Kirat, M. Bartkowski, K. Haupt, Probing the recognition specificity of a protein molecularly imprinted polymer using force spectroscopy, Biosensors and Bioelectronics 24 (8) (2009) 2618–2624. https://doi.org/10.1016/j.bios.2009.01.018.

[39] F.L. Dickert, O. Hayden, R. Bindeus, K.-J. Mann, D. Blaas, E. Waigmann, Bioimprinted QCM sensors for virus detection—screening of plant sap, Analytical and Bioanalytical Chemistry 378 (8) (2004) 1929–1934. https://doi.org/10.1007/S00216-004-2521-5.

[40] D.F. Tai, C.Y. Lin, T.Z. Wu, L.K. Chen, Recognition of dengue virus protein using epitope-mediated molecularly imprinted film, Analytical Chemistry 77 (16) (2005) 5140–5143. https://doi.org/10.1021/ac0504060.

[41] D.F. Tai, C.Y. Lin, T.Z. Wu, J.H. Huang, P.Y. Shu, Artificial receptors in serologic tests for the early diagnosis of dengue virus infection, Clinical Chemistry 52 (8) (2006) 1486–1491. https://doi.org/10.1373/clinchem.2005.064501.

[42] M. Jenik, R. Schirhagl, C. Schirk, O. Hayden, P. Lieberzeit, D. Blaas, et al., Sensing picornaviruses using molecular imprinting techniques on a quartz crystal microbalance, Analytical Chemistry 81 (13) (2009) 5320–5326. https://doi.org/10.1021/ac8019569.

[43] L. Uzun, R. Say, S. Ünal, A. Denizli, Hepatitis B surface antibody purification with hepatitis B surface antibody imprinted poly(hydroxyethyl methacrylate-N-methacryloyl-l-tyrosine methyl ester) particles, Journal of Chromatography B: Analytical Technologies in the Biomedical and Life Sciences 877 (3) (2009) 181–188. https://doi.org/10.1016/j.jchromb.2008.12.004.

[44] J. Zhou, N. Gan, T. Li, F. Hu, X. Li, L. Wang, L. Zheng, A cost-effective sandwich electrochemiluminescence immunosensor for ultrasensitive detection of HIV-1 antibody using magnetic molecularly imprinted polymers as capture probes, Biosensors and Bioelectronics 54 (2014) 199–206. https://doi.org/10.1016/j.bios.2013.10.044.

[45] K. He, C. Chen, C. Liang, C. Liu, B. Yang, X. Chen, et al., Highly selective recognition and fluorescent detection of JEV via virus-imprinted magnetic silicon microspheres, Sensors and Actuators B: Chemical 233 (2016) 607–614. https://doi.org/10.1016/j.snb.2016.04.127.

[46] C. Liang, H. Wang, K. He, C. Chen, X. Chen, H. Gong, et al., A virus-MIPs fluorescent sensor based on FRET for highly sensitive detection of JEV, Talanta 160 (2016) 360–366. https://doi.org/10.1016/j.talanta.2016.06.010.

[47] B. Yang, H. Gong, C. Chen, X. Chen, C. Cai, A virus resonance light scattering sensor based on mussel-inspired molecularly imprinted polymers for high sensitive and high selective detection of Hepatitis A Virus, Biosensors and Bioelectronics 87 (2017) 679–685. https://doi.org/10.1016/j.bios.2016.08.087.

[48] S. Klangprapan, B. Choke-arpornchai, P.A. Lieberzeit, K. Choowongkomon, Sensing the classical swine fever virus with molecularly imprinted polymer on quartz crystal microbalance, Heliyon 6 (6) (2020) e04137. https://doi.org/10.1016/j.heliyon.2020.e04137.

[49] A.A. Malik, C. Nantasenamat, T. Piacham, Molecularly imprinted polymer for human viral pathogen detection, Materials Science and Engineering: C 77 (2017) 1341–1348. https://doi.org/10.1016/j.msec.2017.03.209.

[50] P. Lach, M. Cieplak, K.R. Noworyta, P. Pieta, W. Lisowski, J. Kalecki, et al., Self-reporting molecularly imprinted polymer with the covalently immobilized ferrocene redox probe for selective electrochemical sensing of p-synephrine, Sensors and Actuators: B. Chemical 344 (2021) 1–10 130276. https://doi.org/10.1016/j.snb.2021.130276 .

[51] S. Yang, C. Bai, Y. Teng, J. Zhang, J. Peng, Z. Fang, et al., Study of horseradish peroxidase and hydrogen peroxide bi-analyte sensor with boronate affinity-based molecularly imprinted film, Canadian Journal of Chemistry 97 (12) (2019) 833–839. https://doi.org/10.1139/cjc-2019-0134.

[52] A. Yarman, Development of a molecularly imprinted polymer-based electrochemical sensor for tyrosinase, Turkish Journal of Chemistry 42 (2) (2018) 346–354.

[53] A. Yarman, Electrosynthesized molecularly imprinted polymer for laccase using the inactivated enzyme as the target, Bulletin of the Korean Chemical Society 39 (4) (2018) 483–488. https://doi.org/10.1002/bkcs.11413.

[54] P.S. Sharma, A. Pietrzyk-Le, F. D'Souza, W Kutner, Electrochemically synthesized polymers in molecular imprinting for chemical sensing, Analytical and Bioanalytical Chemistry 402 (10) (2012) 3177–3204. https://doi.org/10.1007/s00216-011-5696-6.

[55] A. Yarman, G. Caserta, X. Zhang, P. Borrero, S. Frielingsdorf, E. Supala, et al., Biosensors 2021, 31th Anniversary World Congress on Biosensors, Potential vs shortcomings of epitope MIPs for the recognition of natural and recombinant proteins, 2021.

[56] P. Singh, SPR biosensors: Historical perspectives and current challenges, Sensors and Actuators B: Chemical 229 (2016) 110–130. https://doi.org/10.1016/J.SNB.2016.01.118.

[57] N. Cennamo, G. D. Agostino, C. Perri, F. Arcadio, G. Chiaretti, E. M. Parisio, et al., Proof of concept for a quick and highly sensitive on-site detection of SARS-CoV-2 by plasmonic optical fibers and molecularly imprinted polymers, Sensors 21 (2021) 1–17.

[58] R. Arshady, & K. Mosbach. Synthesis of Substrate-selective Polymers by Host-Guest Polymerization. Makromolekulare Chemie, 692 (1981) 687–692.

[59] G. Wulff, A. Sarhan. Use of polymers with enzyme-analogous structures for the resolution of racemates. Angew. Chem. Int. Ed. Engl, 11 (1972) 341–344.

[60] M.V. Polyakov, Adsorption properties and structure of silica gel. Zhur Fiz Khim, 2 (1931) 709–805.

[61] C. Alexander, H.S. Andersson, L.I. Andersson, R.J. Ansell, N. Kirsch, I.A. Nicholls, J. O'Mahony, M.J. Whitcombe, et al., Molecular imprinting science and technology: A survey of the literature for the years up to and including 2003. Journal of Molecular Recognition, 19 (2) (2006) 106–180. https://doi.org/10.1002/jmr.760.

[62] B.T.S. Bui, K. Haupt, Molecularly imprinted polymers: synthetic receptors in bioanalysis. Analytical and Bioanalytical Chemistry, 398 (6) (2010), 2481–2492. https://doi.org/10.1007/s00216-010-4158-x.

[63] H.A. Hussein, A. Kandeil, M. Gomaa, R. Mohamed El Nashar, I.M. El-Sherbiny, R.Y.A Hassan, SARS-CoV-2-Impedimetric Biosensor: Virus-Imprinted Chips for Early and Rapid Diagnosis, ACS Sensors *6* (11) (2021) 4098–4107. https://doi.org/10.1021/acssensors.1c01614.

[64] H.F. EL Sharif, S.R. Dennison, M. Tully, S. Crossley, W. Mwangi, D. Bailey, …, S.M. Reddy, Evaluation of electropolymerized molecularly imprinted polymers (E-MIPs) on disposable electrodes for detection of SARS-CoV-2 in saliva, Anal. Chim. Acta (2022) 339777. https://doi.org/10.1016/J.ACA.2022.339777 .

[65] W. Sukjee, A. Thitithanyanont, S. Manopwisedjaroen, S. Seetaha, C. Thepparit, C. Sangma, Virus MIP-composites for SARS-CoV-2 detection in the aquatic environment, Mater. Lett. *315* (September 2021) (2022) 131973. https://doi.org/10.1016/j.matlet.2022.131973 .

[66] T. Zhang, L. Sun, Y. Zhang, Highly sensitive electrochemical determination of the SARS-COV-2 antigen based on a gold/graphene imprinted poly-arginine sensor, Anal. Methods *13* (47) (2021) 5772–5776. https://doi.org/10.1039/d1ay01478a.

[67] V. Ratautaite, R. Boguzaite, E. Brazys, A. Ramanaviciene, E. Ciplys, M. Juozapaitis, …, A. Ramanavicius, Molecularly imprinted polypyrrole based sensor for the detection of SARS-CoV-2 spike glycoprotein, Electrochim. Acta *403* (2022) 139581. https://doi.org/10.1016/j.electacta.2021.139581 .
[68] M. Amouzadeh Tabrizi, J.P. Fernández-Blázquez, D.M. Medina, P. Acedo, An ultrasensitive molecularly imprinted polymer-based electrochemical sensor for the determination of SARS-CoV-2-RBD by using macroporous gold screen-printed electrode, Biosens. Bioelectron. 196 (2021) (2022). https://doi.org/10.1016/j.bios.2021.113729.
[69] A.G. Ayankojo, R. Boroznjak, J. Reut, A. Öpik, V. Syritski, Molecularly imprinted polymer based electrochemical sensor for quantitative detection of SARS-CoV-2 spike protein, Sens. Actuators B *353* (2022). https://doi.org/10.1016/j.snb.2021.131160.
[70] Z. Bognár, E. Supala, A. Yarman, X. Zhang, F.F. Bier, F.W. Scheller, R.E. Gyurcsányi, Peptide epitope-imprinted polymer microarrays for selective protein recognition. Application for SARS-CoV-2 RBD protein, Chem. Sci. *13* (5) (2022) 1263–1269. https://doi.org/10.1039/d1sc04502d.

CHAPTER

2

Lab-on-paper based devices for COVID-19 sensors

Mohd. Rahil Hasan, Shariq Suleman, and Jagriti Narang

Department of Biotechnology, Jamia Hamdard University, New Delhi, India

2.1 Introduction

In December 2019, a disease linked to the coronavirus (CoV) was identified in the capital of China's Wuhan [1]. When seen under an electron microscope, CoVs, which are enveloped positive-sense RNA viruses, appear like crown-shaped viruses. There are four subtypes of CoVs such as (a) alpha, (b) beta, (c) delta and, (d) gamma CoV. Coronavirus disease is caused by the extreme acute respiratory syndrome coronavirus 2, which is caused by a beta coronavirus (-CoVs or Beta-CoVs) (SARS-CoV-2). Infected people may have a fever of 38° C, cough, and shortness of breath [2]. WHO officially called COVID-19, an abbreviated form of coronavirus disease 2019, on February 12, 2020. The original name for the virus was 2019-nCoV that caused the COVID-19 pneumonia epidemic. Coronavirus, which resembles SARS (bat-SL-CoVZC45 and bat-SL-CoVZXC21) and was later renamed severe acute respiratory syndrome coronavirus-2, or SARS-CoV-2, owing to its close genetic relation (88% similarity) with two bat-originated viruses. Besides that, the SARS-CoV-2 has a genetic resemblance of 79% to SARS-CoV and 50% to the Middle East Respiratory Syndrome Coronavirus (MERS-CoV) [3]. COVID-19 was declared a pandemic by the WHO on March 11, 2020, indicating a significant risk of global contamination and the need for precautions. Droplets and direct contact are the leading modes of SARS-CoV-2 transmission among humans, as per reports from affected countries [4]. Relevant genes for viral replication, nucleocapsid formation, and spike formation are located in the ORF1 downstream regions of all coronaviruses. The coronavirus's outer surface

Sensing Tools and Techniques for COVID-19: Developments and Challenges in Analysis and Detection of Coronavirus

DOI: https://doi.org/10.1016/B978-0-323-90280-9.00006-7

glycoprotein spikes are responsible for the virus's attachment and entrance into host cells. The virus can infect many hosts because its RBD, that is, the receptor-binding domain is roughly connected. Coronavirus's entry mechanism is dependent upon cellular proteases such as-human airway trypsin-like protease (HAT), cathepsins, and transmembrane protease serine 2 (TMPRSS2), which split the spike protein and trigger more penetration changes. SARS-CoV-2 has the classic coronavirus structure with spike protein and also expresses RNA polymerase, 3-chymotrypsin-like protease, papain-like protease, helicase, glycoprotein, and accessory proteins, among other polyprotein, nucleoproteins, and membrane proteins [5]. To maintain van der Waals forces, the SARS-CoV-2 spike protein has a three-dimensional structure in the receptor-binding domain.

2.1.1 Life cycle of SARS-CoV-2

As the S protein binds to the cellular receptor ACE2, SARS-CoV-2 begins its life cycle in host cells [6]. After receptor binding, the conformation change in the S protein induces viral envelope fusion with the cell membrane via an endosomal pathway. The SARS-CoV-2 then injects RNA into the host cell. Pp1a and 1ab are viral replicase polyproteins that are encoded into genome RNA and then cleaved into small products by viral proteinases. The polymerase produces a sequence of subgenomic mRNAs, which are then converted into viral proteins, using the discontinuous transcription. Virions are formed in the endoplasmic reticulum and Golgi by combining viral proteins and genome RNA, which are then transferred by vesicles and released from the cell [7]. Diagrammatic representation of SARS-CoV-2 life cycle is shown in Fig. 2.1.

Coronavirus has a high transmitting ability which makes it impossible to deter and control [8]. Since there is no clear vaccine/antiviral therapy for coronavirus, it is important to get a diagnosis as soon as possible to decrease the chance of severe difficulties and mortality [9]. When it comes to detection techniques, the whole world depends on a single conventional tool, that is, RT-PCR, this diagnostic test gives various benefits like-reproducibility, accuracy, sensitivity and it detect the virus at the initial stage and can be used on weak immunity patients, although it has disadvantages in that it takes experience and is expensive due to the cumbersome instrument used. One significant disadvantage of this approach is that diverse temperature is needed for diverse reaction cycles, which make it challenging to use at the point-of-care (POC). As a result, POC instruments for on-site diagnosis have become increasingly important [10]. Improvements in POC diagnosis, chip- or paper-based biosensors have been established for quick analysis of transmittable infections. Based on colorimetric, fluorescent, or electrochemical identification methods, they are commonly used to identify- nucleic acids, antibodies/antigens, in basic samples like- blood, saliva and, sputum. They have a variety of benefits,

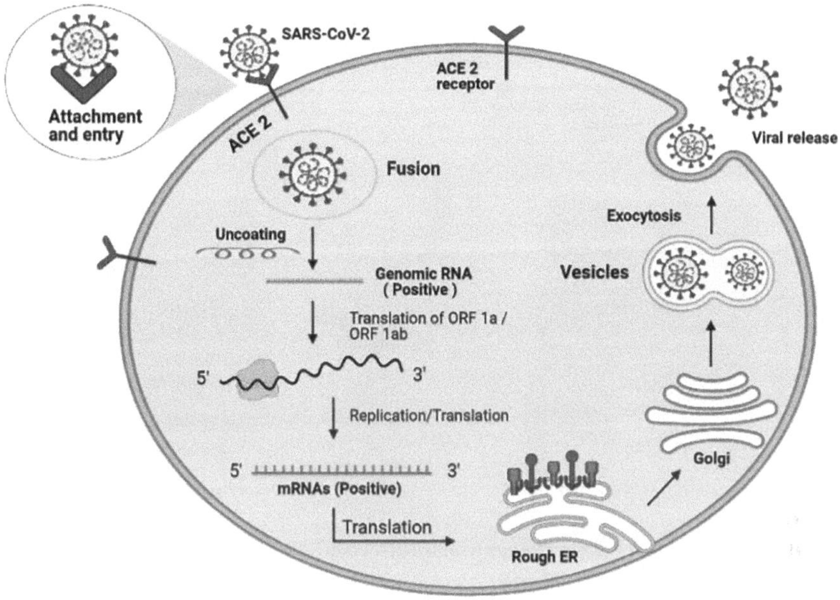

FIGURE 2.1 Entry of SARS-CoV2 inside the cell through receptor, translation, and replication of virus inside the cell, the release of viral components through exocytosis to infect the other cells.

including being cost-effective, responsive, and precise, accessible, fast and durable, less instruments and deliverable to end-users (ASSURED). The outcome can be achieved quickly and easily, allowing for quick decision-making and reducing the chance of human-to-human transmission [11]. The most recent developments in POC biosensors to detect coronavirus are reviewed in this chapter, along with the benefits of lab-on-chip biosensors are also discussed, as well as the commercially available COVID-19 lab-on-chip biosensors. Fig. 2.2 represents the different diagnosis approaches to detect COVID-19 infection.

2.2 LOC biosensor (overview and fabrication)

Amid all analytical methods, the biosensor was confirmed as one of the progressive systems [12]. A biosensor is a bioanalytical instrument that contains a recognition feature as well as a transducing device. The word "biosensor" refers to a common diagnostic method that consists of biocomponents like protein, enzyme, antibody, and so on. Clark and Lyons invented the first biosensor in 1962 [13]. Since then, there has been incredible progress in both science and biosensor use, including emerging technologies including bioelectronics/electrochemistry and nanotechnology. A

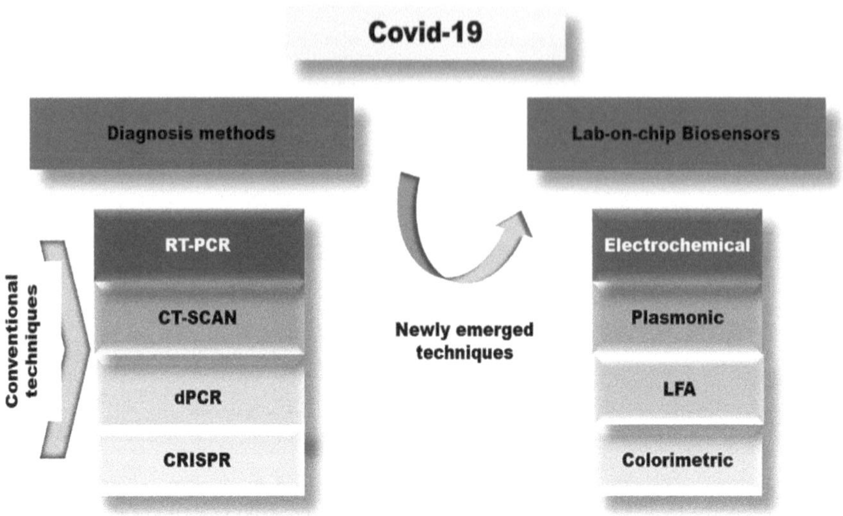

FIGURE 2.2 Different diagnosis approaches of COVID-19.

biosensor is made up of three parts: a recognizing unit, a signal transducer, and a signal amplifier that transmits and demonstrates the data. The recognition element identifies a signal from the surroundings in an analyte form, which is then converted to a measured electric output by the transducer [14]. A well-defined COVID-19 biosensor along with its components is represented in Fig. 2.3. Biosensors should be thought of as an innovative diagnostic instrument with all of the advantages of high sensitivity, accuracy, and low volume. Biology and electronics are combined to make the system more available, digital, and to create a small-scale multiplexed system [15].

Biosensors' enormous promise in medical diagnostics has prompted researchers to evolve this technology and develop novel techniques over time. The popularity of biosensors in medical diagnostics is focused on their ability to identify a wide variety of biomolecules. However, meeting efficiency demands while maintaining simplicity and affordability remains a significant challenge. The ultimate aim is to bring POC research to remote areas around the world, especially in developed countries. It involves continuous technological advancements in multiplexing, fabrication, and miniaturization of biosensor instruments in order to provide lab-on-chip-diagnosis systems to the public [16]. Lab-on-chip technology refers to approaches that conduct numerous laboratory activities on a small scale like-preparation of chemicals and analysis which helps it to make a handheld and portable instrument. To put it another way, a lab-on-chip is a method for scaling one or many laboratory operations down to chip size. This chip can be anything from a few millimeters to a few square centimeters

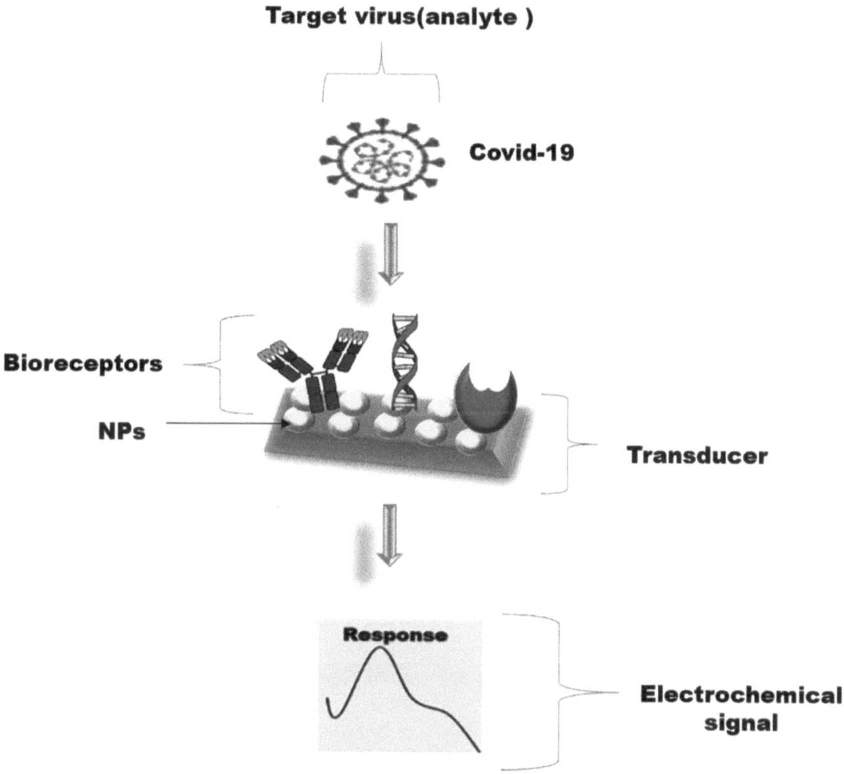

FIGURE 2.3 Diagrammatic representation of COVID-19 biosensor.

in size. The fusion of fluidics, circuitry, optics, and biosensors is known as LOC [17]. Compactness, portability, modularity, reconfigurability, integrated computing, automatic sample handling, low electronic noise, minimal power consumption, and simple component integration are some of the noteworthy technological advantages of lab-on-chip systems. Besides that, lab-on-chip systems can accommodate a broad variety of processes, including sampling, routing, transport, dispensing, and combining, with less moving or rotating parts, extending the device's flexibility and lifetime [18].

2.3 LOC fabrication

Micro and nanoscale LOC applications have been made possible thanks to advancements in microfluidic and microfabrication processes. Based on the liquid propulsion concept, microfluidic-LOC systems are divided into five categories: (a) capillary, (b) pressure-driven, (c) centrifugal, (d) acoustic-driven, and (e) electrokinetic. Various fabrication approaches are used to make lab-on-chip-based devices by using different materials such as polymeric, glass, and, silicon [19]. Similarly, there are numerous fabrication approaches for LOC and paper-based microfluidic tools like-wax printing, alkyl ketene dimer (AKD) printing, flexographic printing, and layer-by-layer 3D affixing, polydimethylsiloxane (PDMS) plotting, wax/inkjet/wax screen printing, wax dipping, plasma treatment, photolithography, deposition, etching, and lift-off lithography [20]. Moreover, a patterning method called soft-lithography is mainly considered for soft materials such as PDMS, Replica molding, microcontact printing, and microtransfer molding are all examples of soft lithography [21]. However, metal lift-off lithography has been generally used for patterning noble metal thin films like- gold, silver, tantalum, nickel, or iron, which are tough to be imprinted by traditional techniques [22]. Another technique used in the fabrication is the screen printing technique. In Fig. 2.4 printing method is described. This method utilizes carbon conductive ink as a substrate.

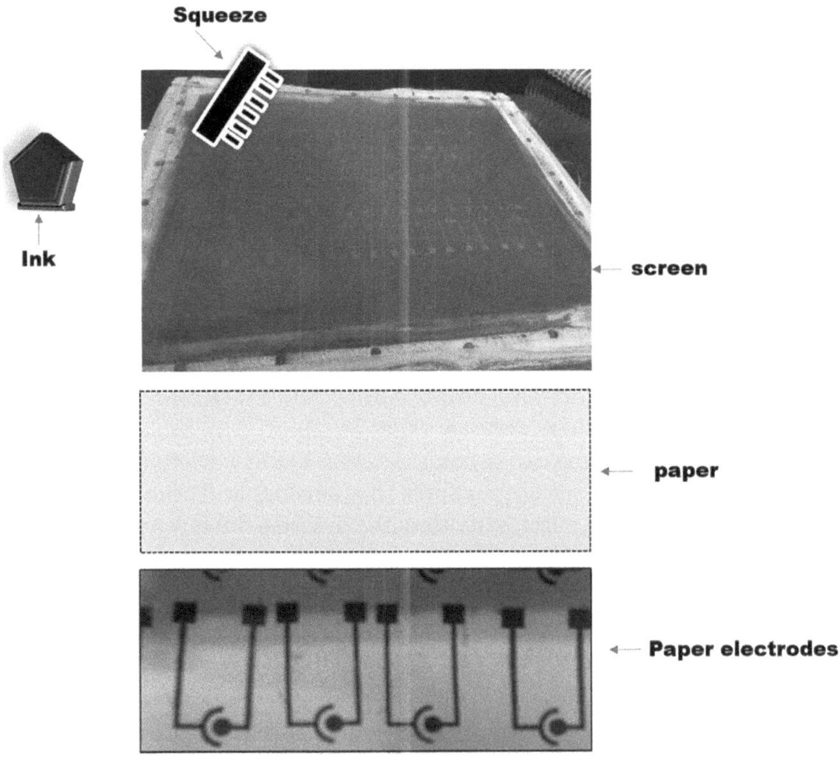

FIGURE 2.4 Fabrication method of lab on chip based biosensor to detect COVID-19.

TABLE 2.1 LOC fabrication materials and methods [12,23,25].

Materials	Techniques of fabrication
Polyurethane	Soft lithography
Polydimethylsiloxane	Soft lithography
Poly(methyl methacrylate)	Simple hot embossing method.
Surface oxidized silicon	Chemical dry etching
Glass	Isotropic wet etching and chemical dry etching
Paper	Screen printing

TABLE 2.2 Commercially available primers/probe of SARS CoV-2.

Gene target	Primer/probe	Sequences (5′-3′)	Developed by	Reference
RdRp	nCoV IP2 Fw	ATGAGCTTAGTCCTGTTG	Institut Pasteur	[22]
	nCoV IP2 Rv	CTCCCTTTGTTGTGTTGT		
	IP2 probe	^{HEX}AGATGTCTTGTGCTGCCGGTA^{BHQ1}		

Carbon conductive ink was squeezed on the screen so that the electrodes were imprinted on the surface of the paper [23]. Table 2.1 represents the fabrication methods along with their materials.

2.4 Conventional method to detect COVID-19

2.4.1 Reverse transcription-PCR

Conventionally, this is a gold standard approach to detect coronavirus [26]. To identify the virus, this molecular detection method evaluates the nucleic acids found in the sample. Real-time RT-PCR reaction is the most widely utilized laboratory identification tool for the clinical detection of viral diseases, like- SARS-CoV and MERS-CoV, which have been diagnosed and monitored using the same technique. Commercially available primers and probes of SARS-CoV-2 for RT-PCR are mentioned in Table 2.2 [27].

2.4.2 Digital PCR (dPCR)

Vogelstein and Kinzler were the first to mention the term "dPCR" in 1999 [29]. The principle of this technique, before amplification its partition the mixture of PCR into various independent subreactions to analyze the original target molecule's numbers by calculating the partitions which show the PCR results (+ve and -ve) and after amplification it determines the Poisson distribution. Commonly, this method can be divided into two terms like- (a) chip-based dPCR (b) droplet-based dPCR [30]. This digital PCR is also suitable for the diagnosis of coronavirus, like- [31] and [32] utilizes d-polymerase chain reaction in context with SARS-CoV-2.

2.4.3 CT-SCAN

This technique is more trustworthy, convenient and faster to classify and assess COVID-19 as compared to PCR. Thorax computer tomography can be used at the initial stage of COVID-19 patients because it is present in almost every clinic. The essential point seen in the images of CT scan is- crazy paving pattern, ground-glass opacities, reticular pattern, and consolidation in COVID-19 patients [33]. Zhao et al. demonstrated the association between the CT-scan outcomes and the medical situations of coronavirus by collecting the data of 101 COVID-19 pneumonia cases from Hunan, China. After collecting, detailing of all the basic imaging features is compared and assessed [34].

2.4.4 CRISPR-based technologies

This method offers an attractive opportunity to permit good alternatives or development to detect COVID-19 by providing a sensitive, specific, and speedy assay. This technique mainly involved the CRISPER-Cas system which used proteins as CRISPR effectors named Cas12/cas13. In this assay these proteins play a very crucial role, which helps in COVID-19 detection [35]. Currently, many researchers used these methods like- Lucia C et al., which stated this CRISPR-Cas 12 assay to diagnose synthetic SARS-Cov-2 RNA sequence [36] (Table 2.3), which comprises all the conventional methods along with their disadvantages.

As a result of the aforementioned drawbacks, these methods are ineffective in situations requiring rapid screening in populated areas where large quantities of samples must be checked.

2.5 Lab-on-chip based biosensors to detect COVID-19

2.5.1 Electrochemical biosensor

Basically, in this device electrochemically transducer is involved along with electrodes that are chemically modified, and the material was used either semiconducting or conducting which is mainly coated with a biochemical film [42]. This device is one of the novel techniques to detect viral antigens/nucleic acids. This electrochemical biosensor mainly focuses on

TABLE 2.3 Limitations of conventional methods.

Conventional methods	Limitations	References
RT-PCR	Requires qualified and educated personnel, as well as facilities, and results take at least 6–24 hours.	[37]
CT-SCAN	Costly, exposes patients to a high dose of radiation, and may be confused, leading to disease-mismanagement.	[38]
dPCR	The major drawbacks of using digital PCR to reliably measuring the proviral DNA or RNA reservoir are false positive results. dPCR assays may be to blame for the sensitivity limitations and cost is high.	[40]
CRISPR	Specificity is very low and a poor selection of delivery tools is all factors that limit its ability to reach its full potential.	[41]

providing a point of care testing (POCT) to the users. There are so many advantages of electrochemical biosensor over conventional as it is cheap, easy to handle, and have time-saving testing. In this current pandemic, this device is really becoming a challenge to detect COVID-19 as it gives a patient –ve or +ve results within minutes or one hour [43]. Many scientists from all over the world used electrochemical biosensors to detect COVID-19 and to make them economically good such as Alafeef et al. used paper in electrochemical biosensor chips to diagnose genetic material SARS-CoV-2 digitally in very little time, that is, 5 minutes. This research is concluded that this paper-based electrochemical chip biosensor is very inexpensive, easily handle, and assessable [44]. To make this testing more advanced Zhao et al. used a smartphone for the detection of SARS-CoV-2 RNA. In this, the author used a super-sandwich type electrochemical biosensor and utilizes calixarene-functionalized GO and the outcome ratios are higher while compare to RT-qPCR, which makes it ultrasensitive testing. In this, author used calixarene-functionalized graphene oxide [45], however, there are so many functionalized materials is used by many authors like-Vadlamani et al. used cobalt-functionalized titanium dioxide nanotubes in an electrochemical biosensor to detect S-receptor binding domain protein (SARS-CoV-2) [46]. To make their research effective and unique scientists try to do new experiments by using different materials and combined with electrochemical biosensors such as Fabiani et al. coupled the magnetic beads with screen printed electrodes which are based on carbon black nanomaterials in an electrochemical immunoassay to detect SARS-CoV-2 (S/N proteins) in the patient saliva [47]. While another author, Chai bun et al. describes an electrochemical biosensor to detect SARS-CoV-2 based on isothermal RCA that is, rolling circle amplification later the amplicons of RCA are hybridized by using probes and functionalized via redox-active labels which can be easily detectable [48]. On the other hand, Funari et al. built an optomicrofluidic chip-based electrochemical biosensor by utilizing gold nano spike for detected antibodies of SARS-CoV-2 spike protein [49].

2.5.2 Plasmonic biosensor

Plasmonic biosensing is one of those methods that have the ability to detect the sample at very low concentrations. In this method, label-free diagnosis of bioanalytical targets can be done in context with a point of care testing [50]. Plasmonic biosensor is also known as SPR that is, surface plasmonic resonance. This device can become an indispensable instrument in the field of medical sciences, which can also be used to diagnose COVID-19 [51]. Using SPR biosensing technology, Masson's research group recently announced the use of human serum samples without dilution for the detection of nucleocapsid antibodies specific for SARS-CoV-2 [52]. This SPR-based sensor was recently identified for detecting nucleocapsid antibodies specific for SARS-CoV-2 in undiluted human serum rather than oropharyngeal swab. Anti-SARS-CoV-2 antibodies were detected in the nM range using an SPR sensor coated with a peptide monolayer and functionalized with SARS-CoV-2 nucleocapsid recombinant protein. As a result, this bioassay is fast and label-free, and it can diagnose samples in as little as 15 minutes after sample or sensor touch [53]. Many scientists work on this plasmonic biosensor to detect COVID-19 as it gives numerous advantages as compared to the RT-PCR such as Peng X et al. developed a near-infrared region plasmonic biosensor to detect SARS-COV-2 along with their S protein and integrated with a telluride-MOS_2 layer with transparent indium tin oxide, which shows great results in term of identification and capturing that helps to detect SARS-CoV-2 and S protein [54]. A plasmonic biosensor is also improved by coupling this biosensor with other methods like Qiu et al., combining these two methods -plasmonic photothermal and localized surface plasmon resonance and term as—dual-functional plasmonic biosensor. The sensitivity is increased by utilizing 2D gold Nis that is, nanoislands, and thermoplasmic heating is done for better results which makes it a novel technique [55]. Pig sera were used by Bong JH et al. to purify antibodies against Coronavirus 2 nucleoprotein (NP) (SARS-CoV-2). After isolation of the antibody fraction using a protein-A axis, the final yield of purified antibodies against SARS-CoV-2 NPs from 1 mL of pig sera was determined to be 0.26–0.05% (absolute total of 143.4 25.2 ng, n = 5). The binding processes of the isolated antibodies were tested using immunoassay and immune-staining. An SPR biosensor with isolated antibodies against SARS-CoV-2 NPs was used to detect SARS-CoV-2, MERS-CoV, and CoV strain 229E in the culture fluid [56].

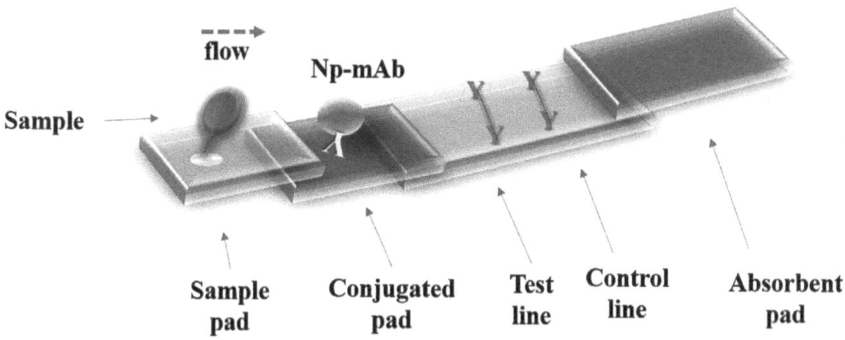

FIGURE 2.5 Diagrammatic representation of lateral flow.

2.5.3 Lateral flow assay (LFA) principle

In 1956, the first time detection (semiquantitative) of glucose in the urine is done by the paper device. One of the popular applications of LFA is the pregnancy kit [57].

Lateral flow tests (LFTs) are an example of a tool that can be used without any special equipment, expertise, or experience [58]. The LFAs/LFTS are well-known POC instruments that have recently emerged as the most striking tools for qualitative, semiquantitative, and even quantitative identification of multiple analytes, particularly biomarkers. The World Health Organization's (WHO) ASSURED guidelines stipulate that all POC devices must meet certain requirements, including being (a) affordable, (b) sensitive, (c) specific, (d) user-friendly, (e) rapid and durable, (f) equipment-free, and (g) delivered to end-users [59]. The cost-effectiveness is achieved primarily by the use of cellulosic materials and plastic supports. Simple process and result explanation, on the other hand, satisfy the requirements of the end-user. Lateral flow assay is very easy for the users. The main protocol to perform this assay is done by adding the sample on the sample pad where it characterized the sample. To make it suitable for the ideal detection after the sample pad, the sample is moved to another pad called a conjugate pad where several interactions take place then the sample is passed to the membrane which mainly consists of two lines that is, test line and control line. At last, the sample reached the absorbent pad where it offers sufficient bed volume to complete the sample flow. After the incubation step, the single line is visible on the pad which defines the +ve/-ve results of this test. A general diagrammatic representation of LFA is shown in Fig. 2.5 [60].

Likewise, this assay is also suitable and simple for the detection of COVID-19 by dropping the COVID patients' blood samples on the pad. Then buffer is used which makes the sample movable. Then the antibodies are bind to the chemicals which are against SARS-CoV-2. As the sample

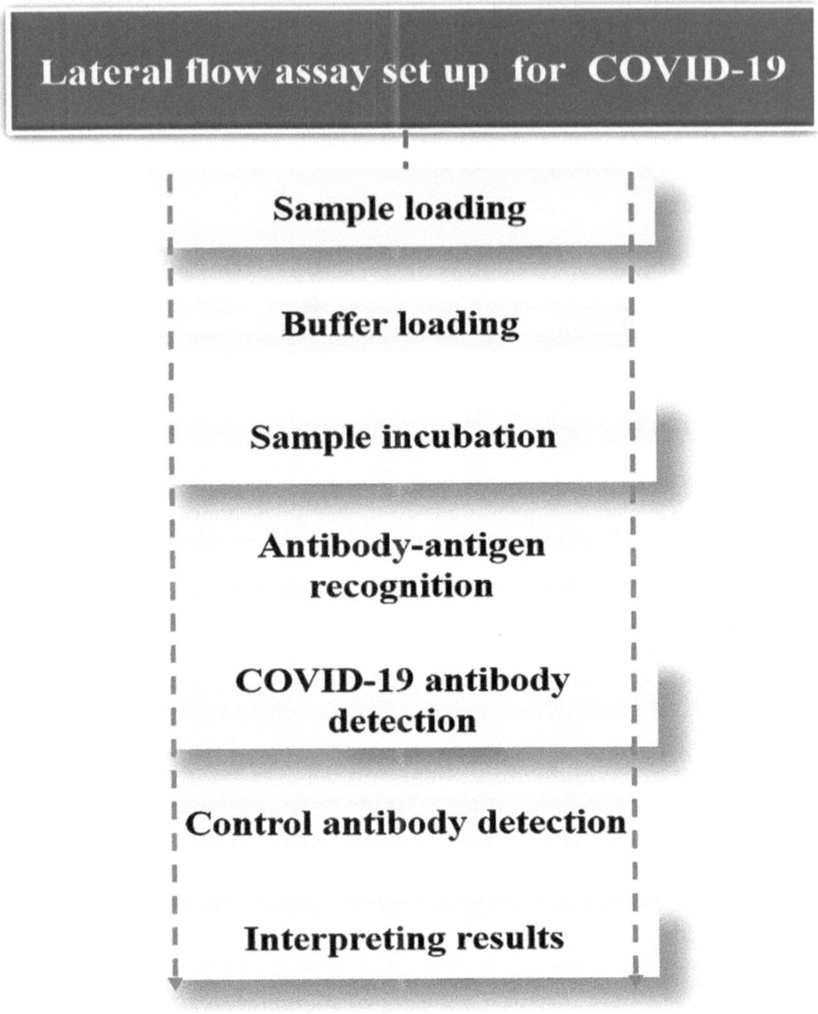

FIGURE 2.6 Flow chart of LFA to detect COVID-19.

moves through the device then the test and control lines (sometimes it may be three lines depending upon the antibodies) capture the antibodies and give color, which interprets the results of this device. The flow chart of LFA set up for the detection of COVID-19 is shown in Fig. 2.6 [61]. Again in 2020, LFA was employed by a group of scientists for detecting COVID-19 antibodies. About 11 LFA were compared and characterized for their performance in a population. Two cohorts were employed—-350 blood donors from past influenza-affected ones and 110 samples from past negative PCR tested ones. Nine out of 11 showed a specificity of less than 88%, while five of them showed sensitivity as same as ELISA. And only one of 11 fulfilled both of the situations. Therefore it was concluded from the study that this needs more negative cohorts and more positive diverse COVID-19 positive samples [62]. The first time a surface-enhanced Raman scattering-based LFA was utilized to diagnose COVID-19 by Liu H et al. As high-performance SERS tags, a new SiO2@Ag nanocomposite labeled with dual layers of DTNB was used. With the SERS-LFIA, anti-SARS-CoV-2 IgM and IgG can be detected simultaneously and with high sensitivity. The SERS-LFIA was shown to be 100% accurate and precise in a study of 68 clinical serum samples [63]. Broughton JP et al. reported the rapid diagnosis of SARS-CoV-2 with the help of CRISPR (clustered regularly interspaced short palindromic repeats)-based DETECTR method followed by LFA assay. Sampling is done by extracting the RNA from the respiratory swab. The RT-PCR method is also used to compare the results from both methods. (SARS-CoV-2 DETECTR, RT-LAMP/Cas12 – LOD – 10 copies/μL input, CDC SARS-CoV-2CoV-2 q RT-PCR–4.2 copies/μL input) [64].

2.5.4 Colorimetric assay

It is one of the simple and cheap methods of detection. Hence, it can make very easy on-site/real-time diagnosis [65]. Chow FW et al. are diagnosed SARS-CoV-2 with the help of a colorimetric assay that leads to color change, and then results can be seen via naked eyes, for this, it established a one-step colorimetric reverse-transcriptional loop-mediated isothermal amplification assay (COVID-19-LAMP) and detect SARS-CoV-2 with the LOD-42 copies per reaction [66]. Lalli MA et al. are demonstrated a rapid and RNA extraction-free detection of SARS CoV-2 in the human samples with the help of colorimetric RT-LAMP (reverse- transcription loop-mediated isothermal amplification) with the limit of detection- 59 particle copies per reaction. This research also highlights the flexibility of LAMP assay implementation using three outcomes- (a) Naked-eye colorimetric, (b) spectrometry, and (c) real-time fluorescence [67]. Garcia et al. introduced a versatile, colorimetric RT-lamp assay for the detection of SARS-CoV-2 RNA in samples by targeting gene N through a specific sequences primer set N15 (sequences are presented in Table 2.4). The results can be diagnosed as soon as 9 minutes after the reaction begin and obtain close to 100% sensitivity within 60 minutes with LOD of 200 copies of SARS-CoV-2 RNA/reaction and study was compared with RT-qPCR [68]. Nawattanapaiboon et al. developed a one-pot visual diagnosis platform to detect coronavirus SARS-CoV-2 with the help of colorimetric RT-LAMP with the LOD- 250 copies of SARS-CoV-2 RNA/reaction and this assay require only 1 hour throughout the practical along with 99.86% accuracy [69]. Gonzalez et al. are developed a colorimetric LAMP assay to detect SARS-CoV-2 nucleic acids in the range of 62-2 $\times$ 10^5 DNA copies. They utilize synthetic samples (with 100%

TABLE 2.4 Primer set for the detection for SARS CoV-2.

Set	Primer[1]	Sequence (5′→3′)	Length (nt)	Gene position	Reference
N15[4]	F3	AGATCACATTGGCACCCG	18	28,703–28,915	[68]
	B3	CCATTGCCAGCCATTCTAGC	20		
	FIP	TGCTCCCTTCTGCGTAGAAGCCA-ATGCTGCAATCGTGCTAC	41		
	BIP	GGCGGCAGTCAAGCCTCTTCCCT-ACTGCTGCCTGGAGTT	39		
	LF	GCAATGTTGTTCCTTGAGGAAGTT	24		
	LB	CCTCATCACGTAGTCGCAACAG	22		

accuracy) and RNA extracts from the COVID patients (sensitivity- 92.85%, specificity- 81.25%) [70]. Another author Moon et al. used this colorimetric assay to detect SARS-CoV-2 NI, N2, N3 genes, and drug-resistance pHINI by utilizing CRISPER/dCAS-9 [70].

Various biosensors based on lab on a chip for the diagnosis of COVID-19 are summarized in Table 2.5, along with their strengths and limitations.

2.6 Conclusion and future aspects

Diagnosis is a big issue in today's world, particularly in heavily populated countries. If the COVID-19 disease has been identified, the disease's treatment and spread can be effectively managed. As a result, there is a crucial need to create a low-cost new technique, such as lab-on-chip biosensors, for a quick point of care diagnosis, so that patients can quickly quarantine/self-isolate, significantly reducing the risk of further infection. While neither traditional methods (qRT-PCR) nor lab-on-chip based biosensors is perfect instruments for COVID-19 diagnosis, these techniques should be used in tandem. Lab-on-chip biosensors, such as LFA, electrochemical, colorimetric approaches, on the other hand, are more compact, allowing for field use since they do not necessitate the use of laboratory equipment. Rather than nucleic acids, these systems seem to rely on detecting antibodies to and antigens upon the virus itself. On a small platform, these innovations demonstrate promise as a replacement for serology studies. On the other hand, plasmonic biosensors require a good laboratory setup but offer greater productivity, permitting the detection of several samples at the same time. The electrochemical and plasmonic assays have the quickest readout among all the lab-on-chip

TABLE 2.5 Various types of labs-on-chip based biosensor along with their strengths and limitations.

	Electrochemical biosensor,	Plasmonic biosensor	Lateral flow assay	Colorimetric assay
Strengths	– Electrochemical biosensors will deliver a more affordable price per analysis than RT-PCR.	– Miniaturization, low cost, and user friendliness are all trends in SPR sensor interface design. It enables label-free, high-reliability, sensitivity, and real-time measurement.	– Very low relative cost, can be done at the point of treatment or at home, is simple to use, and produces fast results.	– Easy operation – Rapid detection. – Highly stable at room temperature.
Limitations	– The manufacturing of biosensors takes a longer time. – In case of multiple sequential reactions, the biosensor response time often increases.	– Small analytes are not appropriate for examination with the SPR. Rather small analytes yield very small responses since the SPR measures the mass of the compounds that bind to the sensor surface.	– Extensive monitoring of cross-reactivity with other immune responses is needed, as is variation in test specificity between manufacturers. – The technology is latest, and the evidence for its accuracy in diagnosing coronaviruses is still being assessed.	– Requires careful cross-reactivity testing with other immune responses, as well as a laboratory setting. – Limited sensitivity

TABLE 2.6 Summary of LOC biosensor based on their principle/transducer for COVID-19 detection.

Lab-on-chip biosensors	Principle/transducer	Detection	Response time	Limit of detection	References
Electrochemical	Paper-based electrochemical sensor chip	SARS-CoV-2	5 minutes	6.9 copies/μL	[68]
	Isothermal rolling circle amplification (RCA)	SARS-COV-2 (N and S genes)	Less than 2 hours	1 copy/μL	[48]
	Cobalt-functionalized TiO2 nanotubes (Co-TNTs)	SARS-CoV-2 (spike -RBD)			[46]
	Electrochemical immunoassay	SARS-CoV-2 (Spike and nucleocapsid)	30 minutes	19 ng/mL for S and 8 ng/mL for N protein	[47]
	Ultrasensitive electrochemical	SARS-CoV-2 RNA		200 copies/mL	[34]
	Optomicrofluidic chip based electrochemical biosensor	Antibody of SARS-CoV-2 spike protein	2 hours	~0.5 p.m	[49]
Plasmonic	A novel NIR plasmonic biosesnor by combining promising 2D tellurene-MoS2-COOH van der Waals heterostructure with plasmonic ITO film	SARS-CoV-2 and its S protein		Linear detection range or S and SARS-CoV-2 –s ~0–301.67 nM and ~0–67.8762 nM respectively.	[54]
	A dual-functional plasmonic biosensor combining the plasmonic photothermal (PPT) effect and localized surface plamon resonance (LSPR) sensing transduction	SARS-CoV-2		0.22 pM	[55]
	SPR biosensor	Coronavirus 2 nucleoprotein		1.02 pM	[56]

(continued on next page)

LFA	Clinical characterization of Eleven LFA	SARS-CoV-2			[47]
	SERS-based LFA	Covid-19		LOD is 800 times higher than that of standard Au NPS-based LFIA for target IgM/IgG. And the LOD of SERS is 1 pg per mL	[63]
	CRISPER-cas12 based LFA	SARS-CoV-2	30 minutes	10 copies per μL input	[64]
Colorimetric	A one step colorimetric reverse-transcriptional loop-mediated isothermal amplification assay	SARS-CoV-2 RNA		42 copies	[66]
	RT-LAMP	SARS-CoV-2		59 copies	[67]
	RT-LAMP	ORF1ab and N genes	60 minutes	200 copies of SARS-CoV-2 RNA/reaction	[68]
	RT-LAMP	RNA- dependent RNA polymerase gene (RdRp)	60 minutes	125 copies per reaction of SARS-CoV-2 RNA	[69]
	LAMP	N protein	50 minutes	62-2 $\times 10^5$ DNA copies	[70]
	CRISPER/d cas-9	N1.n2.n3 genes	~2 hours	30 pM	[71]

biosensor technologies, however, the LFA and colorimetric biosensor takes time to develop a response perception diagnosis.

References

[1] K. Yuki, M. Fujiogi, S. Koutsogiannaki, COVID-19 pathophysiology: A review, Clin. Immunol. 215 (2020) 108427.

[2] S. Shafiee, S. Davaran, A mini-review on the current COVID-19 treatments, Chemical Review and Letters 3 (1) (2020 Jan 1) 19–22.

[3] C.-C. Lai, T.-P. Shih, P.-R. Hsueh, H.-J. Tang, W.-C. Ko, Severe acute respiratory syndrome coronavirus 2 (SARS-CoV-2) and coronavirus disease 2019 (COVID-19): The epidemic and the challenges, Int. J. Antimicrob. Agents 55 (3) (2020) 105924.

[4] F. Wu, H. Yi, Z. Su, S. Zhi-Gang, W. Wen, C. Yan-Mei, A new coronavirus associated with human respiratory disease in China, Nature 579 (7798) (2020) 265–269.

[5] I. Glowacka, et al., Evidence that TMPRSS2 activates the severe acute respiratory syndrome coronavirus spike protein for membrane fusion and reduces viral control by the humoral immune response, J. Virol. 85 (9) (2011) 4122–4134.

[6] P. V'kovski, A. Kratzel, S. Steiner, H. Stalder, V. Thiel, Coronavirus biology and replication: implications for SARS-CoV-2, Nat. Rev. Microbiol. 19 (3) (2021 Mar) 155–170.

[7] M.A. Shereen, et al., COVID-19 infection: Origin, transmission, and characteristics of human coronaviruses, J. Adv. Res. 24 (2020) 91–98.

[8] S. Chen, J. Yang, W. Yang, C. Wang, T. Bärnighausen, COVID-19 control in China during mass population movements at New Year, The Lancet 395 (10226) (2020) 764–766.

[9] C. Zhou, G. Chen, X. Yuanliang, X. Maosheng, et al., COVID-19 with spontaneous pneumomediastinum, Lancet Infect. Dis. 20 (4) (2020) 510.

[10] S. Suleman, S.K. Shukla, N. Malhotra, S.D. Bukkitgar, N.P. Shetti, R. Pilloton, et al., Point of care detection of COVID-19: Advancement in biosensing and diagnostic methods, Chem. Eng. J. 414 (2021) 128759.

[11] J.R. Choi, Development of point-of-care biosensors for COVID-19, Frontiers in chemistry 8 (2020) 517.

[12] M.R. Hasan, P. Sharma, N. Anzar, C.S. Pundir, R. Pilloton, J. Narang, et al., Analytical methods for detection of human cytomegalovirus clinched biosensor a cutting-edge diagnostic tool, Biomedical Engineering Advances 1 (2021) 100006.

[13] P.R. Coulet, What is a biosensor, Biosensor Principles and Applications, Boca Raton, 1991, pp. 1–6. (Book Reference).

[14] N. Anzar, M.R. Hasan, M. Akram, N. Yadav, J. Narang, Systematic and validated techniques for the detection of ovarian cancer emphasizing the electro-analytical approach, Process Biochem. 94 (2020 Apr 13) 126–135.

[15] M.A. Alam, M.R. Hasan, N. Anzar, S. Suleman, J. Narang, Diagnostic approaches for the rapid detection of Zika virus-A review, Process Biochem. 101 (2020 Nov 15) 156–168.

[16] S. Patel, R. Nanda, S. Sahoo, E. Mohapatra, Biosensors in health care: the milestones achieved in their development towards lab-on-chip-analysis, Biochem. Res. Int. 2016 (2016 Mar 3) 1–12.

[17] S. Gupta, K. Ramesh, S. Ahmed, V. Kakkar, Lab-on-chip technology: A review on design trends and future scope in biomedical applications, Int. J. Biosci. 8 (2016) 311–322.

[18] A.T. Giannitsis, Microfabrication of biomedical lab-on-chip devices. A review, Est. J. Eng. 17 (2) (2011 Jun 1) 109.

[19] C. Tymm, J. Zhou, A. Tadimety, A. Burklund, J.X. Zhang, Scalable COVID-19 detection enabled by lab-on-chip biosensors, Cell. Mol. Bioeng. 13 (4) (2020 Aug) 313–329.

[20] Q. Hui, Y. Pan, Z. Yang, Paper based devices for rapid diagnostics and testing sewage for early warning of COVID-19 outbreak, Case Studies in Chemical and Environmental Engineering 2 (2020) 100064.

[21] J.X. Zhang, K. Hoshino, Molecular Sensors and Nanodevices: Principles, Designs and Applications in Biomedical Engineering, Academic Press, United states, 2018 Nov 19.

[22] L.Y. Wing, Z. ShiYun, C. Pang, M. Hazren, S.A. Kumar, W.C. Chung, Optimization studies of lift-off methods and its application in electrochemical biosensors, in: In2014 IEEE 16th Electronics Packaging Technology Conference (EPTC), IEEE, 2014 Dec 3, pp. 768–773.

[23] A. Gautam, C. Singhal, A. Mishra, A.T. John, U.K. Gautam, R. Abolhassani, R. Adelung, D. Kumar Avasthi, Y. Kumar Mishra, Detection of prostate cancer DNA using tetrapods based disposable paper ecofriendly biosensor device, Medical Devices & Sensors 3 (6) (2020 Dec) e10122.

[24] M.R. Hasan, N. Anzar, M. Tyagi, N. Yadav, J. Narang, Lab-on-a-chip devices—Advancement in the designing of biosensors, Functionalized Nanomaterials Based Devices for Environmental Applications, Elsevier, Netherlands, 2021 Jan 1, pp. 175–198.

[25] D.U. Xiaoguang, G. Yanxia, W. Furen, F. Zhaolun, Fabrication of poly (methyl methacrylate)(PMMA) microfluidic chips by a simple hot embossing method, Chemical Journal of Chinese Universities 24 (11) (2003 Jan 1) 1962–1966.

[26] K.I. Kamei, Y. Mashimo, Y. Koyama, C. Fockenberg, M. Nakashima, M. Nakajima, J. Li, Y. Chen, 3D printing of soft lithography mold for rapid production of polydimethylsiloxane-based microfluidic devices for cell stimulation with concentration gradients, Biomed. Microdevices 17 (2) (2015 Apr) 1–8.

[27] A. Yakoh, U. Pimpitak, S. Rengpipat, N. Hirankarn, O. Chailapakul, S. Chaiyo, based electrochemical biosensor for diagnosing COVID-19: Detection of SARS-CoV-2 antibodies and antigen, Biosens. Bioelectron. 176 (2021 Mar 15) 112912.

[28] B. Giri, S. Pandey, R. Shrestha, K. Pokharel, F.S. Ligler, B.B. Neupane, Review of analytical performance of COVID-19 detection methods, Analytical and bioanalytical chemistry 413 (1) (2021 Jan) 35–48.

[29] R.H. Sedlak, K.R. Jerome, Viral diagnostics in the era of digital polymerase chain reaction, Diagn. Microbiol. Infect. Dis. 75 (1) (2013 Jan 1) 1–4.

[30] Tan C, Fan D, Wang N, Wang F, Wang B, Zhu L, Guo Y. Applications of digital PCR in COVID-19 pandemic. View.:20200082.

[31] P. Poggio, P. Songia, C. Vavassori, V. Ricci, C. Banfi, S.S. Barbieri, G. Garoffolo, V.A. Myasoedova, L. Piacentini, A. Raucci, A. Scopece, Digital PCR for high sensitivity viral detection in false-negative SARS-CoV-2 patients, Sci. Rep. 11 (1) (2021 Feb 22) 1–7.

[32] Y.I. Wong, S.Y. Lau, K.K. To, B.W. Mok, P. Wang, S. Deng, et al., Natural transmission of bat-like severe acute respiratory syndrome coronavirus 2 without proline-arginine-arginine-alanine variants in coronavirus disease 2019 patients, Clin. Infect. Dis. 73 (2020) 437–444.

[33] V. Shah, R. Keniya, A. Shridharani, M. Punjabi, J. Shah, N. Mehendale, Diagnosis of COVID-19 using CT scan images and deep learning techniques, Emerg. Radiol. 28 (3) (2021) 1–9.

[34] W. Zhao, Z. Zhong, X. Xie, Q. Yu, J. Liu, Relation between chest CT findings and clinical conditions of coronavirus disease (COVID-19) pneumonia: a multicenter study, Am. J. Roentgenol. 214 (5) (2020 May) 1072–1077.

[35] R. Nouri, Z. Tang, M. Dong, T. Liu, A. Kshirsagar, W. Guan, CRISPR-based detection of SARS-CoV-2: A review from sample to result, Biosens. Bioelectron. 178 (2021) 113012.

[36] C. Lucia, P.B. Federico, G.C. Alejandra, An ultrasensitive, rapid, and portable coronavirus SARS-CoV-2 sequence detection method based on CRISPR-Cas12, BioRxiv (2020 Jan 1) 1–10.

[37] S.D. Bukkitgar, N.P. Shetti, T.M. Aminabhavi, Electrochemical investigations for COVID-19 detection-A comparison with other viral detection methods, Chem. Eng. J. 420 (2020) 127575 Nov 2.

[38] D.L. Lubell, Drawbacks and limitations of computed tomography, Tex. Heart Inst. J. 32 (2) (2005) 250.

[39] A.A. Kojabad, M. Farzanehpour, H.E. Galeh, R. Dorostkar, A. Jafarpour, M. Bolandian, M.M. Nodooshan, Droplet digital PCR of viral DNA/RNA, Current progress, challenges, and future perspectives, J. Med. Virol. 93 (7) (2021 Feb 4) 4182–4197.

[40] Liu H, dwita Lestari S. Application of digital PCR (dPCR) in the detection of Covid-19 in food. InE3S Web of Conferences 2021 May (Vol. 271, p. 02022). EDP Sciences.

[41] N. Munawar, A. Ahmad, CRISPR/Cas System: An Introduction, CRISPR Crops, Springer, Singapore, 2021, pp. 1–35.

[42] M.G. Dastidar, S. Roy, Public health management during COVID-19 and applications of point-of-care based biomolecular detection approaches, Environmental and Health Management of Novel Coronavirus Disease (COVID-19), Academic Press, United States, 2021 Jan 1, pp. 345–378.

[43] J. Kudr, P. Michalek, L. Ilieva, V. Adam, O. Zitka, COVID-19: a challenge for electrochemical biosensors, TrAC Trends Anal. Chem. 136 (2021 Jan 18) 116192 116192.

[44] M. Alafeef, K. Dighe, P. Moitra, D. Pan, Rapid, ultrasensitive, and quantitative detection of SARS-CoV-2 using antisense oligonucleotides directed electrochemical biosensor chip, ACS nano 14 (12) (2020 Oct 20) 17028–17045.

[45] H. Zhao, F. Liu, W. Xie, T.C. Zhou, J. OuYang, L. Jin, H. Li, C.Y. Zhao, L. Zhang, J. Wei, Y.P. Zhang, Ultrasensitive supersandwich-type electrochemical sensor for SARS-CoV-2 from the infected COVID-19 patients using a smartphone, Sens. Actuators B 327 (2021 Jan) 128899.

[46] B.S. Vadlamani, T. Uppal, S.C. Verma, M. Misra, Functionalized TiO2 nanotube-based electrochemical biosensor for rapid detection of SARS-CoV-2, Sensors 20 (20) (2020 Jan) 5871.

[47] L. Fabiani, M. Saroglia, G. Galatà, R. De Santis, S. Fillo, V. Luca, G. Faggioni, N. D'Amore, E. Regalbuto, P. Salvatori, G. Terova, Magnetic beads combined with carbon black-based screen-printed electrodes for COVID-19: A reliable and miniaturized electrochemical immunosensor for SARS-CoV-2 detection in saliva, Biosens. Bioelectron. 171 (2021) 112686.

[48] T. Chaibun, J. Puenpa, T. Ngamdee, N. Boonapatcharoen, P. Athamanolap, A.P. O'Mullane, S. Vongpunsawad, Y. Poovorawan, S.Y. Lee, B. Lertanantawong, Rapid electrochemical detection of coronavirus SARS-CoV-2, Nat. Commun. 12 (1) (2021 Feb 5) 1–10.

[49] R. Funari, K.Y. Chu, A.Q. Shen, Detection of antibodies against SARS-CoV-2 spike protein by gold nanospikes in an opto-microfluidic chip, Biosens. Bioelectron. 169 (2020 Dec 1) 112578.

[50] J.R. Mejía-Salazar, O.N. Oliveira Jr, Plasmonic biosensing: focus review, Chem. Rev. 118 (20) (2018 Sep 24) 10617–10625.

[51] M. Asif, M. Ajmal, G. Ashraf, N. Muhammad, A. Aziz, T. Iftikhar, J. Wang, H. Liu, The role of biosensors in COVID-19 outbreak, Current Opinion in Electrochemistry 23 (2020 Sep 18) 174–184.

[52] A. Djaileb, B. Charron, M.H. Jodaylami, V. Thibault, J. Coutu, K. Stevenson, S. Forest, L.S. Live, D. Boudreau, J.N. Pelletier, J.F. Masson, A rapid and quantitative serum test for SARS-CoV-2 antibodies with portable surface plasmon resonance sensing, ChemRvix 15 (2020 Apr 15) 10–26434 April.

[53] R. Samson, G.R. Navale, M.S. Dharne, Biosensors: frontiers in rapid detection of COVID-19, 3 Biotech 10 (9) (2020 Sep) 1–9.

[54] X. Peng, Y. Zhou, K. Nie, F. Zhou, Y. Yuan, J. Song, J. Qu, Promising near-infrared plasmonic biosensor employed for specific detection of SARS-CoV-2 and its spike glycoprotein, New J. Phys. 22 (10) (2020 Oct 21) 103046.
[55] G. Qiu, Z. Gai, Y. Tao, J. Schmitt, G.A. Kullak-Ublick, J. Wang, Dual-functional plasmonic photothermal biosensors for highly accurate severe acute respiratory syndrome coronavirus 2 detection, ACS nano 14 (5) (2020 Apr 13) 5268–5277.
[56] J.H. Bong, T.H. Kim, J. Jung, S.J. Lee, J.S. Sung, C.K. Lee, M.J. Kang, H.O. Kim, J.C. Pyun, Pig Sera-derived Anti-SARS-CoV-2 Antibodies in Surface Plasmon Resonance Biosensors, Biochip Journal 14 (4) (2020 Dec) 358–368.
[57] D.D. Liana, B. Raguse, J.J. Gooding, E. Chow, Recent advances in paper-based sensors, sensors 12 (9) (2012 Sep) 11505–11526.
[58] M. Pohanka, Point-of-Care Diagnoses and Assays Based on Lateral Flow Test, Int. J. Environ. Anal. Chem. 2021 (2021 Jan 20) 1–9.
[59] T. Mahmoudi, M. de la Guardia, B. Baradaran, Lateral flow assays towards point-of-care cancer detection: A review of current progress and future trends, TrAC Trends Anal. Chem. 125 (2020 Apr 1) 115842.
[60] C. Parolo, A. Sena-Torralba, J.F. Bergua, E. Calucho, C. Fuentes-Chust, L. Hu, L. Rivas, R. Álvarez-Diduk, E.P. Nguyen, S. Cinti, D. Quesada-González, Tutorial: design and fabrication of nanoparticle-based lateral-flow immunoassays, Nat. Protoc. 15 (12) (2020 Dec) 3788–3816.
[61] K. Green, A. Winter, R. Dickinson, S. Graziadio, R. Wolff, S. Mallett, A.J. Allen, What tests could potentially be used for the screening, diagnosis and monitoring of COVID-19 and what are their advantages and disadvantages, CEBM2020 13 (2020 Apr).
[62] F. Rudolf, H.M. Kaltenbach, J. Linnik, M.T. Ruf, C. Niederhauser, B. Nickel, D. Gygax, M. Savic, Clinical characterisation of eleven lateral flow assays for detection of COVID-19 antibodies in a population, MedRxiv (2020 Jan 1).
[63] H. Liu, E. Dai, R. Xiao, Z. Zhou, M. Zhang, Z. Bai, Y. Shao, K. Qi, J. Tu, C. Wang, S. Wang, Development of a SERS-based lateral flow immunoassay for rapid and ultra-sensitive detection of anti-SARS-CoV-2 IgM/IgG in clinical samples, Sens. Actuators B 329 (2021 Feb 15) 129196.
[64] Broughton JP, Deng X, Yu G, Fasching CL, Singh J, Streithorst J, Granados A, Sotomayor-Gonzalez A, Zorn K, Gopez A, Hsu E. Rapid Detection of 2019 Novel Coronavirus SARS-CoV-2 Using a CRISPR-based DETECTR Lateral Flow Assay. 2020 Mar; 10: 1–27.
[65] J. Liu, Y. Lu, A colorimetric lead biosensor using DNAzyme-directed assembly of gold nanoparticles, J. Am. Chem. Soc. 125 (22) (2003 Jun 4) 6642–6643.
[66] F.W. Chow, T.T. Chan, A.R. Tam, S. Zhao, W. Yao, J. Fung, F.K. Cheng, G.C. Lo, S. Chu, K.L. Aw-Yong, J.Y. Tang, A rapid, simple, inexpensive, and mobile colorimetric assay COVID-19-LAMP for mass on-site screening of COVID-19, Int. J. Mol. Sci. 21 (15) (2020 Jan) 5380.
[67] M.A. Lalli, J.S. Langmade, X. Chen, C.C. Fronick, C.S. Sawyer, L.C. Burcea, M.N. Wilkinson, R.S. Fulton, M. Heinz, W.J. Buchser, R.D. Head, Rapid and extraction-free detection of SARS-CoV-2 from saliva by colorimetric reverse-transcription loop-mediated isothermal amplification, Clin. Chem. 67 (2) (2021 Feb) 415–424.
[68] J. García-Bernalt Diego, P. Fernández-Soto, M. Domínguez-Gil, M. Belhassen-García, J.L. Bellido, A. Muro, A Simple, Affordable, Rapid, Stabilized, Colorimetric, Versatile RT-LAMP Assay to Detect SARS-CoV-2, Diagnostics 11 (3) (2021 Mar) 438.
[69] K. Nawattanapaiboon, E. Pasomsub, P. Prombun, A. Wongbunmak, A. Jenjitwanich, P. Mahasupachai, P. Vetcho, C. Chayrach, N. Manatjaroenlap, C. Samphaongern, T. Watthanachockchai, Colorimetric reverse transcription loop-mediated isothermal amplification (RT-LAMP) as a visual diagnostic platform for the detection of the emerging coronavirus SARS-CoV-2, Analyst 146 (2) (2021) 471–477.

[70] E. González-González, I.M. Lara-Mayorga, I.P. Rodríguez-Sánchez, Y.S. Zhang, S.O. Martínez-Chapa, G. Trujillo-de Santiago, M.M. Alvarez, Colorimetric loop-mediated isothermal amplification (LAMP) for cost-effective and quantitative detection of SARS-CoV-2: the change in color in LAMP-based assays quantitatively correlates with viral copy number, Anal. Methods 13 (2) (2021) 169–178.
[71] J. Moon, H.J. Kwon, D. Yong, I.C. Lee, H. Kim, H. Kang, E.K. Lim, K.S. Lee, J. Jung, H.G. Park, T. Kang, Colorimetric detection of sars-cov-2 and drug-resistant ph1n1 using crispr/dcas9, ACS sensors 5 (12) (2020 Dec 3) 4017–4026.
[72] J. Kudr, P. Michalek, L. Ilieva, V. Adam, O. Zitka, COVID-19: a challenge for electrochemical biosensors, TrAC Trends Anal. Chem. (2021 Jan 18) 116192.
[73] I.S. Kucherenko, O.O. Soldatkin, S.V. Dzyadevych, A.P. Soldatkin, Electrochemical biosensors based on multienzyme systems: Main groups, advantages and limitations–A review, Anal. Chim. Acta 1111 (2020 May 15) 114–131.
[74] S.A. Abid, A. Suhail, I.M. Al-Kadmy, A.A. Sattar, A.M. Beshbishy, G.E. Batiha, et al., Biosensors as a future diagnostic approach for COVID-19, Life Sci. 273 (Jan 26) (2021) 119117.
[75] K. Green, A. Winter, R. Dickinson, S. Graziadio, R. Wolff, S. Mallett, A.J. Allen, What tests could potentially be used for the screening, diagnosis and monitoring of COVID-19 and what are their advantages and disadvantages, CEBM2020 13 (2020 Apr).
[76] F. Charbgoo, M. Ramezani, M. Darroudi, Bio-sensing applications of cerium oxide nanoparticles: advantages and disadvantages, Biosens. Bioelectron. 96 (2017 Oct 15) 33–43.

CHAPTER

3

Role of nanotechnology for coronavirus detection

Amit Mall[a], *Trupti R. Das*[a,*], *Prama Bhattacharjee*[b], *Sweety Pal*[c], *Shiva Sharma*[c], *Sudheesh K. Shukla*[c,d], *and Santanu Patra*[e,**]

[a]CIPET: Institute of Petrochemicals Technology (IPT) - Bhubaneswar, Patia, Bhubaneswar, India, [b]Department of Chemistry, Indian Institute of Technology (ISM), Dhanbad, Jharkhand, India, [c]Department of Biomedical Engineering, School of Biological Engineering and Life Science, Shobhit Institute of Engineering & Technology (Deemed-to-be University), Modipuram, Meerut, India, [d]Department of Chemical Sciences, University of Johannesburg, Doornfontein Campus, Johannesburg, 2028, South Africa, [e]Institute of Advanced Materials, IAAM, Ulrika Sweden

*CIPET: Institute of Petrochemicals Technology (IPT) – Bhubaneswar, B-25 CNI Complex, Patia, Bhubaneswar-751024, India

**Institute of Advanced Materials, IAAM, Ulrika Sweden

3.1 Introduction

The world has been challenged by many pandemics regularly from the past. Major pandemics such as flu, plague, smallpox, SARS, MARS, EBOLA, etc. have already grievously troubled humanity. The humanity is now facing novel coronavirus disease pandemic, which is caused by severe acute respiratory syndrome coronavirus-2 (SARS-CoV-2). It has caused more mortality due to its high infectivity rate, lack of specific treatment, drugs and standard healthcare facilities. On January 30, 2020,

Sensing Tools and Techniques for COVID-19: Developments and Challenges in Analysis and Detection of Coronavirus
DOI: https://doi.org/10.1016/B978-0-323-90280-9.00005-5

the novel coronavirus infection was declared as a "public health emergency of international concern" and this disease was named coronavirus disease 2019 (COVID-19).The outbreak of COVID-19 disease was initially noticed in a seafood market in WUHAN city in Hubei province of China in mid-December. The disease has since spread worldwide, leading to an ongoing pandemic. Now it has spread to 221 countries and territories around the world. It was first isolated from three people with pneumonia connected to the cluster of acute respiratory illness. Later, a new strain of coronavirus belongs to the broad family of coronaviruses which was subsequently isolated from bronchialveolar large fluid. The COVID-19 disease is globally resulted in 245,373,039 confirmed cases, including 4,979,421 deaths, reported to World Health Organization (WHO: as of 4:40 pm CEST, October 29, 2021). COVID-19 was proclaimed as pandemic on March 11, 2020 by WHO. It is a deadly infectious disease. The virus mainly transmits from person to person through the droplets from saliva and respiratory secretions of an infected person. Once the virus enters the respiratory tract, it causes upper respiratory infection and subsequently it affects the lower respiratory tract and lungs causing pneumonia [1–5].

The human being can be infected by COVID-19 through different ways. Most infected people would develop mild to moderate illness and recover without hospitalization. The most common symptoms are fever, cough, tiredness, and loss of taste or smell. Less common symptoms are sore throat, head-ache aches, pains, diarrhea, red or irritated eye, etc. As the virus slowly moves toward the bronchial system and lungs, it directly impacts the lungs and damages the alveoli, which leads to organ failure like complications and eventually death. The coronavirus disease is transmitted when people breathe in air contaminated by droplets and small airborne particles containing the virus [6–10]. The COVID-19 disease can also be spread if splashed or sprayed with contaminated fluids in the eyes, nose or mouth and rarely via contaminated surfaces. People remain contagious for up to 20 days and can spread virus even if they do not develop symptoms. The incubation period is ranging from1 to 14 days. The precise interval during which a person with COVID-19 infectious is uncertain. As per current evidence, the period of infectivity starts two days prior to onset of symptoms and declines rapidly within the first week of symptoms onset [11–14]. Currently, there are no effective medication to treat COVID-19. Because clinical manifestation of coronavirus disease ranges from mild flu like symptoms to life threatening pneumonia and acute respiratory illness. It is essential to have proper diagnosis during an early stage of infection for efficient implementation of control measure to slow the spread of COVID-19 [15].

Nanoparticles (NPs) have been widely used in many medical applications, such as biosensing, drug delivery, imaging, and antimicrobial treatment. SARS-CoV-2 is an enveloped virus with particle-like characteristics and a diameter of 60−140 nm. Synthetic NPs can closely mimic the virus

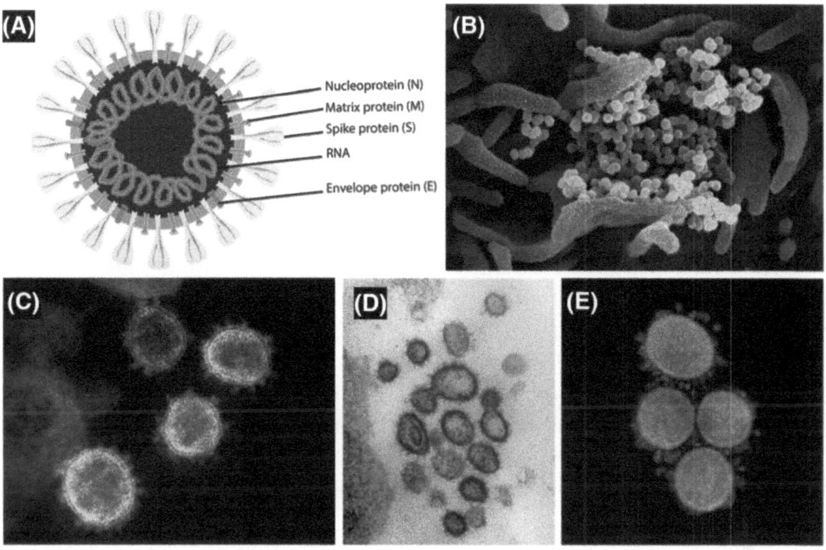

FIGURE 3.1 (A) The SARS-CoV-2 structure is illustrated, with its structural viral proteins indicated. (B) Scanning electron microscopy (SEM) image showing the particulate nature of SARS-CoV-2 (yellow) isolated from a patient in the U.S., emerging from the surface of cells (pink) cultured in the laboratory. Image captured and colorized at NIAID's Rocky Mountain Laboratories (RML) in Hamilton, Montana. Credit: NIAID-RML. (C–E) Transmission electron microscopy (TEM) images show SARS-CoV-2 isolated from a patient in the U.S. Virus particles are shown emerging from the surface of cells cultured in the laboratory. The crown-like spikes on the outer edge of the virus particles give coronaviruses their name [21].

and interact strongly with its proteins due to their morphological similarities. Hence, NP-based strategies for tackling this virus have immense potential. NPs have been previously found to be effective tools against many viruses, especially against those from the *Corona viridae* family [16].

3.2 Structure of COVID-19

SARS-CoV-2 is an enveloped single-stranded positive RNA-virus with a large RNA genome of approximately 30kb with genome characteristics similar to the corona virus (Fig. 3.1). On the basis of genomic organization and phylogenetic relationship, coronaviruses have been classified into subfamily, that is, Coronavirinae that consists of four genera such as Alpha coronavirus, Beta coronavirus, Gamma coronavirus, and delta coronavirus. The coronavirus genomic RNA encodes replication and transcription complexes from a single large open reading frame and structural proteins of virus. The major structural proteins of coronavirus are spike, envelope, membrane, and nucleocapsid.

(a) Spike(S) protein: The spike protein is a highly glycosylated and large type I transmembrane fusion protein that is made up of 1160 to

1400 amino acids depending upon the type of virus. As compared to other proteins that are primarily involved in virus assembly, the spike protein plays a crucial role in penetrating host cells and initiating infection.

(b) Envelope protein-: Envelope protein is contributed to virus assembly and as a membrane protein, also possesses viroporin channel properties that may contribute to epithelial barrier damage, pathogenesis, and disease severity.

(c) Membrane protein: This may contain a T-cell epitope cluster revealed by the immunogenic and structural analysis of a panel of truncated peptides overlapping with Mn2 and Md3. The M-protein of SARS-CoV-2 holds dominant cellular immunogenicity.

(d) Nucleocapsid protein-: N proteins are expressed abundantly during infection. High levels of IgG antibodies against N protein have been detected in sera from SARS patients. The N protein is a representative antigen for the T cell response in a vaccine setting [2].

3.3 Nanotechnology toward the diagnosis of COVID-19

For combating the COVID-19 disease, nanotechnology-based approach should be initiated. This technique can be employed in number of ways toward the betterment of mankind, which includes the development of drugs and vaccine and high responsive testing kits for detecting the infections or immunity. This technique can also be used in developing super fine filters face mask through surface coatings. It will offer higher opposition to the adhesion of virus by inactivating the toxic nature and will improve the tools for contact tracing [17].

Using metallic gold nanoparticles, a colorimetric assay was developed by Moitra et al., through which naked eye detection of SARS-CoV-2 virus was performed. Gold nanoparticles (AuNPs) were capped suitably designed thiol-modified antisense oligonucleotides (ASOs) which is specific for N-gene (nucleocapsid phosphoprotein) of SARS-CoV-2. This experiment resulted with the diagnosis of positive COVID 19 within 10 minutes. Fig. 3.2 shows the schematic representation showing the detail procedure and techniques considered toward the detection of SARS-CoV2 virus. The process includes the collection of viral sample and extraction of viral and then RNA was added to the ASO-capped gold nanoparticles. The thiolated ASO-capped AuNPs are found to be agglomerated and it is observed after five minutes incubation. This agglomeration is because of the target RNA sequence of SARS-CoV-2 which exhibits a change on its surface plasmon resonance. Again, RNaseH was added and incubated for 5 min and at a temperature of 65°C. Finally, precipitation of gold nanoparticles was visually detected [18].

Again, a dual functional plasmonic biosensor was developed by Qiu et al. for rapid clinical diagnosis of COVID-19. The combine behavior

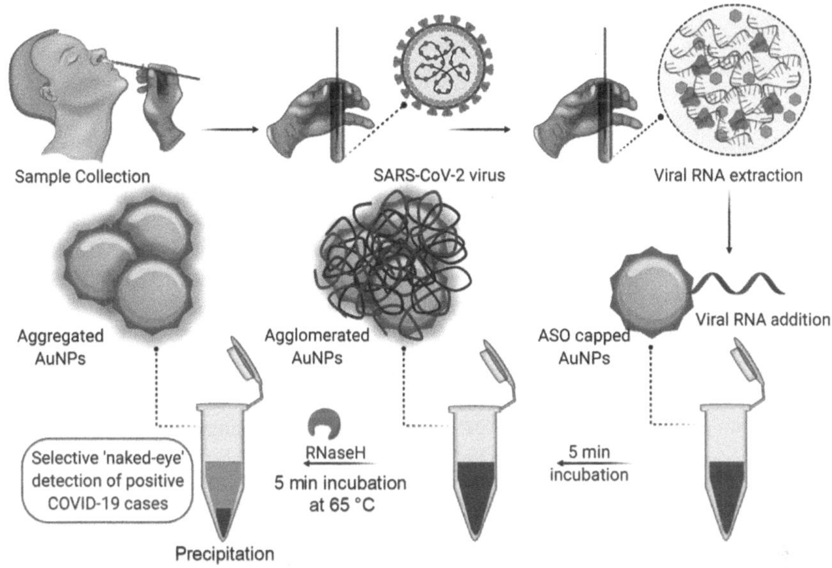

FIGURE 3.2 Schematic representation for the selective naked-eye detection of SARS-CoV-2 RNA mediated by the suitably designed ASO-capped AuNPs. This figure was adopted from Moitra et al. [18].

of plasmonic photothermal (PPT) effect and localized surface plasmon resonance (LSPR), make the biosensor more effective. It uses two-dimensional nanoislands of gold nanoparticles which is functionalized with complementary DNA receptors. Through nucleic acid hybridization, the SARS-CoV-2 is detected. The biosensor exhibits a high sensitivity with a limit of detection of 0.22 pM toward the selected SARS-CoV-2 sequences [19].

Abo-zeit et al. have performed a molecular docking study repurposes US food and drug administration (FDA)-approved iron oxide nanoparticles for combating and controlling the COVID-19 infections. Prior to this iron oxide nanoparticles have been approved by the FDA for curing the anemia treatment. Several studies related to the in vitro antiviral activities have been demonstrated. The docking study was performed to explore the interaction of IONPs (Fe_2O_3 and Fe_3O_4) with the spike protein receptor binding domain (S1-RBD) of SARS-CoV2 that is required for virus attachment to the host cell receptors. These interactions of IONPs are expected to be associated with viral proteins conformational changes and hence, viral inactivation. Therefore, we recommend FDA-approved-IONPs to proceed for COVID-19 treatment clinical trials [20].

Shan et al. have developed a nanomaterial-based sensor array by linking different gold nanoparticles to organic ligands, which has the ability to detect the COVID-19 from the exhaled breath. The array of gold nanoparticles provides excellent sensing layer that can swell and shrink upon exposure to volatile organic compounds (Fig. 3.3). It causes change in

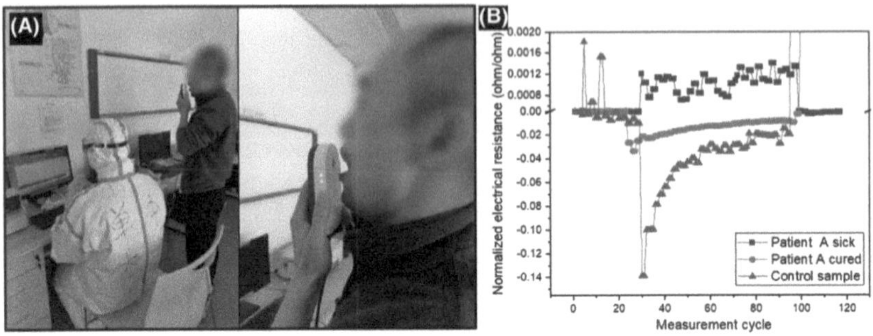

FIGURE 3.3 (A) Example of breath collection with the developed hand-held breathalyzer system from a patient in Wuhan, China. (B) Representative response of a sensor to three different breath samples. The normalized response of sensor 7 of the breathalyzer system to three different samples: patient A, COVID-19, first sample while infected; patient A, second sample after determined as recovered; and a healthy control. The x-axis represents the cycle measurement; each unit is one cycle of the sensor. The infected sample had a positive change response, while the recovered and control showed negative charges [16].

electrical resistance which leads to sensing. They have conducted the study by observing 49 confirmed COVID-19 patients, 58 healthy controls, and 33 non-COVID lung infection controls [16].

An array consisting of eight gold nanoparticles working on the principle described earlier in the text was developed and integrated with electronic circuitry and an advanced apparatus that collects an exhaled breath sample, by blowing into the device for 2–3 s from a distance of 1–2 cm. A built-in sensor technology advised the study subject when the test was complete. As the breath passes through the array, a mixture of COVID-19-related VOCs reacts with the sensors, and these emit a set of electrical resistance signals as a function of time. An example of a sensor response to different samples can be found. If the breath collection was not satisfactory, the subject was asked to repeat the test. Software-based machine learning methods probe the pattern of output signals to get a signature of COVID-19. A more detailed description is provided in the Materials and Methods section, the Supporting Information using this nanotechnology, an exploratory clinical study in Wuhan, China, was conducted during March 2020 on 140 participants. The selection of participants from three distinct groups COVID-19 patients, healthy controls, and non-COVID lung infection/disease controls is described. The characteristics of the 140 participants are tested. Among the patients, more than 60% had no underlying chronic disease, while all the rest suffered at least from hypertension; 10% of those suffered from diabetes mellitus. Most patients were nonsmokers (73%), with a mean age of 59 years, with 57% females. For the control group, 67% were nonsmokers, and the mean age was 52 years, with 46% females. For the lung infection control group, 73% were nonsmokers, and the mean age was 63 years, with 44% females. Statistical analysis was employed on the signals emitted from three binary comparisons: COVID-19 vs. control; COVID-19 vs. other lung infections; and COVID-19 first vs. COVID-19 second sample. Seventy percent of the data were used to calculate the discriminant factor analysis (DFA) models. For each comparison, a receiver-operating-characteristic (ROC) curve and cut-off value were determined. The remaining test data were classified blindly based on the cut-off as presented. Panels A, B, and C show data classification from cumulative sensor responses to breath samples as represented by the canonical variable of the discriminant analysis. Panel D shows ROC curves for the cumulative breath-sensor response. The results showed an area under the curve (AUC) of 0.81 (95% CI, 0.70–0.89) in patients with COVID-19 vs. with controls, 0.97 (95% CI, 0.92–0.99) in COVID-19 vs. other lung infection/conditions, and 0.87 (95% CI, 0.671.00) in COVID-19 first sample vs. COVID-19 second sample. $P <.001$ was found for the comparisons of the training set for each of the binary classifications.

Lew et al. have developed a colorimetric serological assay by conjugating short antigenic epitopes to gold nanoparticles, for detecting the SARS-CoV-2 IgGs in patients' plasma. In SARS-CoV-2, four immunodominant linear B-cell epitopes are present on the spike (S) and nucleocapsid (N) proteins. These epitopes were characterized owing to their IgG binding affinity and they can be used as highly specific biological motifs on the nanoparticle to identify the target antibodies. The binding affinity and the specificity of the four epitopes were determined by using two distinct techniques, that is, surface plasmon resonance (SPR) and fluorescence polarization (FP) techniques (Fig. 3.4A and B). SPR spectroscopy detects binding between immobilized ligands and target analytes at the surface of a thin metal film via changes in the refractive index. SPR sensorgram data confirmed the binding of all four epitopes when titrated against varying concentrations of epitope-specific SARS-CoV-2 IgGs (Fig. 3.4C). For determining the selectivity of these epitopes, an orthogonal FP approach was also pursued and further it confirms their IgG binding affinity. FP detects the binding of a fluorescent ligand (FITC-labeled epitopes, in this case) to a larger molecule (IgG) by measuring the increase in the polarization of fluorescence emission induced by the formation of a larger complex. Compared to SPR spectroscopy, FP is a free solution technique that allows for high-throughput screening and does not require any separation of bound and free ligand. The three S-protein-based epitopes (S14P5, S20P2, and S21P2) exhibited a significant increase in FP values upon the introduction of SARS-CoV-2 IgGs relative to the addition of the closely related SARS-CoVIgGs or the control normal human IgG (Fig. 3.4D).

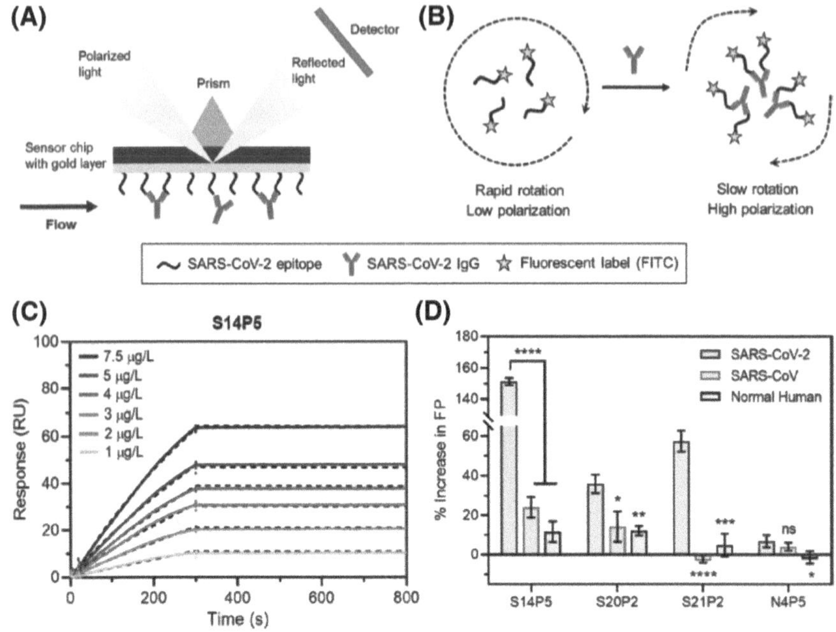

FIGURE 3.4 Characterization of the binding affinity and specificity of four SARS-CoV-2 linear B-cell epitopes. Schematic depiction of (A) SPR spectroscopy and (B) FP working principles to characterize the epitopes' binding affinity and specificity toward SARS-CoV-2 IgG. (C) SPR sensorgram data showing real-time binding of anti-SARS-CoV-2 S1 IgG at varying concentrations to the immobilized S14P5 epitope. Dashed line represents model fit. (D) Relative FP increase when 100 nM peptide epitopes were treated with 200 nM SARS-CoV-2, SARS-CoV, and normal human IgGs [12].

Conclusion

Herein, we have summarized the role of nanotechnology for the COVID treatment. The various detection methods have been discussed in detail. For preventing the COVID-19, nanotechnology-based methodology should be instigated. This method can be utilized in number of ways toward the advancement of manhood, which involves the growth of drugs and vaccine and high responsive assessment kits for detecting the diseases. Although, number of steps have been made, but still, we need to work hard and proceed to use the nanoapproach for future real time application. We need to identify the challenges and need to work on them on a stepwise manner. There is no hurry, but a continuous work should keep as the forefront. Hope everything will be solved and once again nanotechnology will be our savior.

References

[1] P. Bhattacherjee, S. Patra, A. Mishra, T.R. Das, H. Dewangan, R. Ghosh et al., "COVID-19 Diagnostics: Current Approach, Challenges, and," (2021) pp. 23–42.

[2] R.L. Pinals, F. Ledesma, D. Yang, N. Navarro, S. Jeong, J.E. Pak, et al., Rapid SARS-CoV-2 spike protein detection by carbon nanotube-based near-infrared nanosensors, Nano Lett. 21 (5) (2021) 2272–2280, doi:10.1021/acs.nanolett.1c00118.

[3] Y. Shen, T. Wu, Y. Zhang, N. Ling, L. Zheng, S. Zhang, et al., Engineering of a dual-recognition ratiometric fluorescent nanosensor with a remarkably large stokes shift for accurate tracking of pathogenic bacteria at the single-cell level, Anal. Chem. 92 (19) (2020) 13396–13404, doi:10.1021/acs.analchem.0c02762.

[4] K. Wu, R. Saha, D. Su, V.D. Krishna, J. Liu, M.C.J. Cheeran, et al., Magnetic-nanosensor-based virus and pathogen detection strategies before and during covid-19, ACS Appl. Nano Mater. 3 (10) (2020) 9560–9580, doi:10.1021/acsanm.0c02048.

[5] X. Yuan, C. Yang, Q. He, J. Chen, D. Yu, J. Li, et al., Current and perspective diagnostic techniques for COVID-19, ACS Infect. Dis. 6 (8) (2020) 1998–2016, doi:10.1021/acsinfecdis.0c00365.

[6] N. Bhalla, Y. Pan, Z. Yang, A.F. Payam, Opportunities and challenges for biosensors and nanoscale analytical tools for pandemics: COVID-19, ACS Nano 14 (7) (2020) 7783–7807, doi:10.1021/acsnano.0c04421.

[7] E.V.R. Campos, J.L. de Oliveira, L.B. Carvalho, M. Guilger–Casagrande, R. de Lima, et al., How can nanotechnology help to combat COVID-19? Opportunities and urgent need, J. Nanobiotechnology 18 (1) (2020) 1–23, doi:10.1186/s12951-020-00685-4.

[8] G. Chauhan, M.J. Madou, S. Kalra, V. Chopra, D. Ghosh, S.O. Martinez-Chapa, Nanotechnology for COVID-19: Therapeutics and vaccine research, ACS Nano 14 (7) (2020) 7760–7782, doi:10.1021/acsnano.0c04006.

[9] P. Paliwal, S. Sargolzaei, S.K. Bhardwaj, V. Bhardwaj, C. Dixit, A. Kaushik, Grand challenges in bio-nanotechnology to manage the COVID-19 pandemic, Front. Nanotechnol. 2 (2020), doi:10.3389/fnano.2020.571284.

[10] S.M. Russell, A. Alba-Patiño, E. Barón, M. Borges, M. Gonzalez-Freire, R. De La Rica, Biosensors for managing the COVID-19 cytokine storm: Challenges ahead, ACS Sensors 5 (6) (2020) 1506–1513, doi:10.1021/acssensors.0c00979.

[11] E. Bidram, Y. Esmaeili, A. Amini, R. Sartorius, F.R. Tay, L. Shariati, et al., Nanobased platforms for diagnosis and treatment of COVID-19: From benchtop to bedside, ACS Biomater. Sci. Eng. 7 (6) (2021) 2150–2176, doi:10.1021/acsbiomaterials.1c00318.

[12] T.T.S. Lew, K.M.M. Aung, S.Y. Ow, S.N. Amrun, L. Sutarlie, L.F.P. Ng, et al., Epitope-functionalized gold nanoparticles for rapid and selective detection of SARS-CoV-2 IgG antibodies, ACS Nano 15 (7) (2021) 12286–12297, doi:10.1021/acsnano.1c04091.

[13] J. Li, D. Wu, Y. Yu, T. Li, K. Li, M. Xiao, et al., Rapid and unamplified identification of COVID-19 with morpholino-modified graphene field-effect transistor nanosensor, Biosens. Bioelectron. 183 (2020) 2021, doi:10.1016/j.bios.2021.113206.

[14] A. Mishra, M. Choudhary, T.R. Das, P. Saren, P. Bhattacherjee, N. Thakur, et al., Sustainable chemical preventive models in COVID-19: Understanding, innovation, adaptations, and impact, J. Indian Chem. Soc. 98 (2021) 100164.

[15] A.A. Saei, S. Sharifi, M. Mahmoudi, COVID-19: Nanomedicine uncovers blood-clot mystery, J. Proteome Res. 19 (11) (2020) 4364–4373, doi:10.1021/acs.jproteome.0c00425.

[16] B. Shan, Y.Y. Broza, W. Li, Y. Wang, S. Wu, Z. Liu, et al., Multiplexed nanomaterial-based sensor array for detection of COVID-19 in exhaled breath, ACS Nano 14 (9) (2020) 12125–12132, doi:10.1021/acsnano.0c05657.

[17] C. Weiss, M. Carriere, L. Fusco, I. Capua, J.A. Regla-Nava, M. Pasquali, et al., Toward nanotechnology-enabled approaches against the COVID-19 pandemic, ACS Nano 14 (6) (2020) 6383–6406. https://doi.org/10.1021/acsnano.0c03697.

[18] P. Moitra, M. Alafeef, K. Dighe, M.B. Frieman, D. Pan, et al., Selective naked-eye detection of SARS-CoV-2 mediated by N gene targeted antisense oligonucleotide capped plasmonic nanoparticles, ACS Nano 14 (6) (2020) 7617–7627, doi:10.1021/acsnano.0c03822.

[19] G. Qiu, Z. Gai, Y. Tao, J. Schmitt, G.A. Kullak-Ublick, J. Wang, Dual-functional plasmonic photothermal biosensors for highly accurate severe acute respiratory syndrome coronavirus 2 detection, ACS Nano 14 (5) (2020) 5268–5277, doi:10.1021/acsnano.0c02439.

[20] Y. Abo-zeid, N.S. Ismail, G.R. McLean, N.M. Hamdy, A molecular docking study repurposes FDA approved iron oxide nanoparticles to treat and control COVID-19 infection, European Journal of Pharmaceutical Sciences 153 (2020) 105465 July, doi:10.1016/j.ejps.2020.105465.

[21] R. Medhi, et al., Nanoparticle-Based Strategies to Combat COVID-19, ACS Appl. Nano Mater 3 (2020) 8557–8580. https://dx.doi.org/10.1021/acsanm.0c01978.

CHAPTER

4

Polymeric-based interface for the development of COVID-19 biosensor

Zeynep Yilmaz-Sercinoglu, PhD[a], *Cansu İlke Kuru, PhD*[b,c], *and Fulden Ulucan-Karnak, PhD*[b]

[a]Marmara University, Faculty of Engineering, Department of Bioengineering, Istanbul, Turkey, [b]Ege University, Faculty of Science, Department of Biochemistry, Izmir, Turkey, [c]Buca Municipality Kızılçullu Science and Art Center, Izmir, Turkey

4.1 Introduction

Contagious infections contribute to death of millions and have a huge burden on countries' public health and medical care systems. There are numerous spreading diseases such as malaria, tuberculosis, and acquired immunodeficiency syndrome and cholera, which are especially causing the death of thousands in poor countries. Government agencies, research communities, and public agencies have made enormous efforts to avoid, diagnose and treat these infections [99]. Nowadays, there is a huge increment of viral infectious disease outbreaks over the world. In 2009, influenza A H1N1 subtype, in 2014 Ebola virus, and now in 2019 SARS-CoV-2 are the outbreaks in recent years we came across.

Viruses need a living host cell to produce their genetic material and can fastly adapt their genetic material to immune system responses for their continued survival. Therefore, it is an emergency to detect viral pathogens [1].

Enzyme-linked immunosorbent assay (ELISA), polymerase chain reaction (PCR), and microscopy are the current laboratory-based diagnostic

Sensing Tools and Techniques for COVID-19: Developments and Challenges in Analysis and Detection of Coronavirus
DOI: https://doi.org/10.1016/B978-0-323-90280-9.00013-4

methods used in clinical applications. All of these methods have disadvantages such as the need of experienced people, high-cost equipment, and results taking so long. In order to overcome these disadvantages, biosensors are ideal. They have remarkable advantages such as high sensitivity, quick response, user-friendliness, and applicability to point-of-care (POC) devices [2]. Molecularly imprinted polymers are commonly used in biosensors to enhance their selectivity, specificity and reproducibility [3].

Biosensors are utilized for measurement and real-time detection of biological materials such as enzymes, antibodies, nucleic acids, cells, tissue, and microorganisms via a physicochemical detector. The transducer can turn biological signal into an electrical signal. Biosensors can be categorized as immunosensors, electrochemical, optical, and mass-based sensors. Their recognition strategies are relied on nanotechnology, affinity, lab-on-a-chip, and organ-on-chip technologies [4].

Biosensor devices are realistic (such as POC-testing by smartphone-based nanosensors) and are considered to be revolutionary, simple, and efficient solutions against standard detection procedures. These technologies are used to research several viruses such as Zika virus, Human Cytomegalovirus (HCMV), Ebola, influenza, murine norovirus, human immunodeficiency virus (HIV), and recently discovered coronavirus 2 (SARS-CoV-2) [5].

In the literature, recognition of viruses was performed with several biosensors. Some examples of these biosensors were listed in Table 4.1.

In this review, we aim to present comprehensive updates and the latest studies about biosensors for viral pathogen detection in last five years. Also we aim to explain biosensor development of polymeric-based surface coatings and their applications in viral diagnosis. All of these approaches can be realized for development of new COVID-19 biosensing platforms and some of the important studies were explained in detail.

4.2 SARS-CoV-2 Infection (COVID-19)

COVID-19 outbreak in Wuhan is still spreading around the world. The infectious agent of COVID-19 is a phylogenetic sister of severe acute respiratory syndrome coronavirus (SARS-CoV) and thus has been named as SARS-CoV-2 [20].

The initial diagnosis of the virus is of paramount importance for the public health. Computed tomography (CT) is used to examine the patient's lungs in order to determine opacities in the lungs. But this method can only perform the first assessment of the disease. Thus viral load of the patient should be confirmed with further analysis. Common diagnosis methods for COVID-19 are illustrated in Fig. 4.1 [21].

TABLE 4.1 Virus recognition biosensors.

Virus type	Sensor type	Ref.
Zika virus	Graphene based field effect biosensor	Afsahi et al. [6]
Zika virus	Electrochemical immunosensor using a functionalized interdigitated microelectrode of gold (IDE-Au)	Kaushik et al. [7]
Zika virus	Reverse-transcription loop-mediated, isothermal amplification (RT-LAMP)	Song et al. [8]
Zika virus	Electrochemical biosensor based on surface imprinted polymers and graphene oxide composites	Tancharoen et al. [9]
Influenza, norovirus	Binary-NP-decorated carbon nanotubes (bNPCNTs) for electrical resistance biosensor	Lee et al. [10]
Murine norovirus	Gold nanoparticle based aptasensor for colorimetric detection	Weerathunge et al. [11]
Chikungunya virus	2D molybdenum disulfide nanosheets based disposable electrochemical biosensor	Singhal et al. [12]
Dengue virus	Silicon nanowire biosensor.	Nuzaihan et al. [13]
Dengue virus	Graphene-based electrochemical biosensor	Navakul et al. [14]
Human CMV	Cyclic olefin copolymer TOPAS-based electrochemical biosensor	Kiilerich-Pedersen et al. [15]
Hepatitis E virus	Graphene quantum dots and gold-embedded polyaniline nanowires based electrochemical biosensor	Chowdhury et al. [16]
Hepatitis A virus	Virus-magnetic-MIPs (Fe_3O_4 magnetic nanoparticles)-based resonance light scattering (RLS) sensor	Zhang et al. [17]
Papillomavirus	Anthraquinone-labeled pyrrolidinyl peptide nucleic acid (acpcPNA) probe (AQ-PNA) and graphene-polyaniline (G-PANI) modified electrode, electrochemical impedance spectroscopy (EIS)	Teengam et al. [18]
HIV	Coupling of a reagent less electrochemical DNA biosensor with conducting polymer film and nanocomposite as matrices for the detection of the HIV DNA sequences	Fu et al. [19]

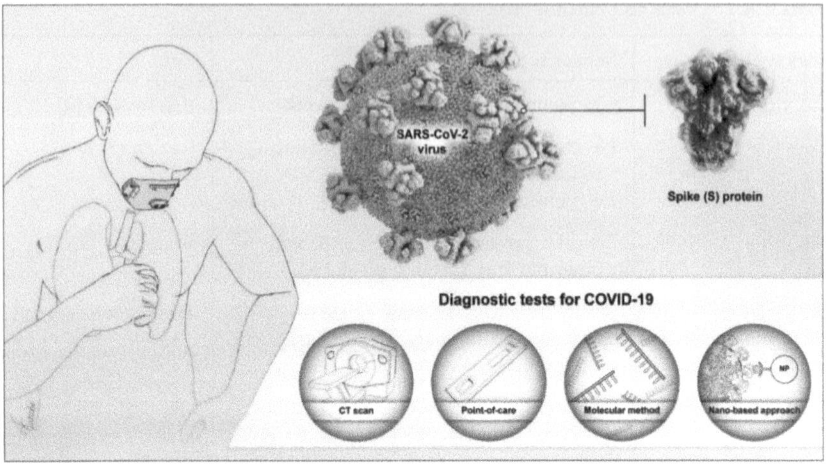

FIGURE 4.1 Common diagnosis methods for SARS-CoV-2 [21]. From Navid R, Mojtaba B, Amir G, et al. Point-of-use rapid detection of SARS-CoV-2: nanotechnology-enabled solutions for the COVID-19 pandemic. International Journal of Molecular Sciences. 2020:5126. doi:10.3390/ijms21145126.

At the beginning of SARS-CoV-2 infection, PCR test is accepted as the gold standard for detection viral nucleic acid. There is a need of additional confirmation test, since PCR test may lead to false negative or positive results. Moreover, PCR test needs trained personnel, expensive equipment and well-equipped laboratory and at least 3 hours to develop results. Nanotechnology can be a great nominee for developing sensitive, selective, rapid and portable screening/diagnostic kits with minimal false positive/negative responses [22].

Serological tests are also used to discover IgG/IgM antibodies against SARS-CoV-2 in the body. They can be either ELISA based or gold (Au-) nanoparticle based colorimetric assays and they generally result in positive, negative or invalid response with naked eye. They are cost-effective and fast, but their results should be analyzed with a holistic approach with other analyses to confirm the infection of SARS-CoV-2 [23].

In this context, development of diagnostic kits is central for fast detection of COVID-19, immediate start of treatment of the infected people and their containment from the community, thus spreading of the disease can be decelerated. As one can comprehend, there is a butterfly effect.

Nasal and throat swab, sputum, saliva and blood test are the samples of potential patients and are used to extraction of RNA, DNA, antibody or antigen. Those extracts can be used in several types of biosensors for the diagnosis of COVID-19 as shown in Fig. 4.2 [24].

Development of fast and user-friendly method for early detection of COVID-19 is of paramount importance. Nanomaterial-based biosensors

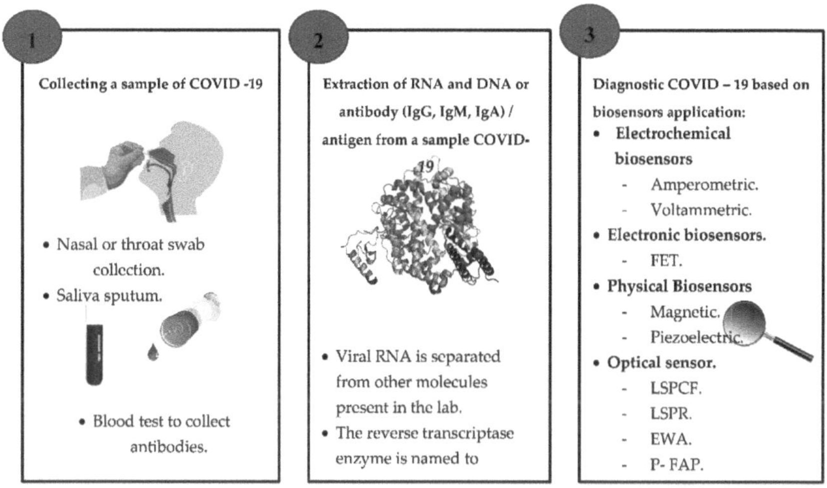

FIGURE 4.2 A schematic represantation of COVID-19 sample collection, extraction, and detection strategies based on biosensor applications (Taha et al., 2020). From Ahmed TB, Yousif AM, Hadri HMM, Saiful DBZM, Norhana A. An analysis review of detection coronavirus disease 2019 (COVID-19) based on biosensor application. Sensors. 2020:6764. doi:10.3390/s20236764.

may fulfill the requirements of the clinical approaches. Different types of nanosized structures were generated and utilized for advanced biosensing of SARS-CoV-2 in different clinical samples, such as urine, blood, saliva, and nasopharyngeal swabs [25]. Current technologies for the detection of SARS-CoV-2 are summarized in the review article of Parihar et al. in 2020 and expanded by us in Table 4.2 [26].

Pathogenic viruses are currently the subject of biosensor technology to detect viral infections. In general, detection techniques currently used for SARS-CoV-2 are time-consuming, labor-intensive, and are inapplicable off the laboratory or on-site testing. These can hinder scientists to obtain real data on SARS-CoV-2 infection and transmission. Polymer-based biosensors have properties such as adequate sensitivity, selectivity, portability, specificity for highly qualified and accurate screening. Novel polymer-based structures are potential candidates to be used in reliable, fast and robust detection and monitorization of pathogenic viruses and obviously they can be helpful to prevent/reduce their spread which is pivotal in pandemics.

4.3 Polymer-based surfaces for biosensor

Several types of nanomaterials (metallic, polymeric, protein-based, carbon-based, composites) were suggested to be used in development of

TABLE 4.2 SARS-CoV-2 sensors and biosensors.

Biomarker	Biological samples	Biorecognition element	Method	LOD/Sensitivity	Ref.
SARS-CoV-2 protein	Gargle solution	-	Mass spectroscopy	10^5–10^6 genome equivalents μL^{-1}	Ihling et al. [27]
nCovid-19 antigen	Saliva	FTO/AuNPs/nCOVID-19 antibody	Differential pulse voltammetry	10 fM	Mahari et al. [28]
SARS-CoV-2 antigen protein	Nasopharyngeal swab	Graphene sheets with SARS-CoV-2 Ab	FET	2.42×10^2 copies mL^{-1}	Seo et al. [29]
Viral sequences including RdRp COVID, ORF1ab-COVID, and E genes	Respiratory sample	Gold nanoislands functionalized with complementary DNA	PPT-LSPR	0.22 pM	Qiu et al. [30]
N-gene	-	Thiolated modified ASO AuNPs	colorimetric	0.18 ng μL^{-1}	Moitra et al. [31]
Nucleocapsid Ab	Serum	Peptide monolayer functionalized with SARS-CoV-2 nucleocapsid recombinant protein	SPR	~1 $\mu g\ mL^{-1}$	Djaileb et al. [32]
IgM/IgG Ab	Blood/serum/plasma sample	Gold nanoparticle conjugated with COVID-19 antigen and rabbit IgG	Colorimetric	88.66%	Li et al.[101]
Anti-SARS-CoV-2 IgG	Serum	Mouse antihuman IgG antibody labeled lanthanide-doped polystyrene NPs	Colorimetric	-	Chen et al. [34]

(continued on next page)

IgM/IgG	Serum	Lanthanide Eu(III) fluorescent microsphere	Colorimetric	98.72%	Feng et al. [35]
IgM Ab	Serum	Antihuman IgM conjugated colloidal gold nanoparticles	Colorimetric	-	Huang et al. [36]
IgG Ab	Serum	NC protein and anti-human IgG modified AuNPs	Colorimetric	69.1%	Wen et al. [37]
IgM/IgG Ab	Serum	SARS-CoV-2 antigen and rabbit IgG antibodies labeled colloidal AuNPs	Colorimetric	85.29%	Zeng et al. [38]
SARS-CoV-2 spike protein	-	Monoclonal antibodies labeled AuNPs	Plasmonic metasensor	~4.2 fM	Ahmadivand et al. [39]
IgM, IgG Ab	Serum	-	Chemiluminescence	≥10 AU mL^{-1}	Hou et al. [40]
SARS-CoV-2 Ab	Blood	Bioconjugate labeled RRBC	Gel card agglutination	-	Alves et al. [41]
S1 Ab	-	Cell-based detection	Membrane engineering	1 fg mL^{-1}	Mavrikou et al. [42]
N-gene	serum	ssDNA aptamers	Aptamer	-	Chen et al. [43]
IgG/IgM/Antigen	Serum/pharyngeal swab	Fluorescent microsphere coated IgG/IgM Ab and SARS-CoV-2 capture Ab	Microfluidic	-	Lin et al. [44]
E and N gene	Respiratory swab	CRISPR-Cas12 based DETECTR	CRISPR	10 copies μL^{-1}	Broughton et al. [45]
orf1ab and N genes	Nasal swab	CRISPR Cas12a/gRNA with fluorescent probe	CRISPR	2 copies mL^{-1}	Huang et al. [46]

(continued on next page)

TABLE 4.2 SARS-CoV-2 sensors and biosensors—cont'd

Biomarker	Biological samples	Biorecognition element	Method	LOD/Sensitivity	Ref.
orf1ab, S and N gene	Throat swab	primer and DNA fragment	RT-LAMP	80 copies viral RNA mL^{-1}	Huang et al. [47]
Nucleocapsid gene	Nasal swab	RT-LAMP primer	RT-LAMP	10^2 RNA copies	Baek et al. [48]
orf1ab and S genes	Swab/bronchoalveolar lavage fluid	Viral RNA extracted using QIAamp Viral RNA Mini Kit	RT-LAMP	2×10^1 copies and 2×10^2 copies/reaction with primer sets orf1ab and S gene	Yan et al. [49]
orf1ab and N genes	Saliva	LAMP primers	RT-LAMP	$\sim 10^2$ viral genome/reaction	Lalli et al. [50]
RNA	Clinical specimens	Au NP and RGO materials	Electrochemical biosensor	200 copies mL^{-1}	Zhao et al. [51]
Surface proteins of SARS-CoV-2	Nasal and throat swabs	Au NP based	Colorimetric	Ct = 36.5	Ventura et al. [52]
1a/b and nucleoprotein gene of SARS-CoV-2	Feces, nasal, pharyngeal and anal swabs	Streptavidin– dye-coated polymer nanoparticles	Reverse transcription multiple cross-displacement amplification (RT-MCDA)	This method: 33.8% while RT-PCR: 30.7%	Li et al. [33]
Signal of the virus was converted to glucose	Clinical samples	CRISPR Cas12a and glucose producing reaction	CRISPR	10 copies μL^{-1}	Huang et al. [53]
Spike (S) protein or nucleocapsid (N) protein	Saliva	Carbon black material and magnetic beads	Electrochemical biosensor	19 ng/mL for S protein and 8 ng/mL for N protein	Fabiani et al. [54]

(continued on next page)

-	-	PLA (PolyLactic Acid) tri-color change filament and thermochromic paint	3D printed thermosensor gear	If T<37°C, the inside of it stays gray (i.e., the color-changing filament is in one stage), but if T > 37°C, it changes to an orange color (i.e., the color-changing filament goes to the second color stage)	Djukanović et al. [55]
Spike subunit 1 protein	Viral stock, bovine serum albumin dilution	Graphene based	Electrochemical biosensor	5.5×10^5 – 13.75×10^5 PFU mL^{-1}	Mojsoska et al. [56]
Pathogen DNA	Synthetic templates of nucleic acid	Nylon mesh attached microfluidic pores and DNA hydrogel	microfludic	0.7 aM	Kim, Abbas, and Shin [57]
SARS-CoV-2 gene N and Orf1ab	Saliva	LAMP primers	RT-Proofman-LAMP	100 copies of gene N RNA mL^{-1}	Ding et al. [58]
SARS-CoV-2 Spike protein or nucleocapsid protein-specific IgG antibodies	Plasma	Biotinylated SARS-CoV-2 spike protein receptor-binding domain (RBD) or the Nucleoprotein (N)	Fluorescent bead-based immunoassay	100% for S-RBD IgG; 100% for S-RBD IgM; 94.26% for S-RBD IgA and 99.13% for Nucleocapsid IgG	Dogan et al. [59]
miR-19b or miR-20a	-	Target-recognition induced collateral nuclease activity of the Cas13a/crRNA-complex	CRISPR/Cas13a-powered electrochemical microfluidic biosensor	1.28 nM	Bruch et al. [60]

biosensors due to their chemical stability, strong electrical conductivity, and surface plasmon effects. Biosensors' performance can be enhanced by this way [61]. Carbon-based nanomaterials (graphene, graphite, nanodiamonds, etc.) are generally used in biosensors thanks to their high thermal and electrical conductivity, stability, corrosion resistance and remarkable strength [62]. But using polymers as a bioactive layer of biosensors may help to modification of surfaces more easily than other nanomaterials, due to existing functional groups.

Polymers are formed by chemical reaction of monomers to generate long polymer chain. They can be natural or synthetic [63]. Polymers have an important place during our daily life in many areas such as agriculture, textile, packaging, automobile, and etc. Their unique characteristics and diverse capabilities as easy fabrication, functionalization and tuneability, makes them an attractive research "playground" to develop new polymers.

Undoubtedly, usage of polymers in the biomedical area, especially in biosensors, is on the march [64].

Finding target analyte in a complex biological environment is challenging. What is more, sustainable biosensors using natural polymers will relieve the impact of excess use of synthetic polymers on the environment [65].

Polymeric interface utilization on biosensors may result in three benefits:

- Porosity of the polymer increases the surface area
- Porosity of the polymer can facilitate the adsorption of the analyte to the surface.
- Interferences can be hindered thanks to the filtering capacity of the polymer.

Polymers can attach on surface with their functional groups by hydrophobic and ionic interaction, hydrogen and chemical bonds [66]. Polymeric substances have considerable advantages such as being light, having good flexibility and high transparency. Therefore, they can be integrated into living systems as smart interfaces and many technologies are purposed in this context [67].

4.4 Polymers for surface coating and biosensor construction

Polymers with conductivity properties are called as "conductive polymers" (CP). They are cheaper than metal electrodes and easy to use. Polyaniline, polyacetylene, polythiophene, polypyrrole, and poly(3,4-ethylenedioxythiophene) are the most commonly used polymers in biosensors, because of their biocompatibility. They can easily be fabricated and modified for targeting biomolecules [68]. Conductive polymers have various properties such as strength, plasticity, elasticity. In addition, due to their charge mobility through polymer chains, they have significant electronic and optical properties [69].

Nonconductive polymers can be used for enzyme immobilization due to their biocompatibility, low toxicity, high affinity, and adsorption ability and inertness. They might also be used as electrode modifiers, electrical insulators, etc. Polystrene, cellulose, chitin, chitosan, gelatin, dextran, nafion, poly(styrene-co-maleimide) are used for developing surface coating of biosensors. They can be natural or synthetic and can be used in nanocomposites with metal, carbon-based materials to enhance electrical conductivity [70, 71].

Most commonly used and well-known conductive and non-conductive polymers are shown in Fig. 4.3 with their chemical structures.

Due their biocompatible, biodegradable nature and possession of multiple active sites, polymers/biopolymers have broad application in biomedical fields. Nonetheless, unique electronic properties of biopolymers are the key point to be used in wearable and biodegradable electronics [100].

Advanced material technologies have given us to use electrospun nanofiber materials to be used in biosensors thanks to their polymer diversity, processability, and enhanced specific surface area. In addition, high responsiveness, sensitivity, and broad detection capability are important aspects of biopolymer-doped/based electrospun materials [72].

4.5 Fabrication of polymeric surface coatings

Nanomaterials or nanocomposites exhibit different functions in biosensing applications such as i) immobilization support for enzymes, ii) mediators, and iii) signal amplifiers. They can bind to electrode surface with electrostatic, covalent, and affinity-based interactions. In Fig. 4.4, these interactions were schematized [73].

Nanomaterials have been deposited on biosensors' electrode surface with several techniques, such as coating, direct deposition, printing, and direct growth [74]. The polymer coatings were also used to inhibit corrosion and provide mechanical strength to surfaces. These techniques are simple solution and dip coatings, doctor-blading, spraying method, and spin and cast coating. There are different types of surface treatment techniques, including electroless plating, electroplating, and chemical conversion coating, to develop surface functionality [75]. Nonionic or charged, chemical, and hydrophilic/hydrophobic coatings have been chosen to improve the efficiency nanoparticles on surfaces. These coatings can enhance the properties of nanoparticles in terms of their therapeutic function,

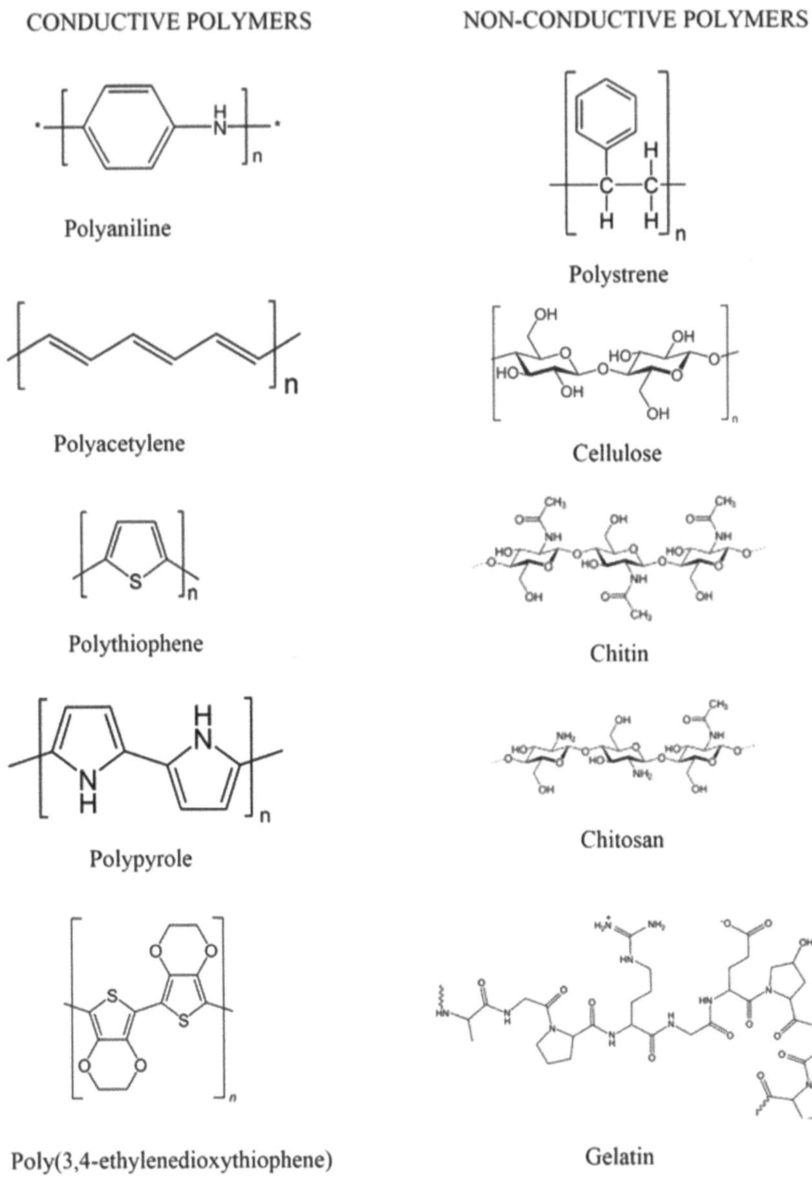

FIGURE 4.3 Most commonly used and well-known polymers. No permission required.

colloidal stability, and biocompatibility [76]. Direct deposition methods include electrospinning, sputtering, electrochemical, electrospray, and vapor depositions. All these methods are multistep fabrication and more expensive than coating techniques. Printing methods can be applicable to

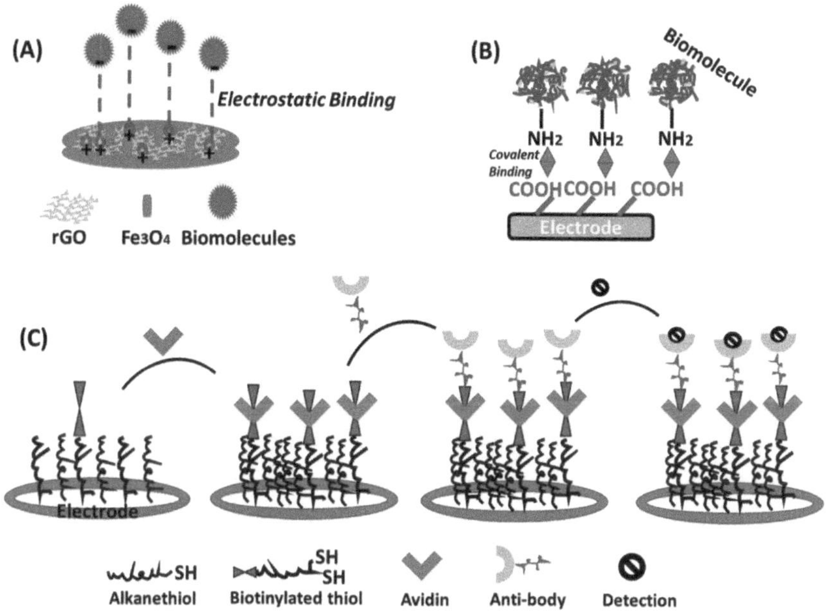

FIGURE 4.4 Interactions between nanomaterials and electrodes (A) Electrostatic (B) Covalent (C) Affinity-based interactions [73]. From Batool R. A Review of the Construction of Nano-Hybrids for Electrochemical Biosensing of Glucose. Biosensors. 2019;9(1).

larger scale and are economical for manufacturing of flexible substrates, especially in biosensor applications. For instance, screen printed electrodes are the most commonly used electrodes in biosensor field, due to their complex structure (counter, reference and working electrodes placed in one electrode). They are disposable and multiple analyses can be conducted with one single screen printed electrode. Direct growth of nanomaterials on the electrode surface can enhance the surface area and increase the mechanical contact which is important for electronic transfer and stability. All of the methods shown in Fig. 4.5 have several advantages and they need to be chosen in accordance with the intended application. Cost-effectiveness and desired biosensor parameters such as stability, reproducibility, and repeatability should also be taken into consideration [74].

Polymers are usually used as surface modificators of sensor's electrodes. In an example study, polydopamine (PDA) was used onto gold chip in order to examine carbohydrate–lectin interactions via quartz crystal microbalance (QCM) biosensor. Target carbohydrates were immobilized onto PDA-coated chips. PDA-coating was achieved by on-surface polymerization onto electrodes. Schematic representation of polymer based surface coating was shown in Fig. 4.6 [77].

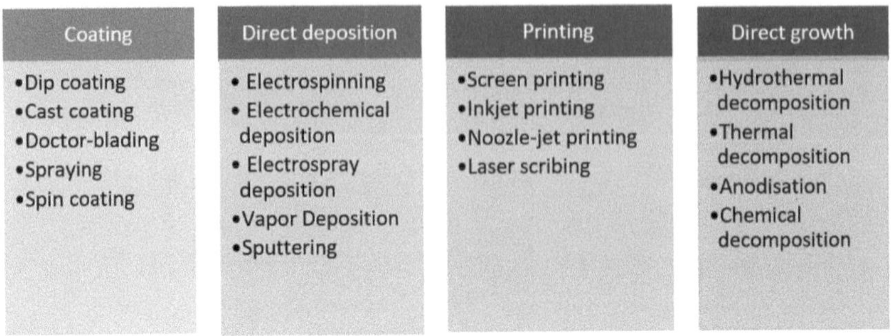

FIGURE 4.5 Techniques of nanomaterial deposition on electrode surfaces [74, 75]. No permission required.

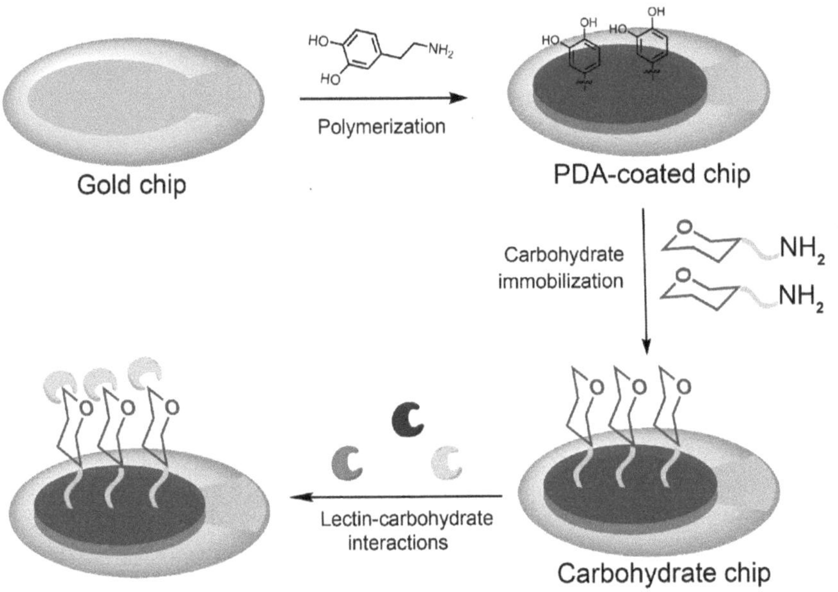

FIGURE 4.6 Schematical representation of carbohydrate-based chip development and its interaction with lectins [77]. From Kun S, Siyu S, Yaping C, et al. Fabrication of carbohydrate chips based on polydopamine for real-time determination of carbohydrate–lectin interactions by QCM biosensor. Polymers. 2018:1275. doi:10.3390/polym10111275.

There are many polymerization methods for preparation of specific polymers as surface coating. Molecular imprinting method (MIP) is generally chosen to generate selective polymers. This method includes a functional monomer, a cross-linker and a template molecule. A pre-complex is generated specific to a template molecule of interest and polymerization

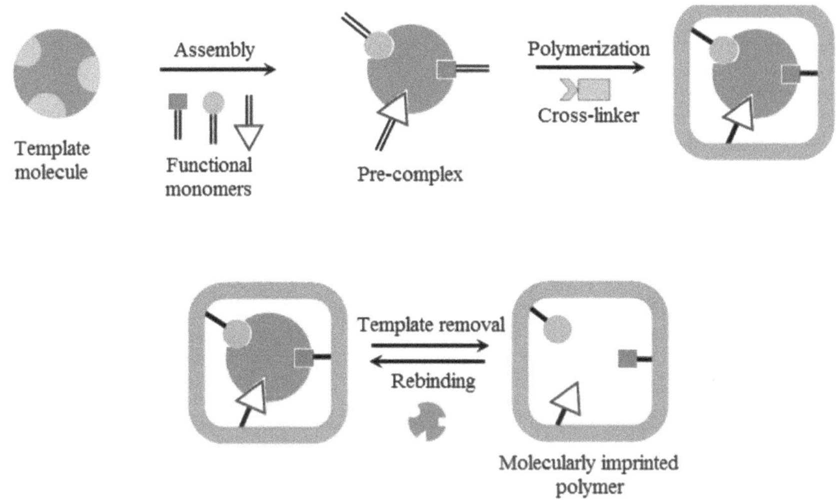

FIGURE 4.7 Step-wise representation of molecularly imprinted polymer (MIP) fabrication [78]. FromYeşeren S, Semra A, Handan Y, Serhat Ü, Adil D. Molecularly imprinted polymer based sensors for medical applications. Sensors. 2019:1279. doi:10.3390/s19061279.

occurs around the pre-complex. This method is easy to control, cost-effective and user-friendly. In Fig. 4.7, step-wise representation of MIP is depicted [78].

MIP based surfaces are superior to existing interfaces due to their operational stability for repeated assays and on-line monitoring, high specificity, and high sensitivity. Aforementioned properties are critical for widespread utilization of biosensors in medicine [79]. There are different molecular imprinting techniques used for biomimetic surfaces, such as bulk, surface and epitope imprinting. All of them have pros and cons, thus proper technique should be chosen in accordance with the desired application [80].

More specific detection of SARS-CoV-2 in early stages of infection can be achieved via surface functionalization, surface coating and modification of nanosized materials. Nanotechnology paves the path of biosensor commercialization since it enables tailor-made fabrications. Moreover, it is easily integrated into sensing devices.

4.6 Polymeric-based SARS-CoV-2 biosensors

Due to their high selectivity, sensitivity and low cost, environmental friendliness and point-of-care diagnostic properties, polymeric based SARS-CoV-2 biosensors focusing on MIP and paper-based sensors are of special interest in the literature.

MIP based sensors can be interpreted as highly selective and sensitive sensors. Thus, for the detection of SARS-CoV-2, a special sensor can also be designed with appropriate polymer and specific aptamer as the recognition element. Polymers containing acrylic group such as acrylamide, acrylic acid, methyl methacrylate, methyl and ethyl acrylate and their mixtures can be selected. They can be modified or combined with conducting metal oxides, graphenes for enhanced conductivity [81].

Puoci and colleagues developed "monoclonal-type" plastic antibodies MIP. This MIP can selectively bind to spike protein of SARS-CoV-2 to block the infection process. Dynamic Light Scattering (DLS) was used for characterization of these nanoparticles and binding experiments using the receptor-binding domain (RBD) of the novel coronavirus and the RBD of SARS-CoV spike protein were conducted to investigate imprinting effect and selectivity, respectively. Finally, the hemocompatibility tests of these antibodies revealed its potential to be used as drug delivery or sensing platform for SARS-CoV-2 [82].

A group of researcher declared that first MIP-based electrochemical sensor detecting the SARS-CoV-2 nucleoprotein (ncovNP). Poly-m-phenylenediamine (PmPD) was used as monomer and resulting polymer modified gold-based thin-film electrodes (Au-TFE). Measurements were performed with a portable potentiostat and results were concluded as 15 fM and 50 fM detection and quantification limit, respectively. This sensor can be a potential rapid COVID-19 diagnostic kit, which determines ncovNP from COVID-19 positive patients' nasopharyngeal swab samples [83].

Rapid and sensitive lateral flow immunoassay (LFIA) was reported by Chen and colleagues. Lanthanide-doped polystyrene nanoparticles (LNPs) synthesized by miniemulsion polymerization were used as the sensing platform for anti-SARV-CoV-2 IgG in human serum. Cut-off value was defined as 0.0666 of 51 normal samples. Only one of the negative samples was turned to be positive by RT-PCR, while others were consistent. Therefore, this system can be regarded as sensitive and fast detection platform for SARS-CoV-2 from human serum [34].

For the disease monitoring and control in COVID-19 pandemic, POC testing systems, such as LFIA, are literally important due to off-lab application and fast response. Despite of being a good alternative for COVID-19 diagnosis, there is still enormous need for investigation and research on colorimetric strips [84].

In another study, Zhao and colleagues reported about a viral RNA extraction method based on poly (amino ester) with carboxyl groups (PC)-coated magnetic nanoparticles (pcMNPs). In this developed system, viral RNA from multiple samples can be purified within 20 min. ORFlab and N gene regions of viral RNA was identified as a 10-copy sensitivity and a strong linear correlation between 10 and 10^5 copies. Automation and simplicity of this new methodcan be effectively reduced the time needed for extraction in current molecular diagnosis of COVID-19, thus leading us to the earlier clinical diagnosis [85].

Cellulose is the main component of plant structure and most widely used biopolymer in Earth. Lignocellulose-based cellulose is the leading source of its industrial use, especially for the production of paper. On the other hand, bacterial cellulose is gathering attention due to its high specific strength, specific physiochemical and surface properties [86] and possible cheap production from different kinds of feedstocks Moreover, bacterial cellulose is free from lignin and hemicellulose, so there is no need to use harsh chemicals to purify it and also dependency of forest ecosystem is decreased. Therefore, overall ecological impact can be reduced [87].

Paper substrates have traditionally been used for analytical and chemical applications. Paper chromatography, in particular, is used to isolate and detect molecules such as proteins, amino acids, small matrix molecules and antibodies [88]. Utilization of paper substrates in POC testing is gaining attention in recent years. Paper-based analytical devices (PADs) are cheaper and simpler to produce, when compared to biosensors using materials, such as plastic and glass. Furthermore, storage and transport PADs are simple and cheap. Their disposal is environment-friendly. Porous nature and intrinsic hydrophilic property of paper help fluid to move by capillary action, resulting in no need for external pumping [89].

Aforementioned advantages make POC analytical devices potential candidates for rapid and reliable testing [90]. Moreover, PADs are of paramount importance in diagnostics, especially in off-lab environments without relying on external reagents and devices [89, 88].

Adrover-Jaume et al. developed a filter paper based and smartphone assisted COVID-19 immunobiosensor which is based on detection IL-6 concentrations. Paper-based signal amplification with polymer-filled reservoirs assists the recognition. This system is resulted with a 10^{-3} pg mL^{-1} LOD and a wide dynamic range between 10^{-3} and 10^{-2} pg mL^{-1} in PBS. The response time of the developed system is stated to be under 10 minutes. This test could be adjusted by shifting the antibodies binded to nanoparticles, in order to be used for several sample types and detect other cytokines [91].

The aggregation of gold and silver nanoparticles was used by Teengam and colleagues to alter the color of their paper wells after MERS-CoV and other pathogens were identified. For other tested pathogens, the observed detection limit for MERS-CoV was 1.53 and 1.3 nM [18].

Yakoh et al. have reported an immunosensor which does not need a specific antibody for detection of SARS-CoV-2 immunoglobulin. This biosensor is a label-free paper-based electrochemical platform in which presence of SARS-CoV-2 antibodies block the conversion of redox indicator that leads a decrease in the control current response. This developed sensor

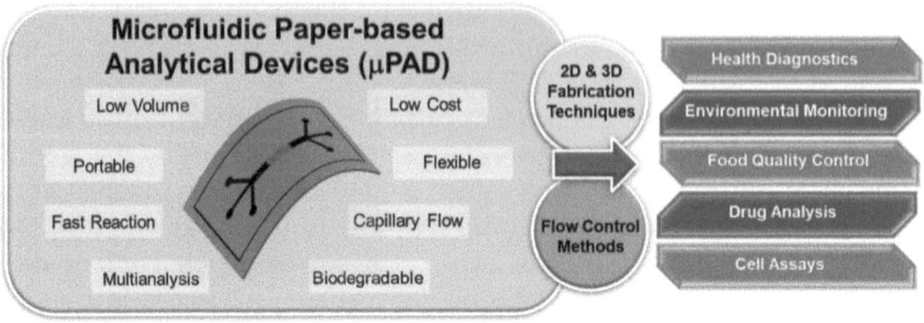

FIGURE 4.8 μPADs, their advantages and application areas [92]. From Lim H, Jafry AT, Lee J. Fabrication, flow control, and applications of microfluidic paper-based analytical devices. Molecules. 2019;24(16). doi:10.3390/molecules24162869.

proved to be effective to detect IgG and IgM in real clinical sera from patients. Square-wave voltammetry (SWV) was used and sensor results were compared with ELISA This system has 30 min response time and 1 ng/mL of LOD, which is 3-fold sensitive than colorimetric LFA. Moreover, it has 100% sensitivity and 90% specificity of detecting targeted antibodies in clinical sera [102].

Resistive pulse sensing (RPS) is another type of label-free detection, in which change in the electric current is the marker for existence of viral particle. Movement of molecules through pores is the mechanism behind the detection, rather than chemical bonding of those molecules with specific labels.

Microfluidic assisted paper-based analytical devices (μPADs) might be used for complex analysis of multiple markers from biochemical samples. In this technique, fluid manipulation is under control of microfluidic system and pumps. Pumps serve as additional pumping source to paper's capillary effect. Limit of detection and required sample volume can be decreased and fast reaction can be induced according to design of interest. These technology can be used in cell assays, drug analysis, environmental monitoring, in addition to biomedical applications. In Fig. 4.8, schematic representation, advantages, and application areas of μPADs are depicted.

In 2017, Xuan Weng and Neethirajan announced their paper-based (nitrocellulose membrane, formed by nitrating cellulose) microfluidic platform as a POC device for successful quantitative detection of norovirus virus-like particles. Recognition elements are labeled aptamer and nanomaterials, multiwalled carbon nanotubes, and graphene oxide. Multiwalled carbon nanotubes and graphene oxide based paper sensors' detection limits were 40 pM and 30 pM, respectively, in a linear range of 0–12.9 μg/mL [93].

In a hot-off-the-press paper, importance of microfluidic systems for diagnosis of RNA viruses is summarized. Integration of this platform with conventional (ELISA, PCR, fluorescence-based assays, aptamers, etc.) and recent (CRISPR, nanotechnology) detection methods showed satisfactory (being user-friendly, highly sensitive, and specific) detection of viral RNA of Zika, HAV, HIV, and norovirus. The study also emphasized the implementation of microfluidic systems on SARS-CoV-2 detection. The reader is kindly directed to this publication [94].

As an effort to monitor the spread of infectious diseases and to prevent further interruption in our social-economic integrity, comprehensive screening of population against COVID-19 has been the main suggestion. Nevertheless, monitoring millions of individuals for a new virus is not easy. The development of successful tests for the new coronavirus and forthcoming possible viruses is and will be a global challenge to confront [95].

Conclusion

Rapid and reliable detection of the viral diseases is the primary objection for developing biosensing platforms. These developments are revolutionary in order to combat several viral infections we are facing with an increasing rate in the last years. Obviously, COVID-19 is not the only and last pandemic. There will be more pandemic challenges to be confronted.

Proposed biosensors are based on different sensing systems. These sensors can include more polymeric interfaces due to unique properties of polymers and reinforced surface coating technologies. These functionalized polymeric interfaces are prepared with special chemical constructions targeting several purposes, such as enzyme immobilization, catalysts and signal amplification. Nanotechnology- and polymer-based developments could be a game changer in development of reliable testing platforms for fast and effective diagnosis and monitorization of diseases resulted from infectious agents, such as Ebola, HIV, Zika, and SARS-CoV-2.

For survival of people, technological evolution and smart technologies will be the leaders of this combat. The current global outbreak has clearly uncovered that nanotechnology assisted novel biosensors platforms, machine learning-based technologies and wearable devices are imperative for health industry now and in the future. Robots, drones, masks, sensors and their combinations are shown to be used for diagnosis of COVID-19 [96]. Machine learning and smart-phone assisted systems should also be thoroughly examined in order to enhance the reproducibility and reliability of biosensors [84]. Since wearable devices provides continuous real-time data, detection of deviations from the "baseline" data of population will be whole story to keep the pulse of population's health and to monitor the spread of the disease (also the health of the population) in a timely manner [97,98].

References

[1] Y. Saylan, Ö. Erdem, S. Ünal, A. Denizli, An alternative medical diagnosis method: biosensors for virus detection, Biosensors 9 (2) (2019) 65. www.mdpi.com/journal/biosensors. (December 14, 2020).

[2] F. Cui, Z. Zhou, H.S Zhou, Molecularly imprinted polymers and surface imprinted polymers based electrochemical biosensor for infectious diseases, Sensors (Switzerland) 20 (4) (2020) 996, doi:10.3390/s20040996.

[3] A.A. Malik, C. Nantasenamat, T. Piacham, Molecularly imprinted polymer for human viral pathogen detection, Materials Science and Engineering C: Materials for Biological Applications 77 (2017) 1341–1348.

[4] A. Mobasheri, Biosensors for the multiplex detection of inflammatory disease biomarkers, Biosensors 11 (1) (2020) 11.

[5] L. Castillo-Henríquez, M. Brenes-Acuña, A. Castro-Rojas, R. Cordero-Salmerón, M. Lopretti-Correa, J.R. Vega-Baudrit, Biosensors for the detection of bacterial and viral clinical pathogens, Sensors 20 (23) (2020) 6926. https://www.mdpi.com/1424-8220/20/23/6926. (December 14, 2020).

[6] S. Afsahi, M.B. Lerner, J. Lee, X. Tang, D.A. Bagarozzi, D. Pan, L. Locascio, et al., Novel graphene-based biosensor for early detection of zika virus infection, Biosens. Bioelectron. 100 (June 2017) 85–88 (2018).

[7] A. Kaushik, A. Yndart, S. Kumar, R.D. Jayant, A. Vashist, A.N. Brown, et al., A sensitive electrochemical immunosensor for label-free detection of zika-virus protein, Sci. Rep. 8 (1) (2018) 3–7.

[8] J. Song, M.G. Mauk, B.A. Hackett, S. Cherry, H.H. Bau, C. Liu, et al., Instrument-free point-of-care molecular detection of zika virus, Anal. Chem. 88 (14) (2016) 7289–7294. 10.1021/acs.analchem.6b01632.

[9] C. Tancharoen, W. Sukjee, C. Thepparit, T. Jaimipuk, P. Auewarakul, A. Thitithanyanont, et al., Electrochemical biosensor based on surface imprinting for zika virus detection in serum, ACS Sensors 4 (1) (2019) 69–75.

[10] J. Lee, M. Morita, K. Takemura, EY. Park, A multi-functional gold/iron-oxide nanoparticle-CNT hybrid nanomaterial as virus DNA sensing platform, Biosens. Bioelectron. 102 (2018) 425–431.

[11] P. Weerathunge, R. Ramanathan, V.A. Torok, K. Hodgson, Y. Xu, R. Goodacre, et al., Ultrasensitive colorimetric detection of murine norovirus using nanozyme aptasensor, Anal. Chem. 91 (5) (2019) 3270–3276.

[12] C. Singhal, M. Khanuja, N. Chaudhary, C.S. Pundir, J. Narang, Detection of chikungunya virus DNA using two-dimensional MoS_2 nanosheets based disposable biosensor, Sci. Rep. 8 (1) (2018) 7734.

[13] M.M.N. Nuzaihan, U. Hashim, M.K. Arshad, S.R. Kasjoo, S.F.A. Rahman, A.R. Ruslinda, et al., Electrical detection of dengue virus (DENV) DNA oligomer using silicon nanowire biosensor with novel molecular gate control, Biosens. Bioelectron. 83 (2016) 106–114.

[14] K. Navakul, C. Warakulwit, P.T. Yenchitsomanus, A. Panya, P.A. Lieberzeit, C. Sangma, A novel method for dengue virus detection and antibody screening using a graphene-polymer based electrochemical biosensor, Nanomedicine: Nanotechnology, Biology, and Medicine 13 (2) (2017) 549–557.

[15] K. Kiilerich-Pedersen, C.R. Poulsen, T. Jain, N. Rozlosnik, Polymer based biosensor for rapid electrochemical detection of virus infection of human cells, Biosens. Bioelectron. 28 (1) (2011) 386–392.

[16] A.D. Chowdhury, K. Takemura, T.C. Li, S. Tetsuro, E.Y. Park, Electrical pulse-induced electrochemical biosensor for hepatitis E virus detection, Nat. Commun. 10 (1) (2019) 4–7.

[17] F. Zhang, L. Luo, H. Gong, C. Chen, C. Changqun, A magnetic molecularly imprinted optical chemical sensor for specific recognition of trace quantities of virus, RSC Adv. 8 (56) (2018) 32262–32268.
[18] P. Teengam, W. Siangproh, A. Tuantranont, T. Vilaivan, O. Chailapakul, C.S. Henry, Multiplex paper-based colorimetric DNA sensor using pyrrolidinyl peptide nucleic acid-induced AgNPs aggregation for detecting MERS-CoV, MTB, and HPV oligonucleotides, Anal. Chem. 89 (10) (2017) 5428–5435.
[19] Y. Fu, R. Yuan, L. Zhou, Y. Zhang, Coupling of a reagentless electrochemical DNA biosensor with conducting polymer film and nanocomposite as matrices for the detection of the HIV DNA sequences, Anal. Lett. 39 (3) (2006) 467–482. https://www.tandfonline.com/doi/abs/10.1080/00032710500536012. (January 30, 2021).
[20] W.-H. Kong, Y. Li, M.-W. Peng, D.-G. Kong, X.-B. Yang, L. Wang, et al., SARS-CoV-2 detection in patients with influenza-like illness, Nat. Microbiol. 5 (5) (2020) 675–678.
[21] N. Rabiee, M. Bargerzadeh, A. Ghasemi, H. Zare, S. Ahmadi, Y. Fatahi, et al., Point-of-use rapid detection of SARS-CoV-2: Nanotechnology-enabled solutions for the COVID-19 pandemic, Int. J. Mol. Sci. 21 (14) (2020) 5126.
[22] S. Talebian, G.G. Wallace, A. Schroeder, F. Stellacci, J. Conde, Nanotechnology-based disinfectants and sensors for SARS-CoV-2, Nat. Nanotechnol. 15 (8) (2020) 618–621.
[23] J. Majumder, T. Minko, Recent developments on therapeutic and diagnostic approaches for COVID-19, The AAPS Journal 23 (1) (2021) 14.
[24] B.A. Taha, Y. Al Mashhadany, M.H. Hafiz Mokhtar, M.S. Dzulkefly Bin Zan, N. Arsad, et al., An analysis review of detection coronavirus disease 2019 (COVID-19) based on biosensor application, Sensors 20 (23) (2020) 6764.
[25] M. Srivastava, N. Srivastava, P.K. Mishra, B.D. Malhotra, Prospects of nanomaterials-enabled biosensors for COVID-19 detection, Sci. Total Environ. 754 (2021) 142363.
[26] A. Parihar, P. Ranjan, S.K. Sanghi, A.K. Srivastava, R. Khan, Point-of-care biosensor-based diagnosis of COVID-19 holds promise to combat current and future pandemics, ACS Applied Bio Mater. (2020).
[27] C. Ihling, D. Tänzler, S. Hagemann, A. Kehlen, S. Hüttelmaier, C. Arlt, et al., Mass spectrometric identification of SARS-CoV-2 proteins from gargle solution samples of COVID-19 patients, J. Proteome Res. (2020).
[28] S. Mahari, A. Roberts, D. Shahdeo, S. Gandhi, ECovSens-Ultrasensitive novel in-house built printed circuit board based electrochemical device for rapid detection of NCovid-19 antigen, a spike protein domain 1 of SARS-CoV-2, bioRxiv (2020) 2020.04.24.059204(October 14, 2020), doi:10.1101/2020.04.24.059204.
[29] G. Seo, G. Lee, M.J. Kim, S.-H. Baek, M. Choi, K.B. Ku, et al., Rapid detection of COVID-19 causative virus (SARS-CoV-2) in human nasopharyngeal swab specimens using field-effect transistor-based biosensor, ACS Nano 14 (4) (2020) 5135–5142. https://pubmed.ncbi.nlm.nih.gov/32293168/. (October 14, 2020).
[30] G. Qiu, Z. Gai, Y. Teo, J. Schmitt, G.A. Kullak-Ublick, J. Wang, Dual-functional plasmonic photothermal biosensors for highly accurate severe acute respiratory syndrome coronavirus 2 detection, ACS Nano 14 (5) (2020) 5268–5277.
[31] P. Moitra, M. Alafeef, K. Dighe, M.B. Frieman, D. Pan, Selective naked-eye detection of SARS-CoV-2 mediated by N gene targeted antisense oligonucleotide capped plasmonic nanoparticles, ACS Nano 14 (6) (2020) 7617–7627.
[32] A. Djaileb, B. Charron, M.H. Jodaylami, V. Thibault, J. Coutu, K. Stevenson, et al., A Rapid and Quantitative Serum Test for SARS-CoV-2 Antibodies with Portable Surface Plasmon Resonance Sensing, ChemRxiv. Cambridge: Cambridge Open Engage (2020), doi:10.26434/chemrxiv.12118914.v1.
[33] S. Li, W. Jiang, J. Huang, Y. Liu, L. Ren, L. Zuang, et al., Highly sensitive and specific diagnosis of coronavirus disease 19 (COVID-19) by reverse transcription multiple

cross displacement amplification-labelled nanoparticles biosensor, Eur. Respir. J. (2020) 2002060.

[34] Z. Chen, Z. Zhang, X. Zhai, Y. Li, L. Lin, H. Zhao, et al., Rapid and sensitive detection of anti-SARS-CoV-2 IgG, using lanthanide-doped nanoparticles-based lateral flow immunoassay, Anal. Chem. 92 (10) (2020) 7226–7231.

[35] M. Feng, J. Chen, J. Xun, R. Dai, W. Zhao, H. Lu, et al., Development of a sensitive immunochromatographic method using lanthanide fluorescent microsphere for rapid serodiagnosis of COVID-19, ACS Sensors 5 (8) (2020) 2331–2337.

[36] W.E. Huang, B. Lim, C.-C. Hsu, D. Xiong, W. Wu, Y. Yu, et al., RT-LAMP for rapid diagnosis of coronavirus SARS-CoV-2, Microb. Biotechnol. 13 (4) (2020) 950–961.

[37] T. Wen, C. Huang, F.-J. Shi, X.-Y. Zeng, T. Lu, S.-N. Ding, et al., Development of a lateral flow immunoassay strip for rapid detection of IgG antibody against SARS-CoV-2 virus, Analyst 145 (15) (2020) 5345–5352.

[38] Lu Zeng, Y. Li, J. Liu, L. Guo, Z. Wang, X. Xu, et al., Rapid, ultrasensitive and highly specific biosensor for the diagnosis of SARS-CoV-2 in clinical blood samples, Mater. Chem. Front. 4 (7) (2020) 2000–2005.

[39] A. Ahmadivand, B. Gerislioglu, Z. Ramezani, A. Kaushik, P. Manickam, S.A. Ghoreishi, Femtomolar-level detection of Sars-Cov-2 spike proteins using toroidal plasmonic metasensors, arXiv (2020) 1–21, doi:10.48550/arxiv.2006.08536.

[40] H. Hou, T. Wang, B. Zhang, Y. Luo, L. Mao, F. Wang, et al., Detection of IgM and IgG antibodies in patients with coronavirus disease 2019, Clinical and Translational Immunology 9 (5) (2020).

[41] D. Alves, R. Curvello, E. Henderson, V. Kesarwani, J.A. Walker, S.C. Leguizamon, et al., Rapid gel card agglutination assays for serological analysis following SARS-CoV-2 infection in humans, ACS Sensors (2020).

[42] S. Mavrikou, G. Moschopoulou, V. Tsekouras, S. Kintzios, Development of a portable, ultra-rapid and ultra-sensitive cell-based biosensor for the direct detection of the SARS-COV-2 S1 spike protein antigen, Sensors (Switzerland) 20 (11) (2020) 3121, doi:10.3390/s20113121.

[43] Z. Chen, Q. Wu, J. Chen, X. Ni, J. Dai, A DNA aptamer based method for detection of SARS-CoV-2 nucleocapsid protein, Virologica Sinica 35 (3) (2020) 351–354.

[44] Q. Lin, D. Wen, J. Wu, L. Liu, W. Wu, X. Fang, et al., Microfluidic immunoassays for sensitive and simultaneous detection of IgG/IgM/antigen of SARS-CoV-2 within 15 min, Anal. Chem. 92 (14) (2020) 9454–9458.

[45] J.P. Broughton, X. Deng, G. Yu, C.L. Fasching, V. Servellita, J. Singh, et al., CRISPR-Cas12-based detection of SARS-CoV-2, Nat. Biotechnol. 38 (7) (2020) 870–874, doi:10.1038/s41587-020-0513-4.

[46] Z. Huang, D. Tian, Y. Liu, Z. Lin, C.J. Lyon, W. Lai, et al., Ultra-sensitive and high-throughput CRISPR-powered COVID-19 diagnosis, Biosens. Bioelectron. 164 (2020) 112316.

[47] C. Huang, T. Wen, F.-J. Shi, X.-Y. Zeng, Y.-J. Jiao, Rapid detection of IgM antibodies against the SARS-CoV-2 virus via colloidal gold nanoparticle-based lateral-flow assay, ACS Omega 5 (21) (2020) 12550–12556.

[48] Y.H. Baek, J. Um, K.J.C. Antigua, J.-H. Park, Y. Kim, S. Oh, et al., Development of a reverse transcription-loop-mediated isothermal amplification as a rapid early-detection method for novel SARS-CoV-2, Emerging Microbes and Infections 9 (1) (2020) 998–1007.

[49] C. Yan, J. Cui, L. Huang, B. Du, L. Chen, G. Xue, et al., Rapid and visual detection of 2019 novel coronavirus (SARS-CoV-2) by a reverse transcription loop-mediated isothermal amplification assay, Clin. Microbiol. Infect. 26 (6) (2020) 773–779.

[50] M.A. Lalli, J.S. Langmade, X. Chen, C.C. Fronick, C.S. Sawyer, L. Burcea, et al., Rapid and extraction-free detection of SARS-CoV-2 from saliva by colorimetric

reverse-transcription loop-mediated isothermal amplification, Clin. Chem. 67 (2) (2020) 1–28, doi:10.1093/clinchem/hvaa267.

[51] H. Zhao, F. Liu, W. Xie, T.C. Zhou, J. OuYang, L. Jin, et al., Ultrasensitive supersandwich-type electrochemical sensor for SARS-CoV-2 from the infected COVID-19 patients using a smartphone, Sensors & Actuators, B: Chemical 327 (2021), doi:10.1016/j.snb.2020.128899.

[52] B.D. Ventura, M. Cennamo, A. Minopoli, R. Campanile, S.B. Censi, D. Terracciano, et al., Colorimetric test for fast detection of SARS-CoV-2 in nasal and throat swabs, ACS Sensors 5 (10) (2020) 3043–3048, doi:10.1021/acssensors.0c01742.

[53] D. Huang, Z. Shi, J. Qian, K. Bi, M. Fang, Z. Xu, et al., A CRISPR-Cas12a-derived biosensor enabling portable personal glucose meter readout for quantitative detection of SARS-CoV-2, Biotechnol. Bioeng. 118 (4) (2021), doi:10.1002/bit.27673.

[54] L. Fabiani, M. Saroglia, G. Galata, R. De Santis, S. Fillo, V. Luca, et al., Magnetic beads combined with carbon black-based screen-printed electrodes for COVID-19: A reliable and miniaturized electrochemical immunosensor for SARS-CoV-2 detection in saliva, Biosensensors and Bioelectronics 171 (2021), doi:10.1016/j.bios.2020.112686.

[55] M. Djukanović, A. Mavric, J. Jovanovic, M. Roganovic, V. Boskovic, et al., Design of 3D printing thermo-sensored medical gear in detecting COVID-19 symptoms, Applied Sciences 11 (1) (2021) 419, doi:10.3390/app11010419.

[56] B. Mojsoska, S. Larsen, D.A. Olsen, J.S. Madsen, I. Brandslund, F.A. Alatrakcthi, et al., Rapid SARS-CoV-2 detection using electrochemical immunosensor, Sensors 21 (2) (2021) 390.

[57] H.-S. Kim, N. Abbas, S. Shin, A rapid diagnosis of SARS-CoV-2 using DNA hydrogel formation onmicrofluidic pores, Biosens. Bioelectron. 177 (2021) 113005.

[58] S. Ding, G. Chen, Y. Wei, J. Dong, F. Du, X. Cui, et al., Sequence-specific and multiplex detection of COVID-19 virus (SARS-CoV-2) using proofreading enzyme-mediated probe cleavage coupled with isothermal amplification, Biosensensor and Bioelectronics 178 (2021). https://linkinghub.elsevier.com/retrieve/pii/S0956566321000774, doi:10.1016/j.bios.2021.113041. (January 30, 2021) 113041.

[59] M. Dogan, L. Kozhaya, L. Placek, C. Gunter, M. Yigit, R. Hardy, et al., SARS-CoV-2 specific antibody and neutralization assays reveal the wide range of the humoral immune response to virus, Communications Biology 4 (1) (2021) 129. http://www.nature.com/articles/s42003-021-01649-6. January 31, 2021.

[60] R. Bruch, M. Johnston, A. Kling, T. Mattmüller, J. Baaske, S. Partel, et al., CRISPR-powered electrochemical microfluidic multiplexed biosensor for target amplification-free MIRNA diagnostics, Biosens. Bioelectron. 177 (2021) 112887.

[61] S.P. Varahachalam, B. Lahooti, M. Chamaneh, S. Bagchi, T. Chhiber, K. Morris, et al., Nanomedicine for the SARS-CoV-2: State-of-the-art and future prospects, Int. J. Nanomed. Vol. 16 (2021) 539–560.

[62] J. Yu, C. Kuwentrai, J.-D. Huang, C. Xu, Carbon-based nanomaterials for viral infection management, Biomicrofluidics 15 (1) (2021) 011501.

[63] H. Namazi, Polymers in our daily life, BioImpacts 7 (2) (2017) 73–74.

[64] N.H. Voelcker, B. Prieto-Simon, A. Shiohara, Porous polymeric membranes: Fabrication techniques and biomedical applications, J. Mater. Chem. B (2020). https://pubs.rsc.org/en/content/articlehtml/2021/tb/d0tb01727b. (December 29, 2020).

[65] R. Pauliukaite, E. Voitechovič, Multisensor systems and arrays for medical applications employing naturally-occurring compounds and materials, Sensors (Switzerland) 20 (12) (2020) 1–38. www.mdpi.com/journal/sensors. (December 29, 2020).

[66] S. Chattopadhyay, E. Heine, H. Keul, M. Moeller, Azetidinium functionalized polytetrahydrofurans: Antimicrobial properties in solution and application to prepare non leaching antimicrobial surfaces, Polymers 6 (5) (2014) 1618–1630.

[67] Ö. Erdem, E. Derin, K. Sagdic, E.G. Yilmaz, F. Inci, Smart materials-integrated sensor technologies for COVID-19 diagnosis, Emergent Mater. (2021).
[68] S. Cherré, N. Rozlosnik, Polymer based biosensors for medical applications, Advanced Polymers in Medicine, Springer International Publishing, 2015, pp. 513–537. https://link.springer.com/content/pdf/10.1007/978-3-319-12478-0.pdf.
[69] B. Lakard, Electrochemical biosensors based on conducting polymers: A review, Applied Sci. 10 (18) (2020) 6614.
[70] J.J. Long, A.M. Benoudjit, F.A. Arris, F. Ali, W.W.A. Wan Salim, et al., Polym. Biosens., Multifaceted Protocol in Biotechnology, Springer, Singapore (2018) 151–165.
[71] X. Wang, S. Uchiyam, Polymers for biosensors construction, State of the Art in Biosensors - General Aspects, InTech (2013).
[72] Y. Liu, M. Hao, Z. Chen, L. Liu, Y. Liu, W. Yang, et al., A review on recent advances in application of electrospun nanofiber materials as biosensors, Current Opinion in Biomedical Engineering 13 (2020) 174–189.
[73] R. Batool, A. Rhouati, M.H. Nawaz, A. Hayat, J.L. Marty, et al., A review of the construction of nano-hybrids for electrochemical biosensing of glucose, Biosensors 9 (1) (2019) 46.
[74] R. Ahmad, O.S. Wolfbeis, Y.B. Hahn, H.N. Alshareef, L. Torsi, K.N. Salama, et al., Deposition of nanomaterials: A crucial step in biosensor fabrication, Mater. Today Communicat. 17 (2018) 289–321.
[75] A. Kausar, Polymer coating technology for high performance applications: Fundamentals and advances, J. Macromolecul. Sci., Part A 55 (5) (2018) 440–448.
[76] D.D. Jurašin, M. Curlin, I. Capjak, T. Crnkovic, M. Lovric, M. Babic, et al., Surface coating affects behavior of metallic nanoparticles in a biological environment, Beilstein J. Nanotechnol. 7 (2016) 246–262.
[77] K. Shang, S. Song, Y. Cheng, L. Guo, Y. Pei, X. Lv, et al., Fabrication of carbohydrate chips based on polydopamine for real-time determination of carbohydrate–lectin interactions by QCM biosensor, Polymers 10 (11) (2018) 1275.
[78] Y. Saylan, S. Akgönüllü, H. Yavuz, S. Ünal, A. Denizli, Molecularly imprinted polymer based sensors for medical applications, Sensors 19 (6) (2019) 1279, doi:10.3390/s19061279.
[79] R.H. Senay, F. Ulucan, C.İ. Kuru, S. Akgol, Biosensors designed by molecularly imprinted polymers, in: A.E. Gursan, M.K. Sezgintürk, A. Denizli (Eds.), Smart Biosensing Technologies: Past, Present and Future, Hacettepe Üniversitesi, Ankara, 2018.
[80] G. Ertürk, Bo Mattiasson, Molecular imprinting techniques used for the preparation of biosensors, Sensors 17 (2) (2017) 288.
[81] T. Nandy Chatterjee, R. Bandyopadhyay, A molecularly imprinted polymer-based technology for rapid testing of COVID-19, Trans. Indian National Academy Eng. 5 (2) (2020) 225–228 /pmc/articles/PMC7293430/?report=abstract (December 20, 2020).
[82] O.I. Parisi, M. Dattilo, F. Patitucci, R. Malivindi, V. Pezzi, I. Perrotta, Monoclonal-type" plastic antibodies for SARS-CoV-2 based on Molecularly Imprinted Polymers, bioRxiv (2020) .05.28.120709, doi:10.1101/2020.05.28.120709.
[83] A. Raziq, A. Kidakova, R. Boroznjak, J. Reut, A. Öpik, V. Syritski, Development of a portable MIP-based electrochemical sensor for detection of SARS-CoV-2 antigen, Biosens. Bioelectron. (2021) 113029.
[84] S. Iravani, Nano- and biosensors for the detection of SARS-CoV-2: Challenges and opportunities, Mater. Adv. 1 (9) (2020) 3092–3103.
[85] Z. Zhao, H. Cui, W. Song, X. Ru, W. Zhou, X. Yu, A simple magnetic nanoparticles-based viral RNA extraction method for efficient detection of SARS-CoV-2, bioRxiv 2020.02.22.961268; doi: 10.1101/2020.02.22.961268
[86] I. Reiniati, A.N. Hrymak, A. Margaritis, Recent developments in the production and applications of bacterial cellulose fibers and nanocrystals, Crit. Rev.

Biotechnol. 37 (4) (2017) 510–524. https://www.tandfonline.com/doi/abs/10.1080/07388551.2016.1189871. (January 31, 2021).

[87] M. Velásquez-Riaño, V. Bojacá, Production of bacterial cellulose from alternative low-cost substrates, Cellulose 24 (7) (2017) 2677–2698. http://bowil.pl/en/. (October 26, 2020).

[88] K. Mahato, A. Srivastava, P. Chandra, Paper based diagnostics for personalized health care: Emerging technologies and commercial aspects, Biosens. Bioelectron. 96 (2017) 246–259.

[89] D. Hristov, C. Rodriguez-Quijada, J. Gomez-Marquez, K. Hamad-Schifferli, Designing paper-based immunoassays for biomedical applications, Sensors 19 (3) (2019) 554.

[90] L. Liu, D. Yang, G. Liu, Signal amplification strategies for paper-based analytical devices, Biosens. Bioelectron. 136 (2019) 60–75.

[91] C. Adrover-Jaume, A. Alba-Patiño, A. Clemente, G. Santopolo, A. Vaquer, S.M. Russell, et al., Paper biosensors for detecting elevated IL-6 levels in blood and respiratory samples from COVID-19 patients, Sens. Actuators B 330 (2021). https://linkinghub.elsevier.com/retrieve/pii/S0925400520316737. December 28, 2020 129333.

[92] J. Lim, Lee, Fabrication, flow control, and applications of microfluidic paper-based analytical devices, Molecules 24 (16) (2019) 2869.

[93] X. Weng, S. Neethirajan, Rapid detection of norovirus using paper-based microfluidic device, bioRxiv (2017) 162396 (January 31, 2021), doi:10.1101/162396.

[94] A. Basiri, A. Heidari, M.F. Nadi, M.T.P. Fallahy, S.S. Nezamabadi, M. Sedighi, N. Rezaei, Microfluidic devices for detection of RNA viruses, Rev. Med. Virol. 31 (1) (2021) 1–11.

[95] J.A. Berkenbrock, R. Grecco-Machado, S. Achenbach, Microfluidic devices for the detection of viruses: aspects of emergency fabrication during the COVID-19 pandemic and other outbreaks, Proceedings of the Royal Society A: Mathematical, Physical and Engineering Sciences 476 (2020) 2243. https://royalsocietypublishing.org/doi/abs/10.1098/rspa.2020.0398. (January 29, 2021).

[96] H. Khan, K.K. Kushwah, S. Singh, H. Urkude, M.R. Maurya, K.K. Sadasivuni, Smart technologies driven approaches to tackle COVID-19 pandemic: A review, 3 Biotech 11 (2) (2021) 50.

[97] M. Al-Emran, J.M. Ehrenfeld, Breaking out of the box: Wearable technology applications for detecting the spread of COVID-19, J Medical Syst. 45 (2) (2021) 20.

[98] H.C. Ates, A.K. Yetisen, F. Güder, C. Dincer, Wearable devices for the detection of COVID-19, Nat. Electron. 4 (1) (2021) 13–14 (January 31, 2021), doi:10.1038/s41928-020-00533-1.

[99] S. Al-Azzam, Y. Ding, J. Liu, P. Pandya, J.P. Ting, S. Afshar, et al., Peptides to combat viral infectious diseases, Peptides 134 (2020), doi:10.1016/j.peptides.2020.170402.

[100] X. Li, C. Ding, X. Li, H. Yang, S. Liu, X. Wang, et al., Electronic biopolymers: From molecular engineering to functional devices, Chem. Eng. J. 397 (2020) 125499, doi:10.1016/j.cej.2020.125499.

[101] Y. Yi, X. Luo, N. Xiong, Y. Liu, S. Li, et al., Development and clinical application of a rapid IgM-IgG combined antibody test for SARS-CoV-2 infection diagnosis, J. Med. Virol. 92 (9) (2020) 1518–1524, doi:10.1002/jmv.25727.

[102] A. Yakoh, U. Pimpitak, S. Rengpipat, N. Hirankarn, O. Chailapakul, S. Chaiyo, Paper-based electrochemical biosensor for diagnosing COVID-19: Detection of SARS-CoV-2 antibodies and antigen, Biosens. Bioelectron. 176 (2021), doi:10.1016/j.bios.2020.112912.

SECTION 2

Developments and Fabrication Strategies of Immunosensors

CHAPTER

5

COVID-19 Immunosensors techniques: Reframing policy and strategic development goals for the post-COVID-19 scenario

Nitika Thakur, and Vanshika AK Dev

Department of Biotechnology, Faculty of Applied Sciences and Biotechnology, Shoolini University of Biotechnology and Management Sciences, Solan (HP), India

5.1 Introduction

The strange is our next typical. This pandemic hugely affects worldwide instruction in a general and enormous sway on our Indian schooling framework specifically. This article manages the utilization of innovation in a pandemic condition. How educators and understudies are utilizing distinctive simultaneous and nonconcurrent mode in instructing and learning. The article will likewise discuss how the pandemic has uncovered the weakness and shortage of both school instruction and advanced education in India. The article will bring out in the open the difficulties the partners are looking during a pandemic. The point of convergence of this article is the ideas to patch up our educational abilities utilizing innovation. In conclusion, to close, we as a showing club in this current circumstance play a part to play. A job to be free, proficient, skillful, taught to hegemonize innovation and not the alternate way round. We can't deny reality; there are numerous chances accessible on the web, more than the reasons we interpret for not profiting it.

Sensing Tools and Techniques for COVID-19: Developments and Challenges in Analysis and Detection of Coronavirus
DOI: https://doi.org/10.1016/B978-0-323-90280-9.00001-8

Managing the anticipated difficulties of the "infection" has negatively affected individuals all throughout the planet. There is an exceptional ascent in the instances of individual influenced by this infection. Indeed, even our policymakers are fighting with the awful trial to choose whether the lockdown will even everything out and save lives. The genuine inquiry that we are battling today is the vulnerability that is sneaking around us. How long do we need to shut down our ordinary daily practice working? The strange is our next ordinary. This pandemic colossally affects worldwide instruction in a general and enormous effect on our Indian instruction framework specifically. With the abrupt lockdown declaration on 22nd March 2020, large numbers of us were intellectually ill-equipped about the term of this lockdown. The month was urgent for every one of the understudies in schools as their scholastic meeting was coming to an end, and they were in the last edge of their year-end tests. This vulnerability has left the understudies and educators in everlasting limbo. Not realizing how to cause the period we to have in our grasp useful. In Higher Education, understudies were associated with their undertakings, tasks, sessional and practicals. All stopped in a second. Numerous outstation understudies passed on the private inns to be in places of refuge of their home. The way they scuttered off, not all the understudies pressed their investigation materials for their home. A significant number of their books, freebees, and notes are locked upon their inns. Educators, as well, have left their scholarly exercises and records in their specializations and schools. As we as a whole are familiar with depend on the printed versions. The end in all the ordinary working and cycles left everybody troubled what's more, destroyed.

Since the beginning of 2020, the world has seen a progression of mind-boggling outlook changes in around each line of business, from gigantic expansions popular in certain ventures to absolute loss of interest in others. The world and the various ages are becoming accustomed to telecommuting or figuring out how to convey items or organizations distantly. This immense extremist shift shows that the future will be unique, and we need to set ourselves up to another distinctive time that needs another world-view and above all, unique ability [1,2]. This exploration paper investigates and premonition the capability and the employability ability needed for the post-COVID-19 time, called the "new ordinary." It is a time loaded with difficulties also, shakiness, where deftness, interest, hazard relief, learning by investigating, learning by doing, yet with center, would be the standard for both endurance and seriousness. The specialists attempt to plan the new capability necessities, with center around the significance of 'request-based learning' since it tends to a considerable lot of the 'new ordinary' prerequisite, for example interest, learning by investigating, and so forth [3,4].

5.2 Characterizing the new normal

Oxford word reference (2020) characterizes the new typical as an occasion, or status, or time that already was new, or abnormal circumstance that has become standard, or regular, or anticipated. A Coronavirus pandemic is an unexpected occasion, yet this new typical can likewise be an outcome of more extensive changes in the cutting-edge world (like environment changes, movement, wars, destitution, hunger); other than the advancement of the innovation. Hence, the new postpandemic reality, the new typical, will be a consequence of all referenced changes. Notwithstanding, the perspectives and practices we needed to follow during COVID-19 pandemic have formed our life, in a couple months, similar to no prior decade in ongoing history. With the endeavor of the nations and the networks for powerful control of this infectious sickness, the world is encountering an immense disturbance to every one of the financial exercises and public activity. The extreme changes that individuals needed to do because of the board of dangers; for example the need to respond, to maintain the new guidelines furthermore, to be ready for the most noticeably awful; have shacked-up the premise of many accepted ideas of life. Indeed, even presently, we changed how we see working together and convey administrations later on [1,2]. Since this pandemic may have a progression of episodes, it is relied upon to have various stages where we need to respond, then, at that point acknowledge and consider the substance of the unexpected significant life challenges. As the world would endeavor to continue its action towards the better usefulness what's more, reformist development, it would have to go through fluffiness and would have to reshape itself to fit the pre-essentials of the following new ordinary [36,37]. The happening to next skylines of the post-COVID-19 time could be described as the criticalness existing apart from everything else' period. Subsequently, the ability here is about the limit of not losing sight of the activities that may be required for later [38]. The new ordinary comes to guarantee the local area and associations quick reaction and the proficient variation to change. This would guarantee that the world would be more prepared for its reappearance with a more grounded position after each shock, or emergency or worldwide test. Buheji [1] referenced that the 'following ordinary' or the post-COVID-19 period would not be like the pre-COVID-19 years. The pandemic would contact the principal issues of life, what we put stock, by the way we think, how we imagine our job throughout everyday life, our cutting-edge basics how we would respond to a coming life emergency. Thusly, we need another method of setting up and overseeing employability skill to meet this difficult time [36,39] noticed that in the new typical, we would observer changes in the speed and nature of preparing for telecommuters. The associations would have as a feature of their bundle, the consideration for the laborer future prosperity, consequently to high employability skill interest. Practically all driving associations would chip away at re-skilling its representatives, or enroll as per the new ordinary necessities [5].

COVID infection (Coronavirus) is a sickness that is brought about by a new serious intense respiratory disorder COVID 2 (SARS-CoV-2). Illnesses like the extreme intense respiratory condition (SARS) and the Center East respiratory disorder (MERS) revealed in 2002 and 2012 separately were brought about by infections like SARS-CoV-2. In any case, SARS-CoV-2 has a more extensive range than the recently detailed comparative infections and along these lines the trouble in the treatment and the board of Coronavirus. Clinical conclusions and explores have shown that people tainted with Coronavirus can be indicative or asymptomatic in the beginning phases of contamination relying upon the individual's resistant framework. It is accounted for that side effects of the sickness incorporate dry hack, fever, sluggishness, windedness, cerebral pain and general body shortcoming because of prompting of different afflictions in the body. The curiosity of SARS-CoV-2 and its obscure hereditary nature result to trouble in treatment of Coronavirus and the improvement of preventive immunizations. Coronavirus was first announced in the city of Wuhan, China in 2019 and has kept on desolating the entire world from that point forward. On eleventh of Walk 2020, the World Wellbeing Association (WHO) proclaimed Coronavirus a pandemic dependent on the rate at which it was spreading and annihilating living souls all throughout the planet. Starting at April 23, 2020, around 2 million individuals have been affirmed to be tainted with the illness across the world with in excess of 182,000 affirmed passing's related to the infection. South Africa detailed its initially affirmed instance of Coronavirus on 5 Walk 2020 and about a month afterward the quantity of affirmed cases expanded to 1585. Starting at April 23, 2020, the number of affirmed cases in South Africa rose to 3953 with 75 related passing's. Scientists accept that COVID is sent through beads from a contaminated individual. When such drops fall on surfaces and somebody interacts with it, for the most part with the hand and from there on contacts the nose or the mouth with the hand, such an individual would be tainted with the infection. Therefore, clinical specialists propose that social separating stays the most ideal approach to contain the spread of the infection. Governments from one side of the planet to the other, including South African government carried out lockdown in their different nations dependent on the seriousness of the pandemic to guarantee that its spreading is contained and the bend leveled.

Exploring Forward the Teaching method of Mixed Learning [6]. The public authority of South Africa has carried out a severe country-wide lockdown to straighten everything out and thus contain the spread of the COVID. Lockdown is compelling in controlling the spread of Coronavirus;

however, it immensely affects the economy. Lockdown involves that organizations are shut and occupations are suspended (aside from those considered as fundamental administrations) and schools are likewise shut. Be that as it may, there is no presently accessible model or experimentally based report that shows positive connection between conclusion of schools and controlling of the spread of Coronavirus. All instructive organizations in South Africa have been briefly closed down to contain the spread of the Coronavirus pandemic. The inquiry has been the means by which students would have the option to make up for the time of lockdown since instructive frameworks in the nation are organized with authoritative schedules. Elective techniques for educating, for example, Mixed Learning can be a valuable intends to guarantee learners" admittance to learning exercises during the time of public lockdown. This paper is intended to basic break down the effect of Coronavirus on the South African instruction area and the chance of coordinating Mixed Learning in the school educational program.

5.3 The effect of Coronavirus on instruction

Coronavirus pandemic has come about to add up to conclusion of schools in around 192 nations all around the world with 91.4% of the complete number of selected students in these nations briefly constrained out of school. It is accounted for that over 1.6 billion students across the world are as of now constrained to avoid schools as friendly removing is being authorized locally and locally all throughout the planet to contain the spread of COVID sickness. Report shows that lockdown of schools is more unmistakable in certain mainlands like Africa, South America what's more, in certain pieces of Europe. It very well may be imagined from situation that Africa and South America are the landmasses which instructive frameworks have been generally influenced by the pandemic as more that 98% of instructing and learning can't be directed because of nationwide lockdown in these mainlands. The conclusion of schools in North America, Australia and Northern Europe is generally lower when contrasted with different pieces of the world and this can be credited to the way that lockdown of schools is just viable in limited regions where the spread of the illness is more pervasive. Exploring Forward the Instructional method of Mixed Learning [6].

Other than drug intercessions and other preventive measures, social removing has been the best method of limiting the spread of Coronavirus. The powerful execution of social separating requests that schools be shut however long every administration is sure that the pandemic has been shortened enough for the wellbeing of students and educators. Significant stretches of learning would be lost however long the terminations endured.

Reports that schools losing extensive stretches of learning because of illness episode can result to both fleeting and long-lasting harm on instructive framework. The fleeting harm incorporate disturbance of educational program which could consume most of the day to be recuperated while the extremely durable harm incorporate the way that a few students may always avoid school in any event, when the sickness flare-up is finished. However the affirmed instances of Coronavirus and related passings in African nations have been low contrasted with different mainlands, it is imagined that the instruction of an African kid would be the most influenced post Coronavirus period because of monetary and innovative backwardness of most African nations.

5.4 Innovation and pandemic

The main beam of expectation in the present circumstance is the "Innovation" we had in our grasp. A similar innovation which we were in a split second utilizing for human correspondence due to social removing. This social separation destitute us of human touch accordingly web and innovation is in effect widely used to restore the human association. Training Area is too one region in which the pandemic made every one of the partners move from their usual range of familiarity and experience what innovation has to bring to the table. However obviously, this innovation might not have similar effect in every one of the financial gatherings of our general public. Youngsters from low financial foundations don't approach what instructive innovation has to bring to the table. Notwithstanding, it very well may be seen as time passes that how instructors, who were once for the most part acclimated with the physical study hall and eye to eye connection, are pushing ahead to take on new strategies for e-educating and e-learning. In truth, the two instructors and understudies are learning together concerning how to make full advantage of functionality of instructive innovation. The more they are getting active experience of it, the more they are propelled to try different things with it. Pandemic has behaved like an impetus and has pushed educators, understudies and chairmen to devise inventive arrangements in a brief time frame. The genuine pith of time might be felt now in this lockdown condition, particularly for an understudy. Hence educators are enticed to foster learning material utilizing an intelligent application, offbeat online instruments like Google homeroom, WhatsApp bunch, Facebook courier or email. Simultaneous eye to eye sound and video communication and learning instruments ZOOM, GoToMeeting, WEBEX, Skype and so forth the pandemic has revealed the weakness and shortage of both school training and advanced education in India. There are many schools and organizations which don't have the offices for e-instructing or e-learning. The employees, particularly the senior workforce, are not educated, and they come up short on the abilities needed for e-instructing. Understanding this, the Human Asset Advancement Center in the College has immediately dispatched on online personnel improvement Projects for every one of the instructors. A few online courses have been started by SWAYAM, UG/PG MOOCs, e-PG Pathshala, and so forth. Another basic change we see in online schooling is the overabundances of sends welcoming the employees, research researchers, and postgraduate understudies to go to online courses. These online courses are resolving the consuming issues of utilizing and acquainting the members with learning devices. The primary target is for the members to shed their hindrance. To make inspired students is essential. We are confronting an advanced split among spurred and noninspired students. We, as students need to foster the interior locus of control with the goal that we are better prepared to control our climate. There will be ramifications for each activity of our own.

5.5 Difficulties ahead

It is for us all to recheck and reevaluate whether the postcoronavirus has adjusted our general public to improve things or in negative ways. Grave fears are being felt in our advanced education framework. Overall, there is the restricted utilization of e_resources, ineffectual conveyance content and the flexibility of assessment which mirrors that we presently can't seem to coordinate with other arising nations who are utilizing instructive innovation to its fullest. Another troubling issue is the monetary downturn we are confronting which enormously affects the employability of our understudies, paying limit of understudies in private foundations and visa limitation of seaward understudies took on colleges abroad. The effect is additionally on the recruiting administrations of the workforce influencing the quality and greatness. Indeed, even the low charge can demonstrate major impediments in dealing with the functioning capital. The internet instructing learning has additionally influenced the in an unexpected way abled understudy. The course evolved online isn't in consistence with the instructive requirement for outwardly and hearing tested students. This is one more key region in which the educators must be prepared to utilize instructive innovation as per their requirements Techno-teaching method as the pandemic has made internet instructing, as the main enduring mode to show the understudies in future. Consequently, that time isn't far away when the best workforce will be looked for across the globe with no political limits. The best workforce will be the one with subject ability and IT abilities. Our showing strategies need extreme patching up and improvement.

Techno-teaching method can be characterized as "Electronically interceded courses that coordinate sound educational standards of instructing/learning with the utilization of innovation." For an instructor, it is without a doubt elating to see their understudies face light up when they are clarifying something baffling and troublesome understand. The productive educator takes convoluted points, reduces them down to their uncovered substance and draws the associations between various ideas. This assists them with comprehension and holds the topic. At the point when instructors are utilizing internet showing teaching method, they ought to realize it isn't assembling the slides and whipping through 40–50 slides in 60 minutes, where everything springs up on the screen. You cannot empty huge course of information into a cup. The thing we teachers ought to comprehend is that we have a greater crowd online that we need to fulfill. And furthermore we need to fulfill our understudies simultaneously. To show someone through some medium (for this situation, is the internet educating learning applications) we better know it well enough to have the option to utilize it to educate them. Utilizing innovation successfully is the thing that we need to dominate now. Every last one of us is prepared to educate and clarify. We need to take on "Learning at the speed of clarification," extremely basic to clarify it to any even out of students. One way is to utilize chalk and talk on the web, which permits us to take into account the distinctive instructive requirements of the understudies. They can record, respite and replay the video. What's more, they can likewise control our speed of show by tapping on 1.5 or 2.0 speed and get the pith. Something else is, utilizing the jargon to get to the crowd and assist them with comprehension. In the event that we can clarify something indeed, then, at that point we ought to have the option to disclose it to pretty much any degree of students, by having a decent supply of words, word usage and manner. The primary concern is, we need to support the inspiration of the understudy with short video clasps of ideas, as it is hard to fold the human mind over some idea or thought with extended recordings or verbiage.

5.6 Changing globalization – The conditions of the postpandemic worldwide monetary reality

The writing features a few ideas concerning the idea of globalization and the quantity of its stages (waves), the World Bank and Oxford College Press (2002) propose three waves, highlight two. As per the three-wave idea, since the turn of the 1980s and 1990s, the worldwide economy has been working in the quick development environment of ICT advancements, worked with communication and a decrease in its expenses. It is a reason for the idea of the demise of distance. The ICT advancement has added to the lessening job of distance while expanding global ventures' significance (MNE). MNEs, having proprietorship benefits, following the OLI worldview by, have taken advantage of the new financial climate, in a circumstance of falling vehicle and media communications and quick data trade, to strengthen worldwide ties inside creation chains and organize exercises in distant areas. The Web has turned into a channel for trading data, a method for conveyance (especially for administrations) and a commercial center for doing exchanges. The hopeful assumptions for an authoritative passing of distance, notwithstanding, have facilitated after some time. As indicated by the new monetary geology (NEG), distance stays fundamental in creating worldwide monetary relations. The inescapable utilization of gravity models in financial aspects affirms that distance is as yet critical in forming exchange and capital relations TRANSNATIONAL Enterprises Survey 75. As of late, we have noticed a lull in the force of globalization measures, which is reflected in a decrease in worldwide FDI streams. The alleged slowbalization, combined with the pandemic, is changing the principles of the game. The interruption of worldwide chains is causing a shift towards provincial financial mix and local economic alliance. The presence of an inward market, a significant degree of improvement, and the elevated place of numerous MS on the worldwide innovation stepping stool make the EU an appealing objective for FDI. The most recent European Advancement Scoreboard results demonstrate that the EU has overwhelmed the USA, thought about the innovative pioneer and other nations, working on its exhibition by 8.9 rate focuses (p.p.) somewhere in the range of 2012 and 2020. The EU nations mirror the least paces of the limitation record, as per OECD. It makes the EU where, for example, financial backers from Asia can take part in mechanical sourcing. Therefore, profoundly aggressive EU organizations, regularly at the mechanical wilderness, may succumb to antagonistic takeovers on the grounds that the pandemic circumstance has debilitated them Worldwide reshuffles, heightened by Coronavirus, will change world areas' situation in the global division of work. The pandemic circumstance will make new local and worldwide settings, and new failures and champs will arise. China and other Asian nations' solid situation as FDI objections might crumble, preferring topographically appealing nations working inside provincial joining groupings. The pandemic emergencies will firmly influence the Asian worldwide worth chains (incl. China), as there is noticed worldwide pressing factor for creation broadening. The shift of MNEs' localization inclination has been seen since somewhere around 2018. The pandemic flare-up fills in as an impetus for these cycles (Joined Countries Meeting on Exchange and Improvement, 2020). It upholds H1, as Poland is a MS of the EU. The normal expanded rivalry for FDI is probably going to influence individual economies in a different way.

5.7 How might COVID-19 change worldwide wellbeing postpandemic? Characterizing exploration and venture openings and needs

Coronavirus addresses the best danger to worldwide general wellbeing and economies in the 21st century. Until now, in excess of 100 million individuals have been tainted and over 2.1 million kicked the bucket; probable a hint of something larger of this staggering pandemic. The effect on different pandemics and wellbeing challenges is veiled by the proceeded with strength of COVID-19 on our lives. The financial and social effect is impressive and developing. In the United States (US) alone, the pandemic's absolute expense is assessed at more than US$16 trillion, or roughly 90% of the country's yearly GDP No nation or locale has been saved; the emergency has uncovered distinct shortcomings in pretty much every medical services framework. Inside and between nations we have seen how currently defenseless and minimized populaces bear a lopsided weight of disease, with inversion of the improvement gains made to date.

The pandemic has additionally featured our association. To accomplish another typical requires a planned, intelligible and firm reaction that is viable universally for manageability. For quite a long time to come, we will without a doubt proceed with basic reflection on why a few nations, including less-resourced nations, have had the option to react all the more successfully to COVID-19, while some well-resourced nations have battled. For the present, a high need in finishing the pandemic, just as speeding up the post-COVID-19 recuperation, should figure out what a successful worldwide reaction to COVID-19 involves. The papers shaping this PLOS Special Collection, Post pandemic: How will COVID-19 change worldwide wellbeing? were considered as a guide to catalyze a worldwide wellbeing examination and strategy plan important to end the pandemic, shape the post-COVID-19 worldwide wellbeing recuperation and establish the frameworks for a stronger worldwide scene to accomplish the objectives of manageable turn of events. While in no way, shape or form extensive, we feature a few basic subjects: interests in better pandemic readiness; more empowered endeavors to address underlying and fundamental disparities inside the reasonable advancement structure and subsidized medical coverage; improving strength of our medical services conveyance frameworks; and accomplishing more noteworthy responsibility for our activities as a danger to one represents a danger to all.

As papers in this Collection diagram, COVID-19 has highlighted the requirement for supported interest in worldwide wellbeing research. The speed and size of immunization improvement endeavors, the multiplication of symptomatic instruments and the advancement of a few demonstrated and promising helpful specialists, are all proof of where worldwide science has prevailed and reflect many years of interests in biomedical examination that could be quickly brought to bear despite this new danger. Regardless, SARS-CoV-2 was an infection we ought to have been prepared for; all things considered, we have been arranging and getting ready since the 1918 flu pandemic. We had effectively experienced the SARS and MERS pandemics: these COVIDS were at that point on the WHO Blueprint program need list, huge scope studies had as of now distinguished an expansive scope of SARS-like COVIDS in the potential supplies and instruments for primary examination and expectation of host-range were prepared So for what reason would we say we were not more ready for this pandemic and ready to contain the worldwide transmission or effectively foresee have range? At the worldwide level, arrangements were insufficient and divided. Also, planned, straightforward and comprehensive worldwide exploration is basic to advise future pandemic readiness. Interpreting and growing our current information on supplies, infections, and drivers for sickness development into a program for early notice and prescient episode hazard is probably going to be testing, yet totally basic to moderate against future pandemics, and will request facilitated, all around subsidized worldwide wellbeing and resourced research endeavors.

Second, COVID-19 has enhanced long-standing foundational and underlying worldwide wellbeing imbalances [6], remembering for destitution, admittance to medical care, race, identity, sexual orientation, and social incohesion. Original copies in this Collection try to feature how tending to these disparities should be both a neighborhood and a worldwide need. While the outcomes of passings from COVID-19 in East Los Angeles are ramifications for California, they are likewise ramifications for the remainder of the world, particularly given the proclivity of this infection for change and transmission. Reestablished consideration should be centered on how worldwide wellbeing endeavors that address neighborhood wellbeing imbalances are vital to tending to worldwide inconsistencies. Indeed, even before the COVID pandemic, social, financial and wellbeing disparities had turned into a predominant worldwide story, with expanding insightful consideration zeroed in on how worldwide wellbeing should have been de-colonized to change power irregular characteristics that sustain these imbalances inside and between countries. Residents universally are all the more noticeably communicating outrage and bigotry of suffering wellbeing, social, and financial shameful acts. Primary imbalances repeated inside the worldwide wellbeing framework itself feature the absence of basic commitment with the political and social determinants of wellbeing differences. While the worldwide reaction to COVID-19 has generally supported these treacheries, post-COVID-19 there is a chance to change worldwide wellbeing through a plan of repoliticizing and rehistoricizing wellbeing, expanding on the recharged and basic mindfulness brought to consideration by the current pandemic.

Third, COVID-19 has filled in as a "stress test" for wellbeing frameworks the world over. Some strong wellbeing frameworks have reacted quickly and successfully to the pandemic, while numerous others have not. The healthy exhibition of East Asian nations in reacting to the current pandemic, clarified partially by engraving and gaining from past episodes, may offer a model for how wellbeing security should be focused on in the post-COVID-19 period. Conversely, the useless presentation of a couple of big league salary countries delineates the risks of under-putting resources into general wellbeing frameworks to adapt to wellbeing emergencies of this scale. Fair admittance to antibodies in the recuperation cycle will likewise be a significant pressure trial of worldwide coordinated effort even with rising immunization patriotism. Creating protected and successful antibodies alone won't be sufficient to end the pandemic, except if those immunizations can be conveyed around the world at a value reasonable to all administrations and dispensed in a manner that amplifies general wellbeing sway and accomplishes value. This Collection will investigate the job of the COVID-19 Vaccine Global Access Facility (COVAX) drive in empowering big time salary countries to take an interest to stay away from a 'misfortune of the halls by extending worldwide antibody supply and conveyance, while likewise creating other positive results [7].

Accomplishing quick pandemic control is conceivable yet is predicated on unequivocal initiative and joint effort for COVID-19 reactions—at worldwide, public and nearby levels—and a pledge to abandon nobody. Worldwide wellbeing strategy plays a basic part to play in catalyzing governments and nonstate entertainers to authorize successful, creative and just approach arrangements. An "each country for itself" approach obviously doesn't hold up in this related world, and we need changes. Specifically, the International Health Regulations, which oversee all nations to have center wellbeing framework abilities to recognize future pandemics, should be revived. More compelling principles of the street are likewise expected to encourage collaboration among nations looking to oversee future episodes and as a worldwide trigger for the United Nations and other global associations to make proper moves. An allaround supported WHO is additionally vital for successful worldwide wellbeing administration and offers a model for worldwide joint effort. The WHO plays a basic part to play in supporting all nations to focus on all inclusive wellbeing frameworks, not just in light of the fact that it will incomprehensibly further develop wellbeing and be a significant rampart against future pandemics, yet in addition to procure checked financial profits.

Despite the fact that we have a lot to find out with regards to SARS-CoV-2, the plague [7] and its results, the infection has made one thing understood: for any emergency that undermines the globe, the issues of any of us are the issues of us all. Worldwide postpandemic recuperation should in this manner be composed and multi-dimensional. Administration frameworks that are comprehensive, responsible and directed by approaches that focus on straightforward, multisectoral dynamic cycles are desperately expected to react successfully. Just an all-encompassing reaction dependent on cross-sectoral cooperation at all degrees of society can construct the important flexibility to react to the prompt and long haul impacts of COVID-19. The COVID-19 pandemic advises us that no nation acting alone can react viably to wellbeing dangers in a globalized world [13]. The emergency has additionally set out an exceptional freedom to rethink and change worldwide wellbeing so future pandemics are not close to as destroying as this one, and that wellbeing gains made to date are maintained and fortified as opposed to turned around.

5.8 Economic heath and COVID

Monetary and social frameworks have been seriously disturbed by the COVID-19 pandemic. Carriers have either shut totally or seriously decreased tasks, with admonitions of corporate disappointment and demands for government monetary help. The car industry has seen markets and creation dive. Shops, cafés and bars have shut. Donning settings and occasions have been delayed or dropped. Colleges and schools have hustled to take on internet learning, while at the same time pondering what the following scholarly year holds. The Association for Petroleum Exporting Countries (OPEC) and its partners have guaranteed a 10% reduction in raw petroleum creation to reinforce costs that were falling beneath US$25 per barrel. Public transport has been disregarded as travelers look for spatial division from one another. After numerous long stretches of starkness following the 2008–2009 monetary emergency, even with the current calamity governments have abruptly had the option to manage the cost of financial salvage bundles that only a couple weeks sooner would have been utter horror. In parallel, the pandemic has uncovered the delicacy of numerous public wellbeing frameworks, the absence of safety in the "gig economy," and the dependence on without a moment to spare least stock worldwide worth chains. For the situation of the Midwestern locale of the United States, for model, gauge between 9,000,000 and 26 million potential employment misfortunes, equivalent to somewhere in the range of 14 and 18% joblessness. Before COVID-19, there was at that point an emergent talk that the globalization of neo-liberal financial aspects was floundering in the midst of numerous strains to social, ecological, and monetary frameworks Pointers of this approaching disappointment incorporate a dial back in worldwide development in gross homegrown product (GDP) and expanding levels of individual, corporate, and government obligation. Solid patriot, protectionist feelings had arisen in the

United States, the European Union and somewhere else, making a pressure or fundamental clash among globalization and patriotism and making some inquiry the corporate rationale of worldwide system. Consequently, COVID-19 might have acted to speed up and concentrate a portion of these eminent strains. The COVID-19 pandemic has as of now started vivacious discussion by means of various media on the job of the state in contemporary social orders, yet generally in monetary terms. Commitment with clients, strategy producers, and other social entertainers can, notwithstanding, be improved by the deduced advancement of scenarios as an instrument to take part in the administration of changes and to energize a zero in on more extensive maintainability issues. Consequently the scenarios proposed in this approach brief are planned for such a job. Situations are helpful appearance gadgets to induce talk around potential fates. Situations are normally utilized to accumulate a multiplicity of viewpoints from various partners on potential fates that might be more, or less, great. In this approach brief the utilization of situations is unique. The situations are presented as elective dreams of the future to invigorate and direct contributions from partners. They permit partners to be more reflexive, to mull over their present arrangement of the circumstance and to participate in a logically educated thought regarding the future from this point. The commitment takes the accompanying structure. To start with, the premise of elective advances is illustrated, drawing "all in all framework" point of view of advances hypothesis. From there on, the four likely situation thumbnails are depicted, and afterward related against the staggered viewpoint (MLP) ideas of landscape, system, and specialty. At long last, the discussion segment considers the ramifications of the contemporary circumstance for the possibilities of the situations happening.

5.9 Elective metachanges

Elective meta-changes Changes hypothesis has normally been utilized to break down explicit parts of contemporary socio-economic frameworks like energy, portability, or food. Nonetheless, the COVID-19 pandemic has had an amazing effect at a more elevated level of total. While explicit frameworks and spots are not all influenced similarly, the COVID episode has been prominent for being worldwide, fast, and astoundingly inescapable in the manner it has upset existing practices. The emergency is, accordingly, in the language of the staggered viewpoint, a meta-progress occasion at the scene level that saturates into numerous systems simultaneously. Advances hypothesis considers and coordinates different framework components into an investigation, including innovation and advancement, markets, business, government, practices and standards, administrative and governance systems, and change pathways. The situations are extended upon in the following segment, the intricacy of the circumstance at worldwide and public levels is quite diminished, and it is possible that particular observational request is expected to research the system and nation level ramifications of the post-COVID-19 period. Notwithstanding, the prevailing conditions for foundational change are characterized in terms of two primary attributes: common society (the state, administration, and guideline), and the economy (business, markets, money, and financial matters). In a sense this double situating mirrors the global national strains implied above, where components of society are scrutinizing the advantages of lower levels of abundance creation and material consumption. This hypothetical position is viable with a "Armageddon" or "calamity" viewpoint, for model, found in apprehensions of an atomic winter in the occasion of a full scale atomic conflict In the event that common society is more grounded in the post-COVID-19 period, there will be improved authenticity to seek after common and social plans. The supremacy stood to business and the economy is disintegrated. Henceforth, the capacity to compel biological weights becomes possible. Once more, there are two variations that can be imagined. With business and the entrepreneur economy still unblemished, and realism for the most part actually upheld, the COVID-19 pandemic could give the conditions to a reestablished and worldwide "green new arrangement" fixated on feasible utilization and the circular material economy. This viewpoint is viable with hypotheses on green development and the de-coupling of environmental weights from monetary action On the other hand, a more extreme vision for cultural prospects would result in the intentional contracting of the material economy as an oversaw withdrawal of monetary movement. This last situation is viable with the hypothetical viewpoint of degrowth. These situations are examined underneath. Globalization. The pandemic has influenced various pieces of the economy differentially. Those personally concerned with globalization have been liable to the most grounded imperatives. Subsequently global air travel, transporting, furthermore, exchange have all endured. Exercises including the assembly of individuals have likewise endured, while those including home conveyance of items and services, or clinical and drug exercises have succeeded. All things considered, private companies and the independently employed are more weak now, while in the future it is likely that capital-escalated ventures that need significant degrees of limit use will likewise be helpless. Nonetheless, as far as the situations, the thought of shortcoming in business and the economy relates more to general full scale monetary conditions where deficiency of capital, feeble interest, and delicate supply chains imply that the suppositions that support everyday business action may no longer apply. On the off chance that common society is more fragile in the post-COVID-19 period, then, at that point, it is feasible to conceive two wide results. In the principal result, when entrepreneur monetary designs stay unblemished and when incitement measures are fruitful there is an expansive get back to "the same old thing," alongside the predominant directions on fossil fuel byproducts, contamination, asset utilization, and different pointers of environmental pressure. This situation can be likened to neoliberal financial aspects and the unregulated economy framework, though with different public varieties. In the subsequent result, a powerless common society combined with slight business and financial frameworks brings the possibility of "turbulent progress" wherein there will be a quick yet, sloppy framework rebalancing around much.

Under "(return to) the same old thing" conditions, environmental weights will keep on ascending as worldwide constructions of creation and utilization return. Governments and global offices are now taking drastic actions with the goal that economies don't breakdown, a re-visitation of "ordinary" life will be conceivable. Recuperation of securities exchanges and other monetary institutions to pre-COVID 19 levels will, best case scenario, take a time of years. Along these lines, the re-visitation of the same old thing is probably going to remember a momentary decrease for carbon dioxide (CO_2) emanation levels (for instance), as likewise happened during the 2008–2009 monetary emergency, before the drawn out pattern line gets back with growing contamination. In addition, the longing to restart economies could bring about the delay of continuous carbon-decrease guideline and other natural measures, remembering for transportation. It is currently evident that the COVID-19 emergency has precipitated cracked connections inside socio-technical frameworks at scene and system levels. The set up specialists of worldwide adjustment, for example, the World Trade Organization, World Bank, United Countries, World Health Organization, and Inter_Governmental Panel on Climate Change had, earlier to the development of this infection, seen a disintegration of their authenticity and suitability. The risk here is that without such balanced governance, the revisitation of the same old thing will speed up the natural emergency. Under the other three situations, it is proposed that there is space for good faith, given the very genuine social and political obstacles can be conquered. It very well might be viewed as that turbulent change and overseen degrowth are more outlandish than overseen progress or a re-visitation of the same old thing. Nonetheless, an occasion of this extent tosses forward different unexpected repercussions. Here, the issue of time is vital. The more extended that the emergency perseveres, the more far off a re-visitation of the same old thing becomes. Nations are differentially situated, being pretty much married to the worldwide economy, pretty much ready to improve independence, and with liking to realism overall. These measurements are probably going to decide the pathway embraced hereon. In the "Oversaw Transition" situation, decoupling asset utilization from GDP, more prominent global political agreement, more grounded guideline and strategy toward business, and consolation of emergent innovations are probably going to prompt a net decrease in natural weights As noted over, this situation requires solid participation between the current system constituents around a type of worldwide "new green arrangement." It is outstanding that interest in green development arose following the worldwide monetary emergency of 2008–2009, with prominent establishments like the Organization for Economic Co-activity and Development (OECD) seeing the system as a twin financial and natural arrangement. Preceding rise of the COVID_19 pandemic there was a developing discernment that green development was neglecting to convey the required speed and size of fossil fuel byproducts decreases at public or worldwide level In the "Tumultuous Transition" situation, which was first proposed in the setting of worldwide asset exhaustion, the environmental, financial, and political frameworks breakdown at a worldwide scale. In result, this would achieve a revisitation of nonintervention, and far and wide business disappointment, with practically zero development, which would in turn bring about critical decreases in natural loads. This situation is obviously profoundly problematic, with conceivably calamitous impacts on social orders all throughout the planet and bunch optional effects like mass relocations and enormous scope clashes. This kind of transdisciplinary situation is portrayed as Cliodynamics. At last, in the "Oversaw Degrowth" situation, worldwide material deficiencies, more grounded guideline at an worldwide level, and development in options to traditional financial development would prompt a fast decrease in biological weights before a consistent state at a lower level happens In this situation, the "new ordinary" is a revolutionary rebalancing of work, creation, and utilization with winning topics of dematerialized ways of life, shared work, green duties, and the disintegration of private riches . Over the long haul, a change in accentuation toward re-use what's more, reusing will drive down net asset utilization.

The travel industry recuperation and the fate of contemporary the travel industry customary way of thinking proposes that the recuperation of the worldwide the travel industry area from an emergency relies upon the scale, typology and extent of the emergency [8]. It is apparent that the sheer scale and effect of the COVID-19 pandemic on the worldwide travel and the travel industry is yet to completely unfurl [9–12]. Meanwhile, endeavoring to demonstrate a reasonable way to deal with the recuperation of the movement and the travel industry from the impacts of the COVID-19 pandemic emergency is predicated on moderating the apparent wellbeing, just as ensuing mental and social danger and its effect on traveler conduct. This has turned into a dire talk inside the surviving of contemporary the travel industry research. The worldwide the travel industry faces a diverse test from both the travel industry interest (saw wellbeing, social, mental danger) and supply-side (gigantic financial shortages, employment misfortunes, business liquidation, human resources exhaustion). Accordingly, the recuperation of the movement and the travel industry requires a purposeful multipartner way to deal with address both the travel industry interest also, supply perspectives. Morethus, the moderation of the intellectual discord related with seen hazard in the travel industry might be predicated on the capacity of the objective or objective nation to oversee both evenhanded and emotional saw hazard factors for the travel industry action to be inside an OK limit to reduce intellectual disharmony [13]. Target apparent danger factors are related with the assessment and the adverse consequence of potential outcomes related with movement. The COVID-19 infection addresses the essential wellbeing hazard that impacts traveler conduct and would be tended to by the viable government and movement intercessions to relieve wellbeing hazard factors. In contrast, abstract apparent danger factors are related with negative sentiments related with the results or adverse consequence of movement. The emotional variables are the ensuing" mental and social dangers related with the travel industry in the COVID-19 time would be moderated by the tertiary mediations to oversee traveler discernment. These intercessions incorporate dealing with the country's media profile, showcasing and homegrown the travel industry. Administration. Becken and Hughey [14] advocate for a multipartner way to deal with better coordinate the travel industry and government structures in the relief of hazard related with the travel industry. To this end, Ritchie and Jiang [8] recognize that the capacity of governments to outfit monetary, human and specialized assets to help the travel industry recuperation measure. For example, post the COVID-19 emergency, governments will be essential to the easing of the negative financial effects related with emergencies by giving the travel industry elements with monetary boost bundles, postponing charges for the travel industry undertakings, just as giving appropriations and low rate advances [8]. Moreover, governments will catalyze objective advertising and advancement through open discretion drives pointed toward opening and overseeing global correspondence channels, rebuilding traveler certainty and recuperating objective picture [15–17]. Nonetheless, above all, legislatures have a basic job in the compelling administration of the infection and mindful re-opening of their economies what's more, nations to the travel industry action. For example, the Spanish Government has started a four_pillar the travel industry recuperation plan that tries to create and uphold the socio-clean particulars to guarantee that vacationer locations are protected and seen to be protected (Association for Economic Co-activity and Development, [2]).

The Spanish Government looks to create and execute new wellbeing focused information what's more, perception models, just as at last actuating the travel industry request when it is fitting [2]. These actions represent how public governments are basic to the relief of the target hazard insight factors identified with the COVID-19 infection. Movement. Public governments regularly acquaint severe measures with confine internal what's more, outbound travel to deal with the spread of sickness during pandemics [18,12]. Rigid measures have overwhelmingly included isolating upon appearance [19] and travel boycotts. This holds in the instance of the current COVID-19 pandemic. Because of the pandemic, essentially 90% of the worldwide populace is right now under cross-line travel limitations including, a ban on inbound and outbound travel by the two local people and unfamiliar nationals, just as a total or then again incomplete hang on visa applications, handling and conceding [20]. As of the first of June 2020, the World Tourism Organization revealed that 156 public legislatures of the travel industry objective nations had forced all out line closers to global the travel industry. To this end, the OECD [2] report on the travel industry strategy reactions to the COVID features that the lifting of movement limitations will go far to reestablishing vacationer certainty. Along these lines, as a basic initial step, nations should open their boundaries and loosen up their rigid impermanent movement strategies to work with more proficient travel and the travel industry once nations are satisfactorily ready for the travel industry movement in a post-COVID-19 infection period.

Media profile. Huynh [21] notices a relationship between hazard insights related with the COVID-19 infection and an expansion in data look via online media. A few investigations [22,23] compare with this perception, tracking down that the higher the apparent danger, the more data sightseers look to help their immoderate the travel industry choices. The data looking for measure is conscious and helps the "heuristic" or "scientific" intellectual course of dynamic [24]. To this end, an objective's postemergency media profile on different stages including the news, narratives, motion pictures, online media destinations and official sites would be a basic touchpoint from which vacationers infer the prompts they might use to advise their decision making [22,25–27]. It follows then that an objective's media profile might be considered as a potential mitigator of saw hazard [28]. For example, Uganda's natural picture was profoundly impacted by cliché factors, for example, the apparent pervasiveness of illnesses in the nation – that brought about its picture as a hazardous the travel industry objective. To relieve these danger insights, a trial concentrate by Lepp et al. [29] found that mediations by Tourism Uganda intended to give data balance through its authority the travel industry site decidedly affected and changed traveler's negative pretest hazard impression of the country as a travel industry objective. In this manner, in any case the worth of postemergency data for the travel industry dynamic in the post-COVID-19 time, it is the validity of the data source that might be basic to the danger data preparing of vacationers [30], related with apparent mental and social danger, post the COVID-19 emergency [33–35].

Promoting. As a postemergency hazard relief methodology, recuperation promoting [31] will be critical to the recuperation of the worldwide travel and the travel industry. Based on Aliperti and Cruz's [30] assurance, postemergency hazard correspondence firmly impacts hazard discernment and at last traveler conduct by advancing mindfulness, persuading person's conative conduct, working with agreement, just as making trust. Therefore, Rodrı'guez-Morales et al. [11] perceive that data balance is basic to overseeing traveler conduct inside the setting of the COVID-19 infection. Proof from Japan recommends that designated promoting efforts assumed a basic part in the moderation of the apparent danger related with movement and the travel industry to Japan post the 2011 quake that brought about the Fukushima debacle [25].

5.10 Journal of tourism futures

Japan and interfacing with Japanese individuals brought about the development of negative picture of Japan as a travel industry objective [25]. To relieve these apparent hazards, the Japanese National Tourism Organization dispatched a serious promoting program that advanced the country in noncustomary business sectors, including Malaysia. By promoting the nation as a protected objective through elite catastrophe the executives abilities, progressed sanitation and medical care capabilities figured out how to impact both the abstract and target apparent danger variables to reestablish and develop its market share in Malaysian outbound the travel industry [25]. In this way, advertising might address a feasible intercession for affecting traveler conduct post the Coronavirus infection and dealing with the emotional parts of apparent danger. Homegrown the travel industry. Hazard avoidance is higher when travelers think about global the travel industry contrasted with when they think about homegrown the travel industry. This is alluded to as "home-is-safer_than-abroad predisposition". This recommends that travel industry professionals might have to zero in on advancing homegrown the travel industry as a transient measure to reviving their country's travel industry and launching the worldwide the travel industry in the medium-term. To this end the OECD [2] reports that as a component of its COVID-19 recuperation system Switzerland has submitted an expected USD$42.2m in government subsidizing for Switzerland Tourism to elevate reasonable homegrown the travel industry to launch the Swiss area by sponsoring homegrown travel and the travel industry. On account of New Zealand, an assessed USD$256.8 m was supported the New Zealand parliament for dispensing through the Tourism Recovery Fund, to among different exercises, reserve a homegrown the travel industry showcasing effort, store a progress program pointed toward "turning" organizations towards homegrown the travel industry, just as help the essential resource insurance of New Zealand's the travel industry contributions and global brand [2]. Such homegrown the travel industry drives advance neighborhood the travel industry and guarantee both homegrown and all the more significantly, global sightseers of the security of taking part in the travel industry, in this way relieving the abstract factors that impact both the mental and social saw hazard.

5.11 End

Fortunately, the pandemic has offered us the chance to utilize this time usefulness. We need to outfit ourselves with the most recent know-hows of various learning applications. Keeping contact with resources in various nations will assist us with breaking down the most ideal arrangements we have in our grasp with restricted assets. This is an experimentation stage for us all. There will be specialized errors and a few issues of safety concern. These also will be managed entry of time. The silver lining in the present circumstance is that lockdown has overall decreased our pressure by dialing our lives back, giving us an opportunity to reflect, think and plan. It has likewise worked on our psychological and actual wellbeing by the changing our way of life and dietary patterns. Instruments like quarantine and disconnection will be utilized as we need to live with this infection. It is imperative to ingrain to us; dislike we learn in first 50% of our life and work in the second 50% of our life. Inclining needs to proceed for the duration of our life. As far as being taught, one ought not take a gander at that point or what amount of time it will require to gain the information multi month, a half year, a year. Regardless of whether it requires years, the years will go back and forth. It's fundamental to engage in teaching ourselves at each front. However we are attempting to instruct understudies on the web, we truly are not putting resources into our schooling. Let not this infection feed our psychological mental soundness and our self-viability. Whatever the procedure is taken on, we in this current circumstance play a part to play. A job to be free, learned, skillful, taught enough to hegemonize innovation and not the alternate way round. You can't deny reality; there are numerous openings accessible on the web, more than the reasons we interpret for not profiting it.

References

[1] M. Buheji, Planning competency in the new normal–Employability competency in post-COVID-19 pandemic, Int. J. Human Res. Stud. 10 (2) (2020) ISSN 2162-3058.

[2] Organisation for Economic Co-operation and Development (OECD) (2020), "Tourism policy responses to the coronavirus (COVID-19)", available at: www.oecd.org/coronavirus/policy-responses/tourism-policy_responses-to-the-coronavirus-covid-19-6466aa20/ (accessed 25 June 2020).
[3] A. Ahmed, Synchronizing pedagogy and technology in post COVID scenario, IJCRT 8 (2020).
[4] A. Mishra, M. Choudhary, T.R. Das, P. Saren, P. Bhattacherjee, N. Thakur, et al., Sustainable chemical preventive models in COVID-19: Understanding, innovation, adaptations, and impact, Journal of Indian Chemical Society 98 (10) (2021) 100164. https://doi.org/10.1016/j.jics.2021.100164.
[5] N. Thakur, R. Sood, C.M. Hussain, S.K. Shukla, Current diagnostic approach for COVID-19. Detection and analysis of SARS Coronavirus: Advanced biosensors for pandemic viruses and related pathogens, (2021) 1–22. https://doi.org/10.1002/9783527832521.ch1.
[6] N.E. Mahaye, The Impact of COVID-19 Pandemic on Education: Navigating Forward the Pedagogy of Blended Learning, (2020).
[7] N. Thakur, B. Mangla, C.M. Hussain, S.K. Shukla, B. Mangla, Restoring quality and sustainability through functionalized nanocatalytic processes, Functionalized Nanomaterials for Catalytic Application (2021) 252–259. https://doi.org/10.1002/9781119809036.ch9.
[8] B.W. Ritchie, Y. Jiang, A review of research on tourism risk, crisis and disaster management launching the annals of tourism research curated collection on tourism risk, crisis and disaster management, Annals Tour. Res. 79 (2019) 102812.
[9] T. McLaughlin (2020), "Coronavirus is devastating Chinese tourism", available at: www.theatlantic.com/international/archive/2020/02/economycoronavirus-myanmar-chinatourism/606715/ (accessed 21 April 2020).
[10] M. Nicola, Z. Alsafi, C. Sohrabi, A. Kerwan, A. Al-Jabir, C. Iosifidis, M. Agha, R. Agha, The socio-economic implications of the coronavirus and COVID-19 pandemic: a review, International Journal of Surgery 78 (2020), doi:10.1016/j.ijsu.2020.04.018.
[11] A.J. Rodrı´guez-Morales, K. MacGregor, S. Kanagarajah, D. Patel, P. Schlagenhauf, Goingglobal – Travel and the 2019 novel coronavirus, Travel Medicine and Infectious Disease 33 (2020) 101578.
[12] Y. Yanga, H. Zhang, X. Chen, Coronavirus pandemic and tourism: Dynamic stochastic general equilibrium modelling of infectious disease outbreak, Annals of Tourism Research 83 (2020), doi:10.1016/j.annals.2020.102913.
[13] F. Cui, Y. Liu, Y. Chang, J. Duan, J. Li, An overview of tourism risk perception, Natural Hazards 82 (1) (2016) 643–658.
[14] S. Becken, K.F.D. Hughey, Linking tourism into emergency management structures to enhance disaster risk reduction, Tourism Management 36 (2013) 77–85.
[15] J.C. Carlsen, M. Hughes, Tourism market recovery in the Maldives after the 2004 Indian ocean tsunami, Journal of Travel & Tourism Marketing 23 (2/4) (2008) 139–149 Nos.
[16] M. Cooper, Japanese tourism and the SARS epidemic of 2003, Journal of Travel & Tourism Marketing 19 (2/3) (2005) 117–131.
[17] H. Gu, G. Wall, SARS in China: Tourism impacts and market rejuvenation, Tourism Analysis 11 (6) (2006) 367–379.
[18] I. Cahyanto, M. Wiblishauser, L. Pennington-Gray, A. Schroeder, The dynamics of travel avoidance: The case of Ebola in the U.S, Tourism Management Perspectives 20 (2016) 195–203.
[19] M. Folayan, B. Brown, (2015), "Ebola and the limited effectiveness of travel restrictions", Disaster Medicine and Public Health Preparedness, p. 1.
[20] Fragomen (2020), "COVID-19 pandemic – immigration situation overview", available at: www.fragomen.com/file/newcoronavirustablefornewspage625v4pdf (accessed 25 June 2020).

[21] T.L.D. Huynh, The COVID-19 risk perception: A survey on socioeconomics and media attention, Economics Bulletin 40 (1) (2020) 758–764.

[22] G. Fuchs, A. Reichel, An exploratory inquiry into destination risk perceptions and riskreduction strategies of first time vs repeat visitors to a highly volatile destination, Tourism Management 32 (2) (2011) 266–276.

[23] J. Swarbrooke, S. Horner, Consumer Behaviour in Tourism, 2nd ed., Butterworth-Heinemann, Amsterdam, 2007.

[24] S. Dunwoody, R.J. Griffin, Risk information seeking and processing model, in: H. Cho, T. Reimer, K.A. McComas (Eds.), The SAGE handbook of risk communication, SAGE, ThousandOaks, CA, 2015, pp. 102–116.

[25] E.Y.T. Chew, S.A. Jahari, Destination image as a mediator between perceived risks and revisit intention: a case of post-disaster Japan, Tourism Management 40 (2014) 382–393.

[26] G. Kapu, B. Richards, News framing effects on destination risk perception, Tourism Management 57 (2016) 234–244.

[27] X. Zhai, D. Zhong, Q. Luo, Turn it around in crisis communication: an ABM approach, Annals of Tourism Research 79 (2019) 102807.

[28] A. Jonas, Y. Mansfeld, S. Paz, I. Potasman (2011), "Determinants of health risk perception among low_risk-taking tour.

[29] A. Lepp, H. Gibson, C. Lane, Image and perceived risk: a study of Uganda and its official tourism website, Tourism Management 32 (3) (2011) 675–684.

[30] G. Aliperti, A.M. Cruz, Investigating tourists' risk information processing, Annals of Tourism Research 79 (2019) 102803.

[31] K. Campiranon, N. Scott, Critical success factors for crisis recovery management: A case study of Phuket hotels, Journal of Travel & Tourism Marketing 31 (3) (2014) 313–326.

[32] K. Wolff, S. Larsen, Flux and permanence of risk perceptions: Tourists' perception of the relative and absolute risk for various destinations, Canadian Journal of Pshycology 57 (6) (2016) 584–590.

[33] P. Wells, W. Abouarghoub, S. Pettit, A. Beresford, A socio-technical transitions perspective for assessing future sustainability following the COVID-19 pandemic, Sustainability: Science, Practice and Policy 16 (1) (2020) 29–36, doi:10.1080/15487733.2020.1763002.

[34] S. Umiński, A. Borowicz, Will multinational enterprises contribute to Poland's economic resilience and recovery during and post COVID-19 pandemic, Transnational Corporations Review 13 (1) (2021) 74–87. https://doi.org/10.1080/19186444.2021.1888638.

[35] N. Thakur, S.K. Shukla, C.M. Hussain. COVID detection via nanotechnology: A promising field in the diagnosis and analysis of novel coronavirus infection,Chaudhery Mustansar Hussain, Sudheesh K. Shukla, Detection and Analysis of SARS Coronavirus: Advanced Biosensors for Pandemic Viruses and Related Pathogens, (2021) 57–73. https://doi.org/10.1002/9783527832521.ch4.

[36] M. Levenson, Scale of China's Wuhan shutdown is believed to Be without precedent[Internet] The New York Times (2020) Jan 22 [cited 2021 Jun 7]; Available from, https://www.nytimes.com/2020/01/22/world/asia/coronavirus-quarantines-history.html.

[37] R. Helyer, D. Lee, The role of work experience in the future employability of higher education graduates. Higher Education Quarterly, 68 (3) (2014), 348–372. https://doi.org/10.1111/hequ.1203.

[38] M. Buheji, D. Ahmed, The Youthineering- 'New Perspectives on Youth Economy', Author House Publishing, U.K.

[39] T.L. Meister, Y. Brüggemann, D. Todt, C. Conzelmann, J.A. Müller, R. Groß, J. Münch, A. Krawczyk, J. Steinmann, J. Steinmann, S. Pfaender, E. Steinmann, Virucidal Efficacy of Different Oral Rinses Against Severe Acute Respiratory Syndrome Coronavirus 2. J Infect Dis. 222 (8) (2020) 1289–1292. doi:10.1093/infdis/jiaa471. Erratum in: J Infect Dis. 223 (3) (2021) 541. PMID: 32726430; PMCID: PMC7454736.

CHAPTER

6

Development of corona sensor

Prama Bhattacherjee[a], *Meenakshi Choudhary*[b], *Trupti R. Das*[c], *Anupama Chaudhary*[d], *Sudheesh K. Shukla*[d,e], *and Santanu Patra*[f]

[a]Department of Chemistry, Indian Institute of Technology (ISM), Dhanbad, Jharkhand, India, [b]The Swiss Institute for Dryland Environmental and Energy Research, The Jacob Blaustein Institutes for Desert Research, Ben-Gurion University of the Negev, Midreshet Ben-Gurion, Israel, [c]CIPET: Institute of Petrochemicals Technology (IPT) – Bhubaneswar, Patia, Bhubaneswar, India, [d]Department of Biomedical Engineering, School of Biological Engineering and Life Science, Shobhit Institute of Engineering & Technology (Deemed-to-be University), Modipuram, Meerut, India, [e]Department of Chemical Sciences, University of Johannesburg, Doornfontein Campus, Johannesburg, 2028, South Africa, [f]Department of Environmental Engineering, Khon Kaen University, Khon Kaen, Thailand

6.1 Introduction

Since the dawn of the twenty-first century, our planet has experienced unprecedented outbreaks of deadly viruses such as Zika, Ebola, SARS, MERS, and others. The advent of novel viral species or the evolution of pre-existing viruses fueled the outbreaks of various viral diseases. Such diseases have already wreaked havoc on civilization. The loss of life was the most tragic, but the ramifications were just as bad: survivors' psychological well-being and the socio-economic fallout were both

Sensing Tools and Techniques for COVID-19: Developments and Challenges in Analysis and Detection of Coronavirus
DOI: https://doi.org/10.1016/B978-0-323-90280-9.00012-2

devastating. In December 2019, the world was struck by a new virus known as SARS-CoV-2 (the disease caused by this virus is known as COVID-19). Even among health specialists, this condition was dismissed as just another strain of influenza. The virus, on the other hand, quickly exhibited its destructive character. If left untreated, the disease quickly spread and overwhelmed society's healthcare infrastructure. The general population became concerned as a result of this. People raced to see doctors even if they had just minor or unrelated symptoms, causing hospitals to overflow. A cluster of pneumonia infections was recorded in Wuhan, China, in December 2019. The virus that caused the disease was identified as SARS-CoV-2 (later renamed COVID-19, Corona Virus Disease 2019, by the WHO) because it shared 80% of its DNA with the SARS-Coronavirus. The first death caused by this virus was reported in China on January 11, 2020. The WHO declared this disease a Public Health Emergency of International Concern (PHEIC) less than a month after the first incidence because it was extremely contagious. COVID-19 was named a pandemic disease by the World Health Organization on March 11 as it began to spread over the world. Sensors are tools to assist detect events or changes in the environment while also sending data to other electronics, most commonly a computer processor. The physical data are gathered and converted into a signal that can be processed, resulting in a readable result. Biosensor is a systematic tool that aids in the recognition of chemical compounds and then merges the organic element with a physicochemical indicator. It has a lot of potential in terms of detecting and diagnosing a variety of diseases, as well as this current pandemic [1–5].

A tool called a sensor is used to determine the occurrences or variations in an environment, as well as assisting in the delivery of data to other electronics, usually a computer processor. Obtaining a physical quantity and converting that quantity into a signal suitable for processing. Currently, an electrical signal is obtained by converting physical phenomena into electrical signals.

Biosensor [3,4,6] is a systematic tool that aids in the recognition of chemical compounds and then combines the organic element with a physicochemical indicator. A biosensor consists of a bioreactor (enzyme/antibody), a transducer (nanomaterials), and an electronic device with a processor and display [6]. In complementary metal oxide semiconductor [7,8], for example, parts like the transducer and electronics can be combined. Biomolecules are used by the recognition substance, which is commonly referred to as a bioreceptor. Biomolecules are used by organisms that are further modeled next to biological systems that interact with an intriguing analyte. The biotransducer calculates this relationship by producing a quantifiable signal that is comparable to the presence of the target analyte in the specimen. Biosensors are designed to enable for quick and accurate testing at the point of interest [9,10]. Biosensors are classified according to their biotransducers. Electrochemical biosensors, optical biosensors, and

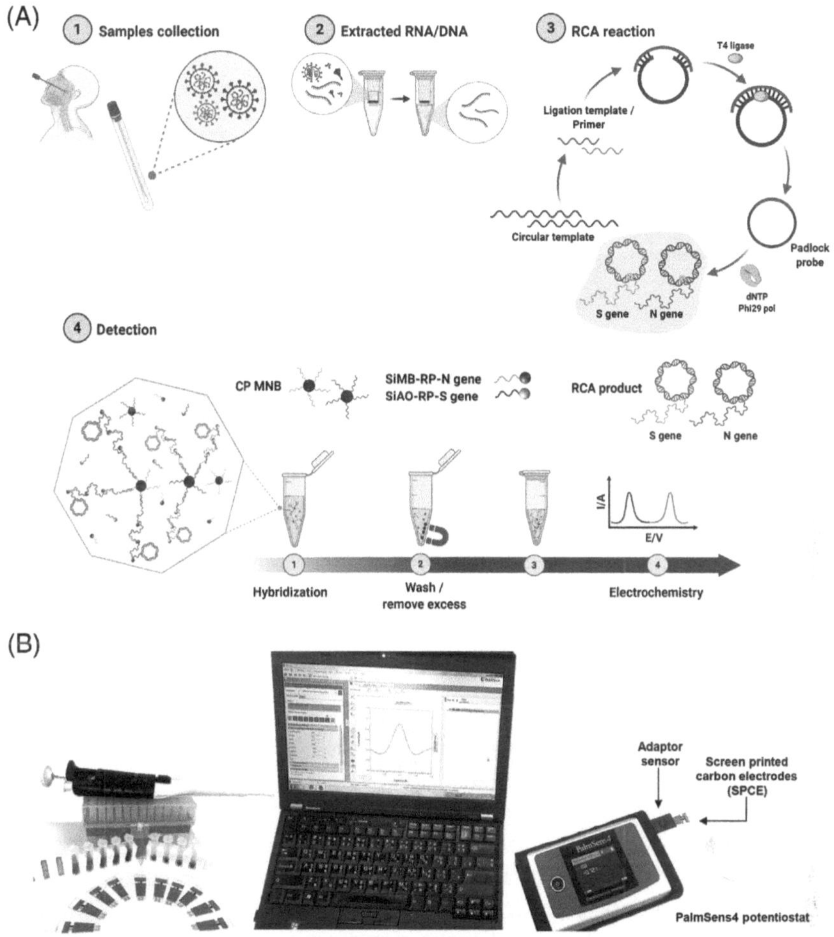

FIGURE 6.1 Rapid electrochemical detection of coronavirus SARS-CoV-2 [9].

other generic types of biotransducers used in chemical canaries include the following: Chemical canaries are categorized as biotechnology, agriculture, food technology, and biomedicine [11], and their optimal organization is determined by their application. Recently one group of authors reported an electrochemical sensor for the detection of SARS-CoV 2 as shown in Fig. 6.1.

6.2 Wearable sensors for COVID-19

COVID-19 wearable sensors COVID-19 is now considered a pandemic by the World Health Organization. COVID-19's physiology and history

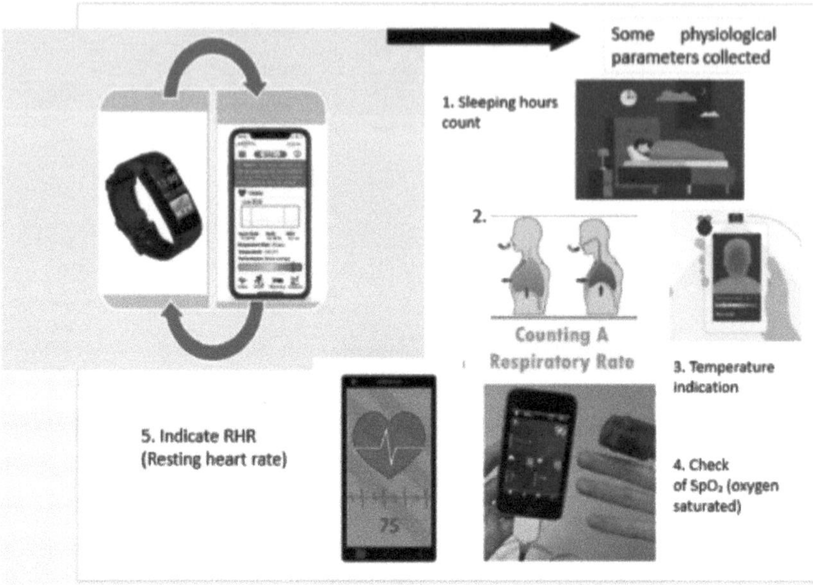

FIGURE 6.2 Schematic detailing of a smart phone application collecting physiological data from a wearable sensor for COVID-19 monitoring [19].

can be distinguished using wearable sensors. About 1.3 billion wearable sensors are utilized to recognize COVID-19. COVID-19 can be detected using modern wearable sensors. Smartphones, Apple Watches, Fit Bits, and Bio Harnesses are among the wearable sensors used to monitor COVID-19. This is the digital level of human health diagnosis instruments [12]. The sensors not only track the COVID-19 virus, but also provide data on the patient's health, disease symptoms, and recovery. Wearable sensors can also detect changes in the body's condition [13]. The most effective technique to detect the presence of COVID-19 is by respiratory monitoring. The respiratory tract is affected by this virus [14]. COVID-19 symptoms are tracked using a remote-like wearable system, which also helps to prevent the disease from spreading [15]. These sensors detected COVID-19 symptoms such as fever and cough, as well as physiological ways for detecting cold. Wearable sensors are being utilized to provide information about people's common health problems [16,17]. The resting heart rate, cardiac rhythm, and fever are all detected by the Huami wearable gadget [18]. Wearable devices are more useful for monitoring the health system and are a fantastic idea for maintaining social distance. In common locations, temperature guns are used to detect the temperature (Fig. 6.2).

6.3 Biosensor on cell for COVID-19

Cell-based biosensors (CBBs) are utilized to identify certain species using analytical information and then amplification into an optical and electrical signal using a processor. In comparison to CBBs, molecular based biosensors are more sensitive to environmental changes; yet, these biosensors may be modified to detect complicated responses inside a living cell using simple genetic engineering techniques [20]. SARS-CoV-2 can transmit through the air and can be found in a sample of air within the aerosol for 3 hours, according to experimental research. SARS-CoV-2 is lethal to humans and causes a variety of diseases. SARS-CoV-2 virus could be detected in the air using cell-based biosensors. However, these biosensors have yet to be demonstrated in real-world applications. A cell-based biosensor is utilized to detect the pollutant in the air, as previously mentioned. When the sensor comes into contact with harmful pollutants, it reports to the cellular response and developing cellular signals, which are then transduced by a secondary transducer, which produces electrical outputs and appropriate readouts.

6.4 Nanosensor for COVID-19

COVID-19 is unquestionably combated by nanotechnology [21]. Sensing technology is critical for the development of an effective sensor [22]. Graphene is a nonmaterial with electrical properties that is employed in sensor technology [23]. Nanomaterials are employed to find nanomedicine in modern technologies. These are used to diagnose, treat, and recover from a wide range of disorders [24] (Fig. 6.3). Chemical sanitization also employs nanomedicines [25]. Hydrogen peroxide in blood samples is measured using dendritic nanochips [26]. Nanomaterials are used by Chinese scientists to regulate SARS-COV-2, and the virus's potency is reduced by 96.5 percent to 99 percent [27]. The identification of RNA is done with nanomaterials, and RNA-based detection is a quick diagnosis method [28]. Iron oxides and gold nanoparticles are now commonly utilized in testing kits [29]. For the detection of viruses and infectious diseases, nanotechnology is critical [30]. 2D materials, graphene, and carbon are used in nanoparticle-based diagnostic tools [31]. A nanomedicine provides information on COVID-19 fast infection [32]. Nanobiosensors are being developed in modern technology to learn more about the advanced identification of COVID-19 [33–35]. Hybrid nanomaterials are utilized to quickly identify Coronavirus [36]. Until now, the medicines hydroxy chloroquine, ribavirin, and lopinavir have been used to treat COVID-19 [37].

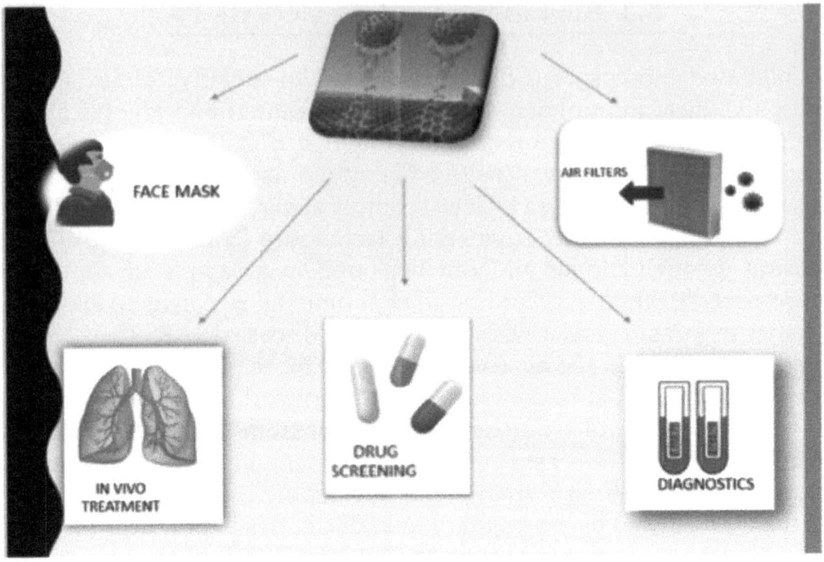

FIGURE 6.3 Different ways of using nanosensors (graphene) [19].

6.5 Immunosensor for COVID-19

Immunosensor is a small, multiplex wireless item that can quickly detect COVID-19. It aids in the detection of nucleocapsid protein, IgM, and IgG antibodies to viral antigens [38]. On the basis of mass-producible laser-engraved graphene electrodes, it is also used for the inflammatory biomarker C-reactive protein [39]. High sensitivity, cheap cost, ultra-fast detection, wireless remote, and multiplexed sensing are some of the primary advantages of this novel approach (Fig. 6.4).

6.6 Recent development of COVID-19 sensors

Many researchers across the world are attempting to discover a possible COVID-19-curing treatment derived from plants, animals, or other microorganisms, or repurposing existing drugs, medications, or vaccines. Bharat Biotech International Limited has demonstrated significant progress in the development of the vaccine, indicating that the month following is the most critical period for researchers. The vaccine is being developed, and the percentage of spread is being reduced, by isolating

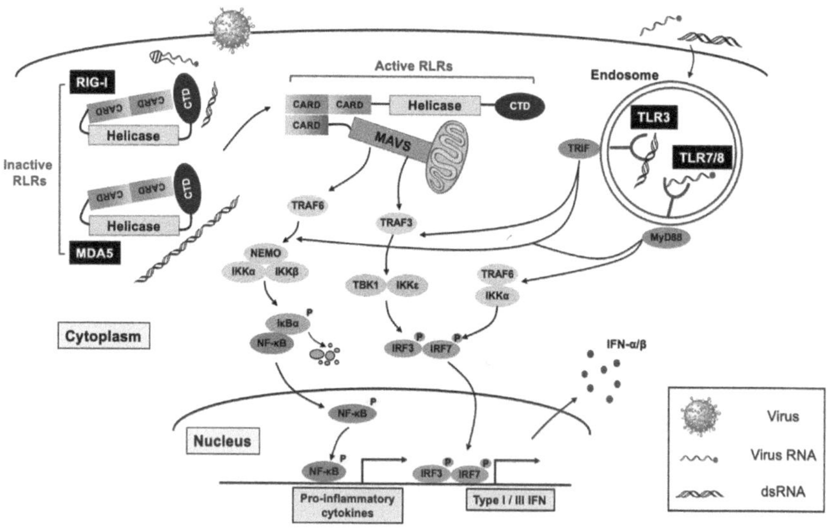

FIGURE 6.4 Immune sensing of coronavirus and viral evasion strategies [40].

the spike protein. India is the third-worst-affected country, with around 1.8 million cases, according to the COVID-19 report. The COVID-19 vaccine is being developed by a collaboration of companies and medical organizations. In which 30 different Indian businesses are working on developing vaccinations to combat the disease. Seven firms are now undergoing testing and investigations, according to the WHO [41]. COVAXIN, ZyCoV-D, Mynvax, and Astra Zeneca are some of the vaccinations that are currently being tested. Russia was the first country to complete human clinical trials of the COVID-19 vaccine [42]. In the month of June 18 [43], Russia permitted two types of possible COVID-19 trials, which were produced by the Gamaleya National Research Centre for Epidemiology and Microbiology. The first vaccine, which is administered intramuscularly, was produced at the Burdenko Military Hospital. After the vaccine, the volunteers were instructed to stay in isolation in the hospital for roughly 28 days [43]. Other hypotheses that place a focus on smartphone-based sensing systems have gotten a lot of attention and have been built within a smartphone (Fig. 6.5). It is built with a technology that can detect any location. This approach could be used as an alternative to more expensive technologies [44]. Practice of Nano bioengineering analytical clinically [45] is another recently established method for COVID-19 diagnosis. Electrochemical biosensors' analytical performance is being improved [46].

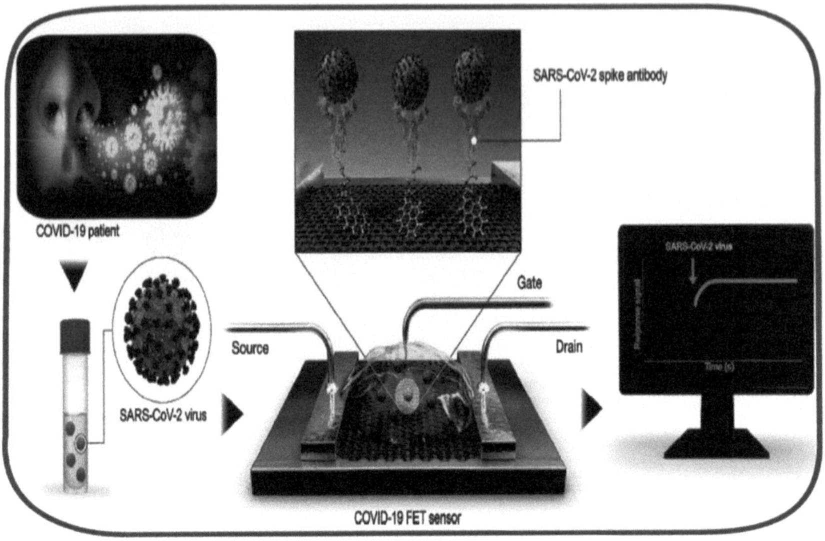

FIGURE 6.5 Point-of-care diagnostics of COVID-19: From current work to future perspectives [25].

Conclusion

COVID-19's resurgence over the world has worried people and drawn more attention to health. It presented a challenge to researchers working on the creation of treatments, vaccines, and new technologies to prevent the virus from spreading. The body temperature is detected using a thermal scanner, but the virus is not detected. RTPCR is also used to identify COVID-19, although it has been deemed ineffective since it produces erroneous results. A biosensing device is also worn in the form of a smart band and has the ability to measure body temperature. To identify the virus, an optical sensor that combines two different types of effects, namely visual and thermal, is used. A powerful biosensor, such as a cell-based biosensor, has the ability to detect and amplify some species, molecular biosensors to monitor virus in the environment, and also plays a critical role in the pharmaceutical sector, environmental monitoring, and food analysis, among other things. In reality, scientists have concentrated their efforts on developing new technology and developing a quick diagnostic sensor that can identify SARS-CoV-2 in less than a minute. Jing Wang and his colleagues created a new optical sensor whose receptor functions as a complimentary sequence for the RNA sequence to identify the virus. A biosensor with plasmonic photothermal effect was also developed, which promotes local temperature and inhibits noncomplementary sequence binding to the sensor's DNA strands. Biosensors based on cells are also used to identify species, which are then amplified into an optical and electrical signal by a processor. It can also be employed in a wide range of conditions, such as pH and temperature. The app "Aarogya Setu" was created to alert users about coming into contact with an infected person. Furthermore, studies have showed that SARS-CoV-2 can transmit through the air and that cell-based biosensors can detect it; however, these sensors have yet to be proved. Many molecular assays were developed to identify SARS-CoV-2 RNA utilizing non-PCR methods, such as nucleic acid sequence-based amplification techniques and loop mediated isothermal amplification. Many more ways for diagnosing the virus at a lower cost should be developed. As a result, instruments like biosensors that can diagnose the virus quickly are required to prevent the virus from spreading quickly. High sensitivity, repeatability, and other aspects of biosensor and other technology should be improved.

References

[1] F. Cui, H.S. Zhou, Diagnostic methods and potential portable biosensors for coronavirus disease 2019, Biosens. Bioelectron. 165 (2020) p. 112349 https://doi.org/10.1016/j.bios.2020.112349.

[2] C. Dincer, R. Bruch, E. Costa-Rama, M.T. Fern_andez-Abedul, A. Merkoçi, A. Manz, et al., Disposable sensors in diagnostics, food, and environmental monitoring, Adv. Mater. 31 (30) (2019) 1806739. https://doi.org/10.1002/adma.201806739.

[3] A. Hierlemann, O. Brand, C. Hagleitner, H. Baltes, Microfabrication techniques for chemical/biosensors, Proc. IEEE 91 (6) (2003) 839–863. https://doi.org/10.1109/JPROC.2003.813583.

[4] H. Kaur, M. Shorie, Nanomaterial based aptasensors for clinical and environmental diagnostic applications, Nanoscale Advances 1 (6) (2019) 2123–2138. https://doi.org/10.1039/C9NA00153K.

[5] A. Turner, I. Karube, G.S. Wilson, Biosensors: Fundamentals and applications, Oxford University Press, United State, 1987.

[6] A. Hierlemann, H. Baltes, CMOS-based chemical microsensors, Analyst 128 (1) (2003) 15–28. https://doi.org/10.1039/B208563C.

[7] R.K. Arora, R.P. Saini, Biosensors: way of diagnosis, Int. J. Pharmaceut. Sci. Res. 4 (7) (2013) 2517–2527. https://doi.org/10.13040/IJPSR.0975-8232.4(7).2517-27.

[8] C. Dincer, R. Bruch, A. Kling, P.S. Dittrich, G.A. Urban, Multiplexed point-of-care testing–xPOCT, Trends Biotechnol. 35 (8) (2017) 728–742. https://doi.org/10.1016/j.tibtech.2017.03.013.

[9] T. Chaibun, J. Puenpa, T. Ngamdee, et al., Rapid electrochemical detection of coronavirus SARS-CoV-2, Nat. Commun. 12 (2021) 802. https://doi.org/10.1038/s41467-021-21121-7.

[10] Tech Desk, Meet the world's first smart wrist band that can detect body temperature, 14, 2020, p. 5. https://indianexpress.com/article/technology/tech-news-technology/meet-the-worlds-first-smart-wrist-band-that-can-detect-body-temperature/.

[11] Y. Khan, A.E. Ostfeld, C.M. Lochner, A. Pierre, A.C. Arias, Monitoring of vital signs with flexible and wearable medical devices, Adv. Mater. 28 (2016) 4373–4395. https://doi.org/10.1002/adma.201504366.

[12] D.R. Seshadri, E.V. Davies, E.R. Harlow, J.J. Hsu, S.C. Knighton, T.A. Walker, et al., Wearable sensors for COVID-19: A call to action to harness our digital infrastructure for remote patient monitoring and virtual assessments, Frontiers in Digital Health 2 (2020) 8. https://doi.org/10.3389/fdgth.2020.00008.
[13] C. Emily, The importance of respiratory rate tracking during the COVID-19 pandemic frontiers. https://www.whoop.com/thelocker/respiratory-rate-tracking-coronavirus/, 2020, 1, 4.
[14] C. Massaroni, A. Nicol_o, E. Schena, M. Sacchetti, Remote respiratory monitoring in the time of COVID-19, Front. Physiol. 11 (2020) 635. https://doi.org/10.3389/fphys.2020.00635.
[15] Z. Wu, J.M. McGoogan, Characteristics of and important lessons from the coronavirus disease 2019 (COVID-19) outbreak in China: summary of a report of 72314 cases from the Chinese Center for Disease Control and Prevention, JAMA 323 (13) (2020) 1239–1242. https://doi.org/10.1001/jama.2020.2648.
[16] G. Ke, Z. Xu, J. Zhang, J. Bian, T.Y. Liu, DeepGBM: A deep learning framework distilled by GBDT for online prediction tasks, in: Proceedings of the 25th ACMSIGKDD International Conference on Knowledge Discovery & Data Mining, 2020, pp. 384–394. https://doi.org/10.1145/3292500.3330858.
[17] B. Purohit, A. Kumar, K. Mahato, P. Chandra, Smartphone-assisted personalized diagnostic devices and wearable sensors, Curr. Opin. Biomed. Eng. 13 (2020) 42–50.
[18] Y. Ranjan, Z. Rashid, C. Stewart, P. Conde, M. Begale, D. Verbeeck, et al., RADAR-base: open source mobile health platform for collecting, monitoring, and analyzing data using sensors, wearables, and mobile devices, JMIR mHealth uHealth 7 (8) (2019) 11734.
[19] S. Behera, G. Rana, S. Satapathy, M. Mohanty, S. Pradhan, M. Kumar Panda, et al., Biosensors in diagnosing COVID-19 and recent development. SI 1 (2020) 100054.
[20] N. Mao, A. Cubillos-Ruiz, D.E. Cameron, J.J. Collins, Probiotic strains detect and suppress cholera in mice, Sci. Transl. Med. 10 (445) (2018). https://doi.org/10.1126/scitranslmed.aao2586. eaao2586.
[21] R.K. Joshi, S. Bhansali, Nanosensor technology, J. Nanomater. 1 (2008). https://doi.org/10.1155/2008/840390. Article ID 840390.
[22] B.J. Kim, H. Jang, S.K. Lee, B.H. Hong, J.H. Ahn, J.H. Cho, High-performance flexible graphene field effect transistors with ion gel gate dielectrics, Nano Lett. 10 (9) (2010) 3464–3466. https://doi.org/10.1021/nl101559n.
[23] A.K. Geim, K.S. Novoselov, The rise of graphene, Nanoscience and technology: A collection of reviews from Nature Journals, World Scientific, USA, 2010, pp. 11–19. https://doi.org/10.1142/9789814287005_0002.
[24] S. Adhikari, U. Adhikari, A. Mishra, B.S. Guragain, Nanomaterials for diagnostic, treatment and prevention of COVID-19, Appl. Sci. Technol. Ann. 1 (1) (2020) 155–164. https://doi.org/10.3126/asta.v1i1.30295.
[25] H.A. Hussein, R.Y.A. Hassan, M. Chino, F. Febbraio, Point-of-care diagnostics of COVID-19: From current work to future perspectives, Sensors 20 (2020) 4289. https://doi.org/10.3390/s20154289.
[26] B. Purohit, A. Kumar, K. Mahato, P. Chandra, Sputtering enhanced peroxidase like activity of a dendritic nanochip for amperometric determination of hydrogen peroxide in blood samples, Microchim. Acta 186 (9) (2019) 658.
[27] Did Chinese scientists already develop nanomaterial to combat coronavirus?, March 29. https://www.dnaindia.com/world/report-did-chinese-scientists-already-develop-nanomaterial-to-combat-coronavirus-2818973, 2020.
[28] Russian scientists to help develop covid-19 antigen standard, April 4. https://russkiymir.ru/en/news/271406/, 2020.
[29] X. Chen, A. Miller, S. Cao, Y. Gan, J. Zhang, Q. He, et al., Rapid Escherichia coli trapping and retrieval from bodily fluids via a three-dimensional bead-stacked

nanodevice, ACS Appl. Mater. Interfaces 12 (7) (2020) 7888–7896. https://doi.org/10.1021/acsami.9b19311.

[30] Advice on the use of point-of-care immunodiagnostic tests for COVID-19, April 8, 2020. https://www.who.int/news-room/commentaries/detail/advice-on-the-use-of-point-of-care-immunodiagnostic-tests-for-covid-19.

[31] P. Rodgers, J. Heath, Nanoscience and technology, Co-Published with Macmillan Publishers Ltd, UK, 2009 https://doi.org/10.1142/7439.

[32] H. Huang, C. Fan, M. Li, H.L. Nie, F.B. Wang, H. Wang, et al., COVID-19: A call for physical scientists and engineers, ACS Nano 14 (4) (2020) 3747–3754. https://doi.org/10.1021/acsnano.0c02618.

[33] P. Chandra, E. Segal, Nanobiosensors for personalized and onsite biomedical diagnosis, The Institution of Engineering and Technology, United State, 2016.

[34] K. Mahato, R.K. Maurya, P. Chandra, Fundamentals and commercial aspects of nanobiosensors in point-of-care clinical diagnostics, 3 Biotech, United State 8 (3) (2018) 149.

[35] S. Ye, K. Shao, Z. Li, N. Guo, Y. Zuo, Q. Li, et al., Antiviral activity of graphene oxide: how sharp edged structure and charge matter, ACS Appl. Mater. Interfaces 7 (38) (2015) 21571–21579. https://doi.org/10.1021/acsami.5b06876.

[36] K. Mahato, A. Prasad, P. Maurya, P. Chandra, Nanobiosensors: next generation point-of-care biomedical devices for personalized diagnosis, J. Anal. Bioanal. Tech. 7 (2016) 125.

[37] R.G. Kerry, S. Malik, Y.T. Redda, S. Sahoo, J.K. Patra, S. Majhi, Nano-based approach to combat emerging viral (NIPAH virus) infection, Nanomed. Nanotechnol. Biol. Med. 18 (2019) 196–220. https://doi.org/10.1016/j.nano.2019.03.004.

[38] P. Bhattacherjee, S. Patra, A. Mishra, T.R. Das, H. Dewangan, R. Ghosh, et al. Detection and analysis of SARS Coronavirus: Advanced biosensors for pandemic viruses and related pathogens; John Wiley & Sons, Ltd., United State. https://doi.org/10.1002/9783527832521.ch2.

[39] R.M. Torrente-Rodríguez, H. Lukas, J. Tu, J. Min, Y. Yang, C. Xu, et al., SARS-CoV-2 RapidPlex: A graphene-based multiplexed telemedicine platform for rapid and low-cost COVID-19 diagnosis and monitoring, Matter, 165, 2020.

[40] A. Kumar, B. Purohit, P.K. Maurya, L.M. Pandey, P. Chandra, Engineered nanomaterial assisted signal-amplification strategies for enhancing analytical performance of electrochemical biosensors, Electroanalysis 31 (9) (2019) 1615–1629.

[41] K. Mahato, S. Kumar, A. Srivastava, P.K. Maurya, R. Singh, P. Chandra, Electrochemical immunosensors: fundamentals and applications in clinical diagnostics, Handbook of immunoassay technologies, Academic Press, 2018, pp. 359–414.

[42] Corona vaccine development in India: here's a list of all vaccine being locally made and developed, July 20, https://timesofindia.indiatimes.com/life-style/health-fitness/health, 2020.

[43] A. Mishra, M. Choudhary, T.R. Das, P. Saren, P. Bhattacherjee, N. Thakur, et al., Biosensors in diagnosing COVID-19 and recent development, J. Indian Chem. Soc. 98 (2021) 100164.

[44] Russia 1st nation to finish human trials for Covid-19 vaccine, July 13, 2020. https://health.economictimes.indiatimes.com/news/industry/russia-1st-nation-to-finish-human-trials-for-covid-19-vaccine/76938659.

[45] P. Chandra, Miniaturized label-free smartphone assisted electrochemical sensing approach for personalized COVID-19 diagnosis, Sensors International 1 (2020) 100019. https://doi.org/10.1016/j.sintl.2020.100019.

[46] S. Mahapatra, P. Chandra, Clinically practiced and commercially viable nanobioengineered analytical methods for COVID-19 diagnosis, Biosens. Bioelectron. 165 (2020) p112361. https://doi.org/10.1016/j.bios.2020.112361.

SECTION 3

Developments and Fabrication Strategies of RNA/DNA Biosensors

CHAPTER

7

Effect of COVID-19: An environmental point-of-view

Brij M Upreti[a], *and Pooja Singh*[b]

[a]Faculty of Agriculture and Veterinary Science, Jayoti Vidyapeeth Women's University, Jaipur, Rajasthan, India, [b]Department of Biotechnology, SRGI Institute of Management and Engineering, Lucknow, Uttar Pradesh, India

7.1 Introduction

The Coronavirus, for example, is a zoonotic virus that jumped from animal to human, wreaking havoc on our modern-day society. The way humanity has reacted to the virus then turns around and affects our immediate surroundings. In this way, COVID-19 affects the environment. However, this relationship also works the other way around: the environment too, has profound effects on the infection rate of COVID-19. Humanity is caught in the middle of this web, and even now, half a year after the epidemic went global, the general public still has only vague ideas about how this interconnectivity functions.

Perhaps the most important issue about COVID-19 and the environment is the question of how COVID-19 is affecting the present environment. As people all over the world settle into quarantine, traffic levels decrease and nonessential businesses close temporarily. In addition, lockdown will decrease overcrowding and the concentration of human activity, particularly in urban areas. The decrease in human activity is changing the air we breathe as CO_2 and other emission levels go down. Decrease in travel will also play quite a large role in changing air quality and emission of levels of air pollutants. In short, COVID-19's impact on the human population's lifestyle choices will cause changes in every aspect of the environment that

Sensing Tools and Techniques for COVID-19: Developments and Challenges in Analysis and Detection of Coronavirus
DOI: https://doi.org/10.1016/B978-0-323-90280-9.00009-2

human activity is tied to. This includes not only air pollution levels, but also water quality and wildlife biodiversity.

The memory of the fear associated with the virus, as well as the habits that people picked up during the shift to quarantine will linger long after the pandemic abates. Humanity's outlook of the future is rapidly changing as the pandemic continues, and there seems to be more good than bad to this difference.

Perhaps the most important issue about COVID-19 and the environment is the question of how COVID-19 is affecting the present environment.

Environmentally, the COVID-19 outbreak also has constructive effects. Studies started in the home environment during the outbreak increased Home Office applications. This will have an effect that will reduce carbon dioxide emissions by reducing the mobility of people in their external environment. Many countries in the world have switched to distance education due to corona virus (Zhou et al., 2020). Social mobility, which causes many carbon emissions, such as student services, has also decreased with distance education. In addition, since the scientific events, meetings and political events started to be held in the form of a remote conference, environmental emissions have decreased globally. In this context, those who are accustomed to the efficiency of teleconferencing and distance education are expected to increase the number of these activities after the outbreak [1]. In most countries, there has been a lockdown with people not allowed to move around leading to a reduction in greenhouse gas emissions. In the course of the Covid-19 outbreak, the greenhouse gas emission reduction is observed during the restrictions of the states, such as China and Italy [2]. A reduction of approximately 25% of carbon emission is reported in China [3]. It is also reported that air pollution (such as nitrogen dioxide and carbon dioxide emission) is reduced in many regions [4].

8.2 Observations

8.2.1 Positive impact of COVID-19 on the environment

Get better air quality through reducing CO_2 emissions by reducing the mobility of vehicles by individuals. A worldwide reduction of nearly 25% of carbon (C) emission is reported.

A decline in greenhouse gas emissions such as nitrogen dioxide (NO_2) and carbon monoxide (CO) is reported.

Increase in ozone (O_3) level (probably due to lower titration of O_3 by NO) has been reported.

Decline air pollution but there were significant differences between the pollutants. It has been observed that the most significant reduction amounts are in black carbon and nitrogen dioxide (NO_2), while a low decline occurs in the particulate matter with a diameter of less than 10 (PM10).

Ecosystems are being significantly recovered. In many big cities the residents are experiencing a clear sky for the first time in their lives.

8.2.2 Negative impact of COVID-19 on the environment

- Increased medical wastes resulting from the increased medical activity: It was stated that for the period when the outbreak peaked in Wuhan, an average of 240 tons of medical waste was produced daily in hospitals and this value was 6 times higher than normal value.
- Masks made of polypropylene, a type of plastic, and it is very difficult to get lost in nature.
- Increase demand for online shopping for home delivery. Consequently, organic and inorganic waste generated by households has increased.
- Reduction in waste recycling

Often, there has been an increased demand for personal protective equipment (PPE). The demand has forced companies to work overtime to keep up. The increased factory use has increased the pollution they create. As well, the materials the masks are made out of are harmful for the environment. Many people have chosen to use single use masks over cotton and reusable ones which have added to the clinical waste in our oceans. This has a negative impact on our marine ecosystems, endangering the animals and plants that live in them. There also has been more water usage, people have started to wash their hands more often, take longer showers, and wash their clothes/dishes more often. Water is a finite resource and it is important to use it responsibly. It is predicted that water usage will return to normal once the lockdown rules are lifted.

Perhaps the most important effect of the pandemic on human activity is how the pandemic is changing industries. There has been an increase in robotics use and automotive equipment. As many workers were forced to work from home and as factories were shut down companies started to look for alternatives and replacements. Many have turned toward Industry 4.0 technology; this includes 3D printing and artificial intelligence. Before the pandemic the shift toward using more robots had already begun but COVID-19 has acted as a catalyst pushing it forward faster, One industry making major changes is the mining industry, companies such as Resolute Mining have started to shift their entire production to automated machines controlling them through a central control center. Another industry is the medical industry, making hospitals have looked into using robots in order to maintain sanitation and decrease the spread of the virus.

References

[1] M. Rohwer-Kahlmann, Will the coronavirus change the way we work from home?, DW, 2020 Retrieved April 20, 2020.

[2] CDPThe relationship between climate change and coronavirus (Covid-19), CDP, Turkey, 2020 Retrieved April 20, 2020.
[3] L. Myllyvirta, Coronavirus has temporarily reduced China's CO2 emissions by a quarter. Carbon Brief. Retrieved April 20, 2020 (2020).
[4] J. McMahon, Coronavirus lockdown likely saved 77,000 lives in China just by reducing pollution. Forbes. Retrieved April 20, 2020 (2020).

CHAPTER

8

Nanotechnology-based bio-tools and techniques for COVID-19 management

Shiva Sharma[a], *Alpana Joshi*[b], *Subrata K. Das*[a], and *Sudheesh K. Shukla*[a,c]

[a]Department of Biomedical Engineering, School of Biological Engineering and Life Science, Shobhit Institute of Engineering & Technology (Deemed-to-be University), Modipuram, Meerut, India, [b]School of Agriculture Technology and Agri-informatics, Shobhit Institute of Engineering & Technology (Deemed-to-be University), Modipuram, Meerut, India, [c]Department of Chemical Sciences, University of Johannesburg, Doornfontein Campus, Johannesburg, 2028, South Africa

8.1 Introduction to COVID-19

COVID-19, a most serious issue and the threat for the human life, has affected millions of people worldwide. It was observed as unknown cases of pneumonia in Wuhan, China and claimed unknown till January 10, 2020 and led to the corona virus disease 2019 (COVID-19) therefore worldwide pandemic. The Director-General-WHO declared the outbreak of COVID-19 and constituted a Public Health Emergency of International Concern (PHEIC) on January 30, 2020 with the recommendations of the Emergency Committee. This outbreak originated from Wuhan, China in 2019 named as COVID-19 approached 115 countries, with 119,239 cases of infection spread and 4287 deaths by March 11, 2020.

There were around 4,995,996 confirmed cases of SARS-CoV-2 including 327,821 deaths in 216 countries by May 22, 2020 [1]. Virus belongs to Coronaviridae family, Nidovirales order, and Cornidovirineae suborder. This

Sensing Tools and Techniques for COVID-19: Developments and Challenges in Analysis and Detection of Coronavirus
DOI: https://doi.org/10.1016/B978-0-323-90280-9.00008-0

Coronaviridae classified into Orthocoronavirinae and Letovirinae two subfamilies. Letovirinae have one genera named as Alphaletovirus whereas Coronaviridae have four genera named as αCoV (alpha-coronavirus), βCoV (beta-coronavirus), γCoV (gamma-coronavirus), and δCoV (delta-coronavirus).

COVID-19 virus belongs to the β coronaviridae family, seventh virus of its kind that infected the human being HKU1, 229E, OC43, and NL63 corona viruses mediates the slight cold or common cold. Whereas SARS-CoV, MERS-CoV, and SARS-CoV-2 are life-threatening viruses this may be the responsible cause of death [2]. Research shows coronavirus has 79.5% of its genome with SARS-CoV (severe acute respiratory syndrome coronavirus) 50% with MERS-CoV (Middle East respiratory syndrome coronavirus), and 96% with TG13 (bat coronavirus).

The International Virus Taxonomy Committee (IVTC) gave the formal name to a new coronavirus, SARS-CoV-2 on February 11, 2020 and named as COVID-19 [3]. The most general pathway of respiratory ailment diffusion is airborne droplets, physical contact, and fomites. The physical contact transmission refers to direct transmission and fomites transmission refers to the indirect transmission through intermediate objects. Fever, dry cough, and fatigue are the most generalized symptoms of COVID-19 patients.

Few patients reported conjunctivitis, loss of taste and smell, nasal blocking, pain in muscles and joint pain, painful throat, diarrhea, vomiting and queasiness, chills and dizziness, rashes on skin, and headache also. The severity of COVID-19 generally includes high temperature, constant pain and pressure in the chest, loss of hunger, breathlessness, and confusion in all ages' peoples. Few COVID-19 severe cases were found with nervousness, hopelessness, sleeping disorders, confusion, irritability, and reduced consciousness.

Neurological difficulties such as strokes, inflammation of brain, delirium, and damaging of nerve were present with the severe cases of COVID-19 [1]. COVID-19, a serious threat to life, has affected millions of propels worldwide. It was observed as unknown cases of pneumonia in Wuhan, China and claimed unknown till January 10, 2020 and led to the corona virus disease 2019 (COVID-19) therefore worldwide pandemic. The Director-General-WHO declared the outbreak of COVID-19 and constituted a Public Health Emergency of International Concern (PHEIC) on January 30, 2020 on the recommendations of the Emergency Committee. This outbreak originated from Wuhan, China in 2019 named as COVID-19 approached 115 countries, with 119,239 cases and 4287 deaths by March 11, 2020. There were around 4,995,996 confirmed cases of SARS-CoV-2 including 327,821 deaths in 216 countries by May 22, 2020 [4].

Virus belongs to Coronaviridae family, Nidovirales order, and Cornidovirineae suborder. This Coronaviridae family is further classified into the two subfamilies Letovirinae and Orthocoronavirinae. Letovirinae have one genera named as Alphaletovirus and Coronaviridae consists of four genera αCoV (alpha-coronavirus), βCoV (beta-coronavirus), γCoV (gamma-coronavirus), and δCoV (delta-coronavirus). COVID-19 virus belongs to the β coronaviridae family, it is the seventh virus of its kind that infected the human HKU1, 229E, OC43 and NL63 corona viruses mediates the slight cold or common cold. Whereas SARS-CoV, MERS-CoV, and SARS-CoV-2 are life-threatening viruses and may cause death [2]. Research shows coronavirus has 79.5% of its genome with SARS-CoV (severe acute respiratory syndrome coronavirus), 50% with MERS-CoV (Middle East respiratory syndrome coronavirus), and 96% with TG13 (bat coronavirus). The International Virus Taxonomy Committee (IVTC) gave the formal name to a new coronavirus, SARS-CoV-2 that is severe acute respiratory syndrome coronavirus-2 on February 11, 2020 and was named as COVID-19 [3].

The most common pathway of respiratory disease transmission is airborne droplets, physical contact, and fomites. The physical contact transmission refers to direct transmission and fomites transmission refers to the indirect transmission through intermediate objects. Fever, dry cough, and fatigue are the most common symptoms of COVID-19 patients. Few patients reported conjunctivitis, loss of taste or smell, nasal congestion, muscle or joint pain, sore throat, diarrhea, nausea or vomiting, chills or dizziness, different types of skin rashes, and headache also. The severity of COVID-19 most commonly includes high temperature, persistent pain or pressure in the chest, loss of appetite, shortness of breath, and confusion in the peoples of all ages. Few severe COVID-19 cases were found with anxiety, depression, sleep disorders, confusion, irritability, and reduced consciousness. More severe and rare neurological complications such as strokes, brain inflammation, and delirium, and nerve damage were also present [1].

8.2 Structure of COVID-19 virus

COVID-19 (SARS-CoV-2) virus structure shows it with single-stranded RNA, enveloped virus covered with high affinity protein glycosylated (S) for angiotensin-converting enzyme-2 receptor (ACE2). SARS-CoV-2 showed 10–20 times more likeness to receptor enzyme AEC-2 in comparison to SARS-CoV. It connects to protein receptor ACE2 of host cell membrane and supports the administration of the virus within the human cell [5]. SARS-CoV-2 composed of nearly 30,000 nucleotides forming a single stranded positive RNA contains 27 proteins, including four structural proteins named as spike and denoted by S, enveloped protein denoted as E, membrane protein denoted as M, and nucleocapsid protein denoted as N. The nucleocapsid protein clutches the RNA, whereas spike

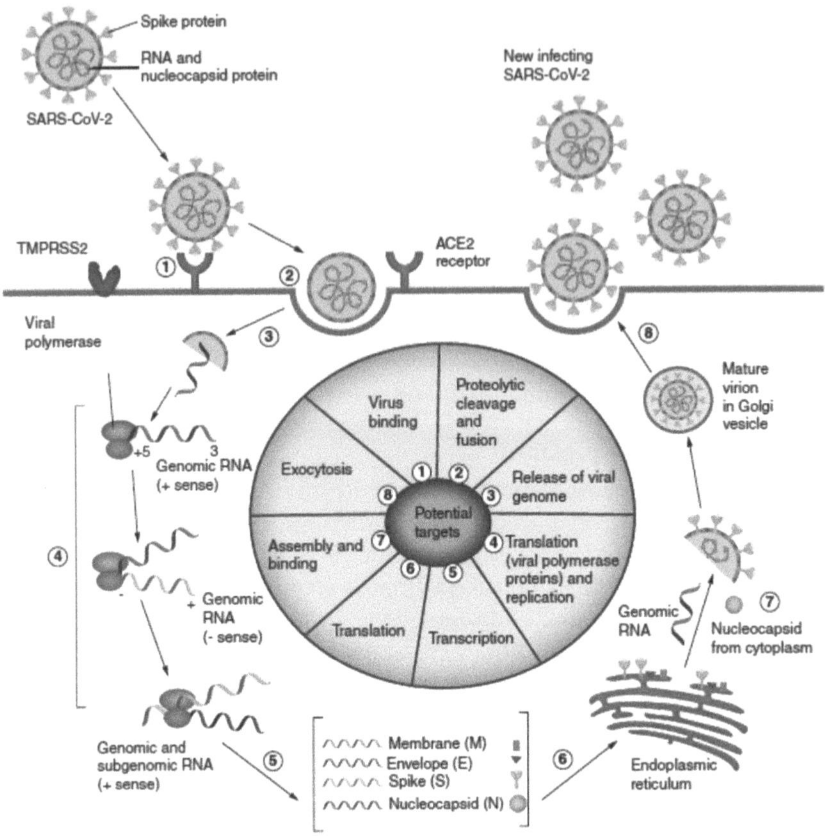

FIGURE 8.1 Targeting by SARS-CoV-2 COVID-19 to host cell Resource [69].

S. enveloped E, and the membrane M protein created the envelope for virus.

Previous studies indicate bat as the possible host for the SARS-CoV-2, and Malayan pangolin as the middle host [3,6–8]. Presence of the type-2 TM serineon on cell membrane promoted the administration of the COVID-19 virus into the host cell membrane with the activation of S protein. Upon entering from the membrane to host cell the virus released its genetic material RNA into the host cell. Further translation take place and translated the polyproteins with the help of genome RNA followed by replication and transcription. Thereafter, synthesis of structural protein, assembling, packaging taken place and particles of the viral released within the host cell [9] as the mechanism of the transmission indicated in Fig. 8.1. Moreover, the presence of ACE2 receptors at most of the organs and cells as endothelial cell, muscular cell and its maximum expression at renal,

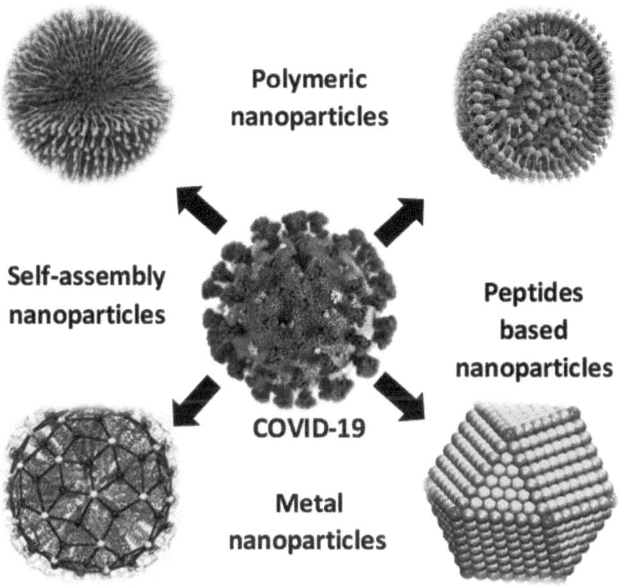

FIGURE 8.2 Nanotechnology as a diagnostic tool for COVID-19.

cardiovascular, and gastrointestinal tissues increases the survival domain and provides larger surface area for the interaction and transmission of SARS-CoV-2 through the membrane into host cell. Therefore, isolation of virus can be done through urine, blood samples, bronchoalveolar lavage fluid, cerebrospinal fluid, sputum, nasal swab, throat and anal swabs, and tracheal aspirates [10–12] (Fig. 8.2).

8.3 Management of COVID-19

The movement and transmission of COVID-19 can be constrained by making mindfulness and taking preventive measures, examples includes face covers, social separating, hand cleanliness, and isolating themselves from affected people [13]. Management strategies of SARS-CoV-2 COVID-19 also required an effective and precise diagnostic, identification, and separation method.

SARS-CoV-2 diagnosis is divided into four categories based on nucleic acid amplification, immunological assay and sequencing methods such as RT-PCR, immunoassay technique based on antibodies IgG or IgM, antigens, and other relevant diagnostic approaches. COVID-19 manifestation can also be verified using chest computed tomography (CT) and magnetic resonance imaging (MRI) [14–16]. There are four primary techniques for nucleic acid-based diagnostic technologies: real-time RT-qPCR, Digital dPCR, multiplex mPCR, and loop-mediated isothermal amplification method (LAMP); each has advantages and limitations.

RT-qPCR offers a high sensitivity and the ability to quantify absolute and relative values with no communication error. It specifically targets RdRP, envelope, and nucleocapsid genes; the overall reaction time is around 80 minutes. The ORF1ab and nucleocapsid genes are targeted using digital PCR, and the entire reaction time is about 30–60 minutes. It is the most sensitive technique, capable of perfect identification. Multiplex PCR has the capacity to address several targets, primarily spike and nucleocapsid genes, with greater efficiency and takes around 2 hours to accomplish. The LAMP technique specifically targets the ORF1ab and nucleocapsid genes.

It does not require any extensive sample processing or thermal cycler. It takes around 30 minutes for complete analysis [17–23]). Immunological tests include the enzyme-linked immunosorbent assay (ELISA), the fluorescence-labeled immunochromatographic assay (FICA), the colloidal gold immunochromatographic assay (GLCA), and the chemiluminescence immunoassay (CMIA). The ELISA targets IgA, IgG, and IgM antibodies; it is a viable approach that requires no viral contact and produces findings in 1 hour.

GLCA is a quick and easy method that uses IgG and IgM antibodies to get results in 15 minutes. FICA is a very sensitive approach for detecting infection quickly and early. It specifically targets the nucleocapsid protein and produces effects in 10–30 minutes. CMIA is a highly sensitive and fully automated method that detects IgA, IgG, and IgM antibodies and produces findings in 15–40 minutes [24–27].

Sequencing-based detection technologies include the two most often used sequencing methods, mNGS (meta genomics next generation sequencing) and NTS (nonopore third generation sequencing). mNGS focuses on the whole gene sequence, whereas NTS focuses on specific genes such as ORF3a, ORF6, ORF7a, ORF8, ORF10, surface, envelope, and nucleocapsid genes. The study takes around 24 hours and is capable of detecting any portion of the genome as well as diagnosing viral mutations, making it a very sensitive and impartial testing tool [12,28–30].

Over the number of advantages, the high relativity of analysis in RT-qPCR gives false positive results. Efficiency of primers and reactions can also affect the testing outcomes. The expensiveness of dPCR limits its uses, mPCR requires multiple primers and optimization makes the process tedious, the complexity of the sample, primer design, and lack of data limits the usability of LAMP techniques. Moreover, all the above mentioned techniques take half an hour to 2 hours times for the analysis that also limits the uses and demands for high speed, sensitive, less complex, and accurate method [17–23].

Among the many benefits, the high relativity of analysis in RT-qPCR produces false positive findings. The efficiency of primers and reactions might also have an impact on test results. The cost of dPCR restricts its application; mPCR needs numerous primers and optimization, making the procedure time-consuming; and the complexity of the sample, primer design, and lack of data limit the applicability of LAMP methods. Furthermore, all of the above-mentioned approaches need a half-hour to 2-hour analysis time, which restricts the applications and demands for high-speed, sensitive, less complicated, and accurate methods [17–23].

Sequencing-based detection methods have drawbacks in terms of complexity, length of procedure, and length of time [12,28–30]. A few additional methods, such as chest computed tomography (CT) or magnetic resonance imaging, are also available for the diagnosis of COVID-19 (MRI). These imaging modalities are extremely expensive, and early diagnosis is not achievable with CT or MRI.

Overall, diagnostic and detection techniques play an essential role in controlling and managing COVID-19 and allowing health care specialists to focus resources to prevent the spread of infections and lower the death ratio due to COVID-19.

8.4 Nanotechnology: A diagnostic tool for COVID-19

A COVID-19 diagnostic instrument nanotechnology is the creation and deployment of nanoparticles that has enormous promise in many disciplines of research, including materials science, and biological science [31]. Nanotechnology is widely used in medical sciences due to its unique properties such as small size, large surface area, surface adaptability, and enhanced solubility, which aid in the development of safer and more efficient drug candidates, tissue-targeted therapies, personalized medicines, and early diagnostic devices. Despite the fact that nanomedicine techniques are being used to grow antibody transporters, nanotechnology approaches to dealing with generating vaccinations and therapy for the ongoing devastation remain missing. The COVID-19 pandemic necessitates an examination of current nanotechnologies [18,32].

COVID-19 instances are rising all the time, and millions of individuals are already infected, therefore it is critical to create specialized sensors to detect the infection for quick point-of-care (POC) diagnosis, observation, and monitoring of viral illness [33]. In the current situation, efficient, simple, and quick detection approaches for COVID-19 are required.

As a result, quick, sensitive, and exact diagnostic techniques, as well as point-of-care testing, are necessary to combat the COVID-19 epidemic. Nucleic acid-based testing was utilized as an effective detection technique for SARS-CoV-2, with nasal or oral swabs used to determine viral presence using the real-time PCR method. This testing procedure was discovered to be costly, time-consuming, and labor-intensive.

Because of its tiny size, high surface to volume ratio, adsorption, and high reactivity, nanotechnology can greatly improve the sensitivity of detection methods such as real-time PCR and immunofluorescence tests. By using metal-based nanoparticles such as gold, copper, zinc, and others, nanoparticles can shorten detection time, improve selectivity and specificity, and aid in the detection of low-magnitude signals. Because of their excellent electric characteristics when coupled with antibodies, RNA aptamers, and single-stranded DNA, inorganic nanoparticles have been utilized in the development of a wide range of viral detection instruments [34].

Furthermore, nanoparticles can be utilized to prevent the transmission of viral illness, produce efficient sanitizers, and self-sterilizing personal protective equipment (PPE) such as masks and gloves [35].

The synthesis and separation of nanomaterials using bioengineering ideas in order to create nanoparticles/nano vaccines to combat COVID-19. In general, synthetic biology is involved in the creation of vaccines, the study of genetic sequences, and the development of antiviral medicines for detecting, treating, and preventing the spread of Coronavirus at the community levels [36].

Nanoparticles are also used in the development of Coronavirus detection devices that are supported by digital technologies and integrated with virus testing facilities for accurate and quick diagnosis. Coronaviruses are composed of RNA enclosed in proteins such as envelope proteins, nucleocapsid proteins, and glycoproteins. Nucleocapsid proteins are degraded by UV exposure, alcohol, and surfactant treatment. Nanoparticles, which are used as coatings and protective layers in nanodrugs and vaccines to combat viruses and break their DNA, will likewise degrade the outer layer of spike protein [37–39].

It has been suggested that monoclonal antibodies or vaccinations based on nanotechnology will be a potential technique for rapid diagnosis and successful therapy. Novel nanomaterials can boost the penetration of polymeric nanoparticles in the lung, increasing the efficacy of COVID-19 therapy. In addition, the researchers created stable, nontoxic, biodegradable nanoparticles that may be employed successfully in the lung.

Aside from that, nanotechnology may be utilized to create innovative antiviral medicines, promote the co-delivery of several treatments, lengthen the circulation duration, and release a consistent amount of pharmaceuticals (Figs. 8.3–8.5). Furthermore, it can be utilized for medication targeting and decreases drug-related adverse effects. Furthermore, by shielding mRNA and DNA vaccines from enzymatic breakdown, nanotechnology can be employed as an efficient delivery method [35,40].

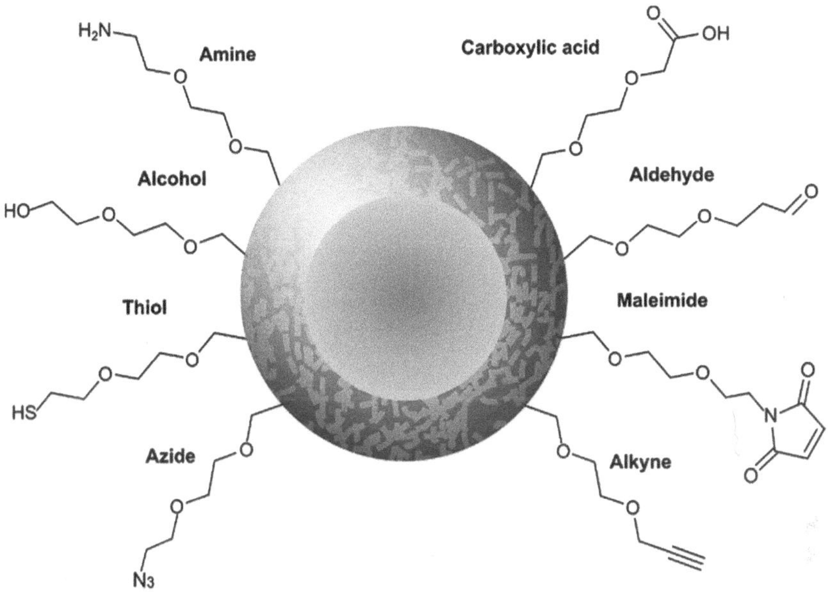

FIGURE 8.3 Representation of functional ligands (amine, carboxylic, aldehyde, maleimide, alkyne, azide, thiol, alcohol, and amine) attached to nanoparticle surfaces.

Anti-COVID-19 agents colloidal polymers composed of nanoparticles, with particle sizes ranging from 10 nm to 100 nm, are extremely efficient as gene carriers. Metal-based nanoparticles can impede viral attachment, replication, and deformation of viral walls and membranes.

Because of their unique characteristics, such as smaller particle sizes, higher surface-volume ratios, and surface charges, nanoparticles are effective tools in viral treatment. Antiviral effects of nanoparticles have been demonstrated against influenza, HIV, and rabies (Jiang et al., 2002). A more cost-effective, faster, and safer type of antiviral therapy has been developed using molecular imprinting with polymeric viral catchers.

Polymers, by sticking to infectious viruses and preventing them from entering cells, inhibit them from replicating and infecting cells. Natural (cyclodextrin, chitosan, and carrageenan) and synthetic polymers such as polyethyleneimine (PEI), poly (dl-lactide-co-glycolide) (PLGA), dendrimers, and metal and metal oxide NPs (silver, copper, titanium, zinc, and gold) have been introduced as effective antiviral agents due to their biodegradability, outstanding biocompatibility, and safety profiles in the human body [41–46]) (Table 8.1).

Cyclodextrins are nontoxic oligosaccharides with cyclic α-D-glucopyranose units of (α-1, 4)-linkage. It can help with formulation development by increasing the solubility of water-insoluble medicines

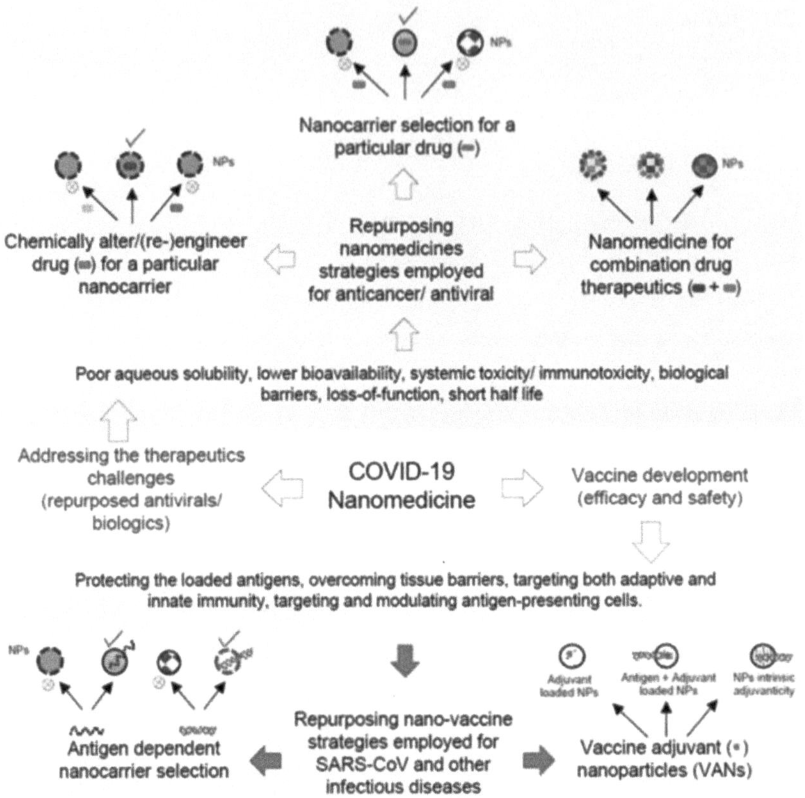

FIGURE 8.4 Nanomedicine approaches for production therapeutics of COVID-19 vaccines [18]. Polymeric-based nanoparticles against COVID-19.

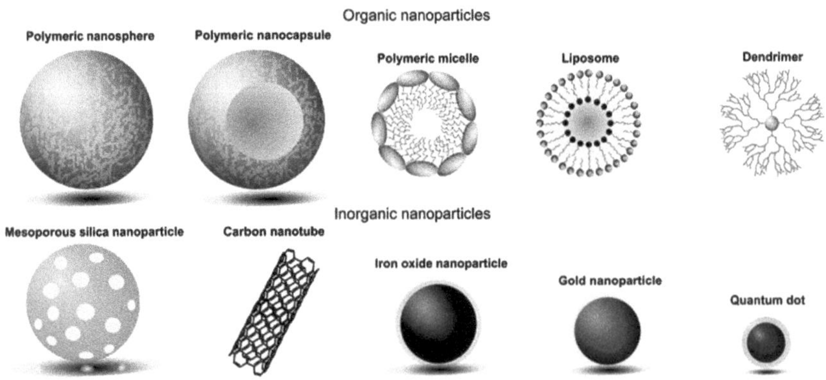

FIGURE 8.5 Representation of different types of nanoparticles used in several biomedical applications.

TABLE 8.1 Nanoparticle drug delivery systems with antiviral properties.

Nanocarriers	Composition	Size (nm)	Loaded molecules
Liposomes	Phospholipids	50–900	Hydrophilic, hydrophobic
Solid lipid nanoparticles	Solid lipid, surfactant	100–500	Hydrophobic
Nanostructured lipid carriers	Solid and liquid lipids, surfactant	100–500	Hydrophobic
Nanoemulsions	Liquid lipid, cosurfactant, surfactant	20–200	Hydrophilic, hydrophobic
Polymer nanoparticles	Polymer crosslinking agent	100–900	Hydrophilic
Cyclodextrins	Cyclic oligosaccharides	5–50	Hydrophilic, hydrophobic
Dendrimers	Oligosaccharides	3–20	Hydrophilic, hydrophobic

by inclusion complexation. The advantages of cyclodextrin complexed medicines include fewer adverse effects, more abundance, higher stability, a less volatile nature, and an enhanced drug release method. Chitosan is a common polysaccharide that is made up of randomly distributed deacetylated and acetylated units of D-glucosamine.

Chitosan, a deacetylated derivative of chitin, a natural polymer found in crab and shrimp shells, is utilized in a variety of our goods. Chitosan has gained popularity because to its antiviral, antibacterial, and anti-inflammatory properties, as well as its biocompatibility, biodegradability, nontoxicity, and high moldability. Researchers looked at the possible applications of chitosan in agriculture, biotechnology, cosmetics, environmental protection, gene therapy, materials science, the food industry, and wastewater treatment [47–49].

Carrageenans (CGNs), which are linear sulfated polysaccharides, are mostly found in marine seaweeds. Natural polymers have an ionic character and an ester sulfate concentration ranging from 15% to 40%. Various molecular weight CGNs polymers have been widely employed in drug discovery and development [50,51]. Dendrimers offer a greater biocompatibility, are less expensive, and provide a carefully regulated large macromolecular surface. One of the most distinctive properties of dendrimers is their ability to typify small size components such as medicines, metals, and imaging moieties that can fit inside their depressions and interact through charge connections, hydrogen holding, and lipophilicity [52].

Poly(lactic-co-glycolic corrosive) (PLGA) and polyethyl glycol (PEG) are FDA-approved polymers that can be used to combat COVID-19.

Furthermore, PLGA can be used in healthcare devices and other drug conveyance applications. A new nanovaccine based on PLGA and DEPE-PEG polymers can deliver subunit viral antigens, STING agonists, and an adjuvant in an infection-like fashion as a safe and effective anti-COVID strategy. The created capsid-like hollow nanostructured polymer exhibited a number of distinct and advantageous characteristics, including reduced systemic reactogenicity, a pH-responsive release profile, and substantial local immune activation. According to the study, this technique allowed the development of safe and efficient vaccinations for combating viral infections [53].

Polyethylenimine (PEI) is a kind of polycationic natural polyamine polymer. PEI is a water-soluble, fanned polymer that is frequently used in vitro and in vivo as a nonviral vector for DNA/RNA transfection and quality control. Covalent restriction of PEG or moieties with PEI was used to improve reliability and decrease poisonousness [54,55].

8.4.1 Self-assembling protein-based nanoparticles (SANPs) against COVID-19

SARS-CoV-2 RBD/spike antigens were developed and optimized for display on self-assembling protein nanoparticles (SApNPs) as possible COVID-19 vaccines. The creation of SApNPs have resulted in the production of a new nanoparticle by protein oligomerization (SANPs). In an animal model, nanoparticles made from protein monomers were able to give full protection against influenza [56].

SANPs have proven a variety of medicinal applications, and their size ranges in the viral size range (20–100 nm) making these particles effective and promising candidates for vaccine development against respiratory viruses such as COVID-19 [57,58]. In animal models, SANPS based on nucleoprotein from the respiratory syncytial virus (RSV) nucleocapsid were discovered to be a promising vaccine antigen, and vaccinated mice were reported to be protected against RSV replication and had reduced viral load in the lungs. Mucosal vaccination with nanovaccines induced both systemic and local immunity, with substantial levels of IgA anti-N, IgG2a and IgG1 antibodies, CD8+, and CD4+ T cells [59]. Furthermore, nanovaccines containing a palivizumab-targeted epitope to the N protein improved immune response against RSV [60].

Virus-like particles (VLPs) are spherical supramolecular assemblies produced by the self-assembly of viral capsid proteins with diameters ranging from 20 to 200 nm. VLPs are supramolecular assemblies with spherical shapes and dimensions ranging from 20 to 200 nm that are formed by the self-assembly of viral capsid proteins. VLPs are thought to be a promising method for vaccine development since they are capable of easily replicating antigenic epitopes despite the fact that they lack genetic components.

Intranasal administration of VLPs induced both humoral and cellular immune responses, and produced vaccines were found to be efficient in reducing viral loads [61–63].

8.4.2 Peptide-based nanoparticles against COVID-19

Peptide-based vaccinations contain peptides in addition to adjuvant mixes, or the peptides can be delivered by a nanocarrier or encoded by nucleic acid corrosive antibody designs. Nanoparticle antibody efficacies can be enhanced and their invulnerable profiles tailored to target specific infections by combining methods for concentrating on lymph hubs (LNs) or cell subsets and subcellular regions. Transformations in the amino corrosive and short peptide inhibitors are predicted to be anti-SARS-CoV apparatuses. Antibodies based on peptides express the C-terminal heptad rehash region, which is an excellent beneficial way for the treatment of SARS-CoV associated contaminations, and this technique is effectively used as peptide-based nanoparticles (PBNPs). Despite advancements in peptide-based COVID-19 vaccinations, industrial and academic initiatives impact expected B- and T-cell epitopes in their subunit antibodies against SARS-CoV-2 [64,65].

Peptide inhibitors derived from angiotensin-converting enzyme 2 (ACE2) are a potential technique for blocking COVID-19 receptor binding domains. For multivalent binding and inhibition of the COVID-19 receptors, peptides should be attached to the surfaces of nanostructured materials. It is possible by employing the PBNPs that have the capacity to neutralize COVID-19 infections. Furthermore, it was proven that by supplying numerous bindings of SPI onto the nanocarriers, the binding effectiveness may be increased [66,67].

8.4.3 Organic/inorganic and metal nanoparticles against COVID-19

Inorganic nanoparticles have a wide range of uses in medicine. Nanoparticles' appealing characteristics, including as biocompatibility, adjustable size, and unique physicochemical properties, make them appropriate for a variety of biomedical applications. The produced nanoparticle must be inert, stable under physiological circumstances, and have a surface that can be readily conjugated to the metal in order to travel freely through the body. The most efficient treatments technique is injecting nanoparticles with encapsulated cargo or magnetic capabilities, and surface targeting ligands assist nanoparticles in delivering to target locations without generating significant harm. Natural and inorganic NPs can be utilized to target the lung in order to counteract the side effects of conventional organization's high serum groupings. Carbon-based NPs, metal oxides

(Fe_2O_3, TiO_2, ZnO_2,) and transition metal NPs (Ag, Cu, Zn) exhibit inherent antipathogenic properties by interfering with at least one viral life-cycle phase.

Natural nanoparticles (NPs) such as liposomes, dendrimers, micelles, and polymers have also been demonstrated to be effective against SARS-CoV-2 (Table 8.2). These particles can be created on their own or in conjunction with inorganic NPs to create a target nanovaccine mixture. Because of their small molecular size, nanovaccine codelivery can also reduce health problems in the lung and respiratory systems ([68]; Shrivastava et al., 2021; [69]).

Nanolabeling can be accomplished by imbuing specific probes with metal nanoparticles such as silver (AgNPs) or gold nanoparticles (AuNPs) or quantum dots, resulting in signal amplification. Because of their regulated intracellular delivery, improved stability, and resistance to degradation, silver nanoparticles (AgNPs) are superior drug carriers for nucleic acid-based delivery. Among inorganic nanoparticles, gold nanoparticles (AuNPs) are suitable for vaccine production because they are easily absorbed by both dendritic cells and macrophages. AuNPs are polymeric NPs and vesicular nanocarriers that depending on their size, shape, and surface science, have the potential to activate cytokine and anti-inflammatory agent responses.

Because of the strong interaction between thiol groups and gold, large-scale synthesis of AuNPs is possible, and they can also be easily functionalized. In addition, because inert carriers, such as AuNPs, do not elicit an immune response, they have emerged as a platform for the development of nanovaccines via antigen fictionalization [15,63,70–73] (Fig. 8.5).

8.5 Conclusions and future perspective

Novel technologies and developments in nanoscience can be utilized to diagnose and cure viral respiratory infections. Because of their unique size, shape, surface area, and capacity to functionalize, nanoparticle therapeutic methods can overcome the limits of traditional vaccinations, resulting in immunogenicity and increased antigen presentation. Novel nanoparticles exhibited simple penetration, biodegradability, and toxicity. Furthermore, nanotechnology has enabled the creation of a number of biosensors, nanovaccines, and antiviral composites against SARS-CoV2.

Polymeric nanoparticles are safe to use, while inorganic nanoparticles based on gold, silver, copper, and zinc are among the finest prospects for use against viral illness and other medicinal uses. Furthermore, the protein-isolated peptide inhibitor inhibits SARS-CoV-2 because its binding effectiveness enhanced with repeated binding of nanocarrier linked peptides. However, the toxicity of Nanoparticles should be carefully studied

TABLE 8.2 Application of nanoparticles for prevention and detection COVID-19.

S.N.	Name of product/manufacturer	Type of nanoparticles	Function
1	Nano Silver sanitizer/SHEPROS	Silver nanoparticle	Hand sanitizer (kills 99% of germs and bacteria)
2	Silvo clean spray/Weinnovate biosolutions	Silver nanoparticle	Sanitizer and disinfectant
3	NanoSepti/NanoTouch Materials, LLC	Mineral nanocrystal	Surface disinfectant
4	TeqAir 200 air ionizer/TEQOYA	–	Air purifier
5	AAVI Leaf/AAVI Technologies Co.	–	Air purifier
6	Mack Antonoff HVAC/Mack Antonoff HVAC	–	Air purifier
7	Air Decontamination Units/Genano Ltd.	–	Air purifier
8	Graphene mask/Flextrapower, Inc.	Graphene nanomaterial	Virus protective respiratory mask
9	Guardian G-Volt (LIGC Applications Ltd)	Graphene nanomaterial	Virus protective respiratory mask
10	G + Fibrics/Directa Plus PLC	Graphene nanomaterial	Antiviral fabric is used in the production of masks, gloves and gowns
11	Antiviral fabrics/Promethean Particles Ltd	Copper nanoparticle	PPE (personal protective equipment)
12	MVX Nano Mask/MVX Prime Ltd.	–	Self-sanitizing surgical mask kills 99.9% of all viruses and bacteria
12	ReSpimask VK/RESPILON	Copper oxide nanoparticles	Face mask with a 99.9% filtration efficiency for viruses
13	Nanofiber mask/YAMASHIN-FILTER CORP.	Nanofibers made from synthetic polymers	Respiratory protective mask
14	NANOHACK (Copper 3D Antibacterial Innovations)	Copper oxide nanoparticles	Protective respiratory mask

(continued on next page)

TABLE 8.2 Application of nanoparticles for prevention and detection COVID-19—cont'd

S.N.	Name of product/manufacturer	Type of nanoparticles	Function
15	COVID-19 Rapid POC CE-IVD Test/NanoComposix	Gold nanoparticles	Detection kit used in point-of-care tests
16	COVID-19 Rapid Test Cassette/SureScreen Diagnostics Ltd	Gold nanoparticles	Detection kit
17	COVID-19 point-of-need diagnostic test/Mologic Ltd	Gold nanoparticles	Detection kit
18	SAFER-sample Kit/Lucence Diagnostics Pte Ltd	–	Sample collection kit (stabilizes viral RNA at room temperature for 1 week)
19	Lateral flow/Sona Nanotech, Inc.	Gold nanorod	Detection kit
20	mRNA-1273	Lipid nanoparticle	RNA vaccine; act as mRNA carrier for safe and efficient transport in vivo
21	NVX-CoV2373	Virus-like nanoparticle	Protein subunit vaccine; Thermostable and has a higher binding affinity for human ACE2 receptor and neutralizes virus infection
22	BNT162	Lipid nanoparticle	mRNA vaccine; acts as mRNA carrier for safe and efficient transport in vivo
23	ARCT-021	Lipid nanoparticle	RNA vaccine; acts as mRNA carrier for safe and efficient transport in vivo
24	ChulaCov19	Novel lipid nanoparticles	RNA vaccine; acts as mRNA carrier for safe and efficient transport in vivo
25	CVnCoV mRNA vaccine	Lipid nanoparticles	Acts as mRNA carrier for safe and efficient transport in vivo

and managed when creating Nanoparticle-based treatments for respiratory illnesses.

Furthermore, knowing the molecular processes of COVID-19, as well as a full understanding of viral transmission, is still necessary to deal with the COVID-19. Many nanovaccines have already been repurposed to treat the new SARS-CoV-2 disease, demonstrating the promise of nanotechnology. Furthermore, research should concentrate on the development of efficient models for assessing the toxicity profile of these nanovaccines, their in vitro and in vivo efficacy, and large-scale commercial production.

References

[1] World Health OrganizationCoronavirus disease (COVID-19) pandemic, WHO, Geneva, 2020 Available at: https://www.who.int/emergencies/diseases/novel-coronavirus-2019 [accessed 25 April 2020].

[2] A.A. Ismail (2020). Serological tests for COVID-19 antibodies: Limitations must be recognized, 57, 274–276.

[3] P. Zhou, X.L. Yang, X.G. Wang, B. Hu, L. Zhang, W. Zhang, et al., A pneumonia outbreak associated with a new coronavirus of probable bat origin, Nature 579 (7798) (2020) 270–273.

[4] D. Wrapp, N. Wang, K.S. Corbett, J.A. Goldsmith, C.L. Hsieh, O. Abiona, et al., Cryo-EM structure of the 2019-nCoV spike in the prefusion conformation, Science 367 (6483) (2020) 1260–1263.

[5] F. Wu, S. Zhao, B. Yu, Y.M. Chen, W. Wang, Z.G. Song, et al., A new coronavirus associated with human respiratory disease in China, Nature 579 (7798) (2020) 265–269.

[6] R. Lu, X. Zhao, J. Li, P. Niu, B. Yang, H. Wu, et al., Genomic characterisation and epidemiology of 2019 novel coronavirus: Implications for virus origins and receptor binding, Lancet North Am. Ed. 395 (10224) (2020) 565–574.

[7] P. Muralidharan, M. Malapit, E. Mallory, D. Hayes Jr, H.M Mansour, Inhalable nanoparticulate powders for respiratory delivery, Nanomedicine: Nanotechnology, Biology, and Medicine 11 (5) (2015) 1189–1199.

[8] D. Yang, Application of nanotechnology in the COVID-19 pandemic, Int J Nanomedicine 16 (2021) 623–649.

[9] D.E. Gordon, G.M. Jang, M. Bouhaddou, J. Xu, K. Obernier, K.M. White, et al., A SARS-CoV-2 protein interaction map reveals targets for drug repurposing, Nature 583 (7816) (2020) 459–468.

[10] Y. Han, P. Král, Computational Design of ACE2-Based Peptide Inhibitors of SARS-CoV-2, ACS Nano 14 (4) (2020) 5143–5147.

[11] B. Udugama, P. Kadhiresan, H.N. Kozlowski, A. Malekjahani, M. Osborne, V.Y. Li, et al., Diagnosing COVID-19: The disease and tools for detection, ACS Nano 14 (4) (2020) 3822–3835.

[12] M. Wang, A. Fu, B. Hu, Y. Tong, R. Liu, Z. Liu, et al., Nanopore targeted sequencing for the accurate and comprehensive detection of SARS-CoV-2 and other respiratory viruses, Small 16 (32) (2020) 2002169.

[13] World Health Organization.Transmission of SARS-CoV-2: Implications for infection prevention precautions: Scientific brief, 09 July 2020 (No. WHO/2019-nCoV/Sci_Brief/Transmission_modes/2020.3) b, World Health Organization, 2020.

[14] M. Mahmoudi, Emerging biomolecular testing to assess the risk of mortality from COVID-19 infection, Mol. Pharmaceutics 18 (2) (2021) 476–482.

[15] US Food and Drug Administration. 2021 Coronavirus (COVID-19) update: FDA issues emergency use authorization for potential COVID-19 treatment. www.fda.gov/news-events/press-announcements/coronavirus-COVID-19-update-fda-issues-emergency-use-authorization-potential-COVID-19-treatment)

[16] D. Wang, One-pot detection of COVID-19 with real-time reverse-transcription loop-mediated isothermal amplification (RT-LAMP) assay and visual RT-LAMP assay, bioRxiv (2020), doi:10.1101/2020.04.21.052530.

[17] B. Diao, K. Wen, J. Chen, Y. Liu, Z. Yuan, C. Han, et al., Diagnosis of acute respiratory syndrome coronavirus 2 infection by detection of nucleocapsid protein, MedRxiv (2020), doi:10.1101/2020.03.07.20032524.

[18] L. Dong, J. Zhou, C. Niu, Q. Wang, Y. Pan, S. Sheng, et al., Highly accurate and sensitive diagnostic detection of SARS-CoV-2 by digital PCR, Talanta 224 (2021) 121726.

[19] I. Hamming, W. Timens, M.L.C. Bulthuis, A.T. Lely, G.V. Navis, H. van Goor, Tissue distribution of ACE2 protein, the functional receptor for SARS coronavirus. A first step in understanding SARS pathogenesis, The Journal of Pathology: A Journal of the Pathological Society of Great Britain and Ireland 203 (2) (2004) 631–637.

[20] G.B. Ramaiah, A. Tegegne, B. Melese, Developments in nano-materials and analysing its role in fighting COVID-19, in: Materials today. Proceedings, 2021, doi:10.1016/j.matpr.2021.05.020.

[21] D. Wang, B. Hu, C. Hu, F. Zhu, X. Liu, J. Zhang, et al., Clinical characteristics of 138 hospitalized patients with 2019 novel coronavirus–infected pneumonia in Wuhan, China, JAMA 323 (11) (2020) 1061–1069.

[22] World Health Organization. Coronavirus disease (COVID-19) (2019). https://www.who.int/emergencies/diseases/novel-coronavirus-2019/question-and-answers-hub/q-a-detail/coronavirus-disease-covid-19

[23] W. Zhang, X. Yu, Y. Li, Z. Su, K.D. Jandt, G. Wei, Protein-mimetic peptide nanofibers: Motif design, self-assembly synthesis, and sequence-specific biomedical applications, Prog. Polymer Sci. 80 (2018) 94–124.

[24] F. Amanat, D. Stadlbauer, S. Strohmeier, T.H. Nguyen, V. Chromikova, M. McMahon, et al., A serological assay to detect SARS-CoV-2 seroconversion in humans, Nat. Med. 26 (7) (2020) 1033–1036.

[25] X.F. Cai, J. Chen, J. Li Hu, Q.X. Long, H.J. Deng, P. Liu, et al., A peptide-based magnetic chemiluminescence enzyme immunoassay for serological diagnosis of coronavirus disease 2019, J. Infect. Dis. 222 (2) (2020) 189–193.

[26] S. Kumar, R. Nyodu, V.K. Maurya, S.K. Saxena, Morphology, genome organization, replication, and pathogenesis of severe acute respiratory syndrome coronavirus 2 (SARS-CoV-2), Coronavirus Disease 2019 (COVID-19) (2020) 23.

[27] P. Singh, D. Singh, P. Sa, P. Mohapatra, A. Khuntia, S. K Sahoo, Insights from nanotechnology in COVID-19: Prevention, detection, therapy and immunomodulation, Nanomedicine 16 (14) (2021) 1219–1235.

[28] L. Chen, W. Liu, Q. Zhang, K. Xu, G. Ye, W. Wu, et al., RNA based mNGS approach identifies a novel human coronavirus from two individual pneumonia cases in 2019 Wuhan outbreak, Emerging Microbes & Infections 9 (1) (2020) 313–319.

[29] R.C. Cheung, T.B. Ng, J.H. Wong, W.Y. Chan, Chitosan: An update on potential biomedical and pharmaceutical applications, Marine Drugs 13 (8) (2015) 5156–5186.

[30] X. Roux, C. Dubuquoy, G. Durand, T.L. Tran-Tolla, N. Castagné, J. Bernard, et al., Sub-nucleocapsid nanoparticles: A nasal vaccine against respiratory syncytial virus, PLoS One 3 (3) (2008) e1766.

[31] A. Salleh, R. Naomi, N.D. Utami, A.W. Mohammad, E. Mahmoudi, N. Mustafa, et al., The potential of silver nanoparticles for antiviral and antibacterial applications: A mechanism of action, Nanomaterials (Basel, Switzerland) 10 (8) (2020) 1566.

[32] N.S. Malik, M. Ahmad, M.U. Minhas, R. Tulain, K. Barkat, I. Khalid, et al., Chitosan/xanthan gum based hydrogels as potential carrier for an antiviral drug: Fabrication, characterization, and safety evaluation, Frontiers in Chemistry 8 (2020) 50.
[33] Y.C. Wong, S.Y. Lau, K.K. Wang To, B.W.Y. Mok, X. Li, P. Wang, et al., Natural transmission of bat-like severe acute respiratory syndrome coronavirus 2 without proline-arginine-arginine-alanine variants in coronavirus disease 2019 patients, Clin. Infect. Dis. 73 (2) (2021) e437–e444.
[34] I. Ali, O. Alharbi, COVID-19: Disease, management, treatment, and social impact, Sci. Total Environ. 728 (2020) 138861.
[35] E. Campos, A. Pereira, J.L. de Oliveira, L.B. Carvalho, M. Guilger-Casagrande, R. de Lima, et al., How can nanotechnology help to combat COVID-19? Opportunities and urgent need, Journal of Nanobiotechnology 18 (1) (2020) 125.
[36] H. Rashidzadeh, H. Danafar, H. Rahimi, F. Mozafari, M. Salehiabar, M.A. Rahmati, et al., Nanotechnology against the novel coronavirus (severe acute respiratory syndrome coronavirus 2): Diagnosis, treatment, therapy and future perspectives, Nanomedicine (London, England) 16 (6) (2021) 497–516.
[37] J.F.W. Chan, S. Yuan, K.H. Kok, K.K.W. To, H. Chu, J. Yang, et al., A familial cluster of pneumonia associated with the 2019 novel coronavirus indicating person-to-person transmission: a study of a family cluster, Lancet North Am. Ed. 395 (10223) (2020) 514–523.
[38] A.R. Fehr, S. Perlman, Coronaviruses: An overview of their replication and pathogenesis, Coronaviruses (2015) 1–23.
[39] N. Sankarakumar, Y.W. Tong, Preventing viral infections with polymeric virus catchers: A novel nanotechnological approach to anti-viral therapy, Journal of Materials Chemistry. B 1 (15) (2013) 2031–2037.
[40] F. Amanat, F. Krammer, SARS-CoV-2 vaccines: Status report, Immunity 52 (4) (2020) 583–589.
[41] M. Aghajanzadeh, M. Zamani, H. Rashidzadeh, K. Rostamizadeh, A. Sharafi, H. Danafar, Amphiphilic Y shaped miktoarm star copolymer for anticancer hydrophobic and hydrophilic drugs codelivery: Synthesis, characterization, in vitro, and in vivo biocompatibility study, J. Biomed. Mater. Res. A 106 (11) (2018) 2817–2826.
[42] S. Amjadi, H. Hamishehkar, M. Ghorbani, A novel smart PEGylated gelatin nanoparticle for co-delivery of doxorubicin and betanin: A strategy for enhancing the therapeutic efficacy of chemotherapy, Materials Science & Engineering. C, Materials for Biological Applications 97 (2019) 833–841.
[43] A. Cohen, K. S., P. S., J.A. Whitsett, R. Goldbart, T. Traitel, J. Kost, SP-D loaded PLGA nanoparticles as a drug delivery system for prevention and treatment of premature infant's lung diseases, Int. J. Pharm. 585 (2020) 119387.
[44] P.L. Hervé, C. Deloizy, D. Descamps, M.A. Rameix-Welti, J. Fix, J.S. McLellan, et al., RSV N-nanorings fused to palivizumab-targeted neutralizing epitope as a nanoparticle RSV vaccine, Nanomedicine: Nanotechnology, Biology, and Medicine 13 (2) (2017) 411–420.
[45] H. Rashidzadeh, S. Rezaei, S. Zamani, E. Sarijloo, A. Ramazani, pH-sensitive curcumin conjugated micelles for tumor triggered drug delivery, Journal of Biomaterials Science. Polymer edition 32 (3) (2021) 320–336.
[46] P. Singh, D. Singh, P. Sa, P. Mohapatra, A. Khuntia, S. K Sahoo, Insights from nanotechnology in COVID-19: Prevention, detection, therapy and immunomodulation, Nanomedicine (London, England) 16 (14) (2021) 1219–1235.
[47] V.M. Corman, O. Landt, M. Kaiser, R. Molenkamp, A. Meijer, D.K. Chu, et al., Detection of 2019 novel coronavirus (2019-nCoV) by real-time RT-PCR, Eurosurveillance 25 (3) (2020) 2000045.

[48] S. McReynolds, S. Jiang, Y. Guo, J. Celigoy, C. Schar, L. Rong, et al., Characterization of the prefusion and transition states of severe acute respiratory syndrome coronavirus S2-HR2, Biochemistry 47 (26) (2008) 6802–6808.
[49] G. Qiu, Z. Gai, Y. Tao, J. Schmitt, G.A. Kullak-Ublick, J. Wang, Dual-functional plasmonic photothermal biosensors for highly accurate severe acute respiratory syndrome coronavirus 2 detection, ACS Nano 14 (5) (2020) 5268–5277.
[50] S.A. Kaba, C.P. Karch, L. Seth, K. Ferlez, C.K. Storme, D.M. Pesavento, et al., Self-assembling protein nanoparticles with built-in flagellin domains increases protective efficacy of a Plasmodium falciparum based vaccine, Vaccine 36 (6) (2018) 906–914.
[51] C. Weiss, M. Carriere, L. Fusco, I. Capua, J.A. Regla-Nava, M. Pasquali, et al., Toward nanotechnology-enabled approaches against the COVID-19 pandemic, ACS Nano 14 (6) (2020) 6383–6406.
[52] E. Abbasi, S.F. Aval, A. Akbarzadeh, M. Milani, H.T. Nasrabadi, S.W. Joo, et al., Dendrimers: Synthesis, applications, and properties, Nanoscale Res. Lett. 9 (1) (2014) 247.
[53] W. Lin, A.G. Coombes, M.C. Davies, S.S. Davis, L. Illum, Preparation of sub-100 nm human serum albumin nanospheres using a pH-coacervation method, J. Drug Targeting 1 (3) (1993) 237–243.
[54] A. Kraegeloh, B. Suarez-Merino, T. Sluijters, C. Micheletti, Implementation of safe-by-design for nanomaterial development and safe innovation: Why we need a comprehensive approach, Nanomaterials (Basel, Switzerland) 8 (4) (2018) 239.
[55] J. Machhi, J. Herskovitz, A.M. Senan, D. Dutta, B. Nath, M.D. Oleynikov, et al., The natural history, pathobiology, and clinical manifestations of SARS-CoV-2 infections, Journal of Neuroimmune Pharmacology 15 (2020) 359–386.
[56] J. Kim, M.J. Biondi, J.J. Feld, W.C. Chan, Clinical validation of quantum dot barcode diagnostic technology, ACS Nano 10 (4) (2016) 4742–4753.
[57] S. Al-Halifa, L. Gauthier, D. Arpin, S. Bourgault, D. Archambault, Nanoparticle-based vaccines against respiratory viruses, Front. Immunol. 10 (2019) 22.
[58] J. Zhao, X. Liao, H. Wang, L. Wei, M. Xing, L. Liu, et al., Early virus clearance and delayed antibody response in a case of coronavirus disease 2019 (COVID-19) with a history of coinfection with human immunodeficiency virus type 1 and hepatitis C virus, Clin. Infect. Dis. 71 (16) (2020) 2233–2235.
[59] S.K. Sahoo, S. Parveen, J.J. Panda, The present and future of nanotechnology in human health care, Nanomedicine: Nanotechnology, Biology, and Medicine 3 (1) (2007) 20–31.
[60] N. Hossein-Khannazer, B. Shokoohian, A. Shpichka, H.A. Aghdaei, P. Timashev, M. Vosough, Novel therapeutic approaches for treatment of COVID-19, J. Mol. Med. 98 (6) (2020) 789–803.
[61] I. Balke, A. Zeltins, Recent advances in the use of plant virus-like particles as vaccines, Viruses 12 (3) (2020) 270.
[62] L.C. Lin, C.Y. Huang, B.Y. Yao, J.C. Lin, A. Agrawal, A. Algaissi, et al., Viromimetic STING agonist-loaded hollow polymeric nanoparticles for safe and effective vaccination against Middle East respiratory syndrome coronavirus, Adv. Funct. Mater. 29 (28) (2019) 1807616.
[63] Nanotechnology Products Database. COVID-19 Rapid POC CE-IVD Test. 2021
[64] M.O. Mohsen, L. Zha, G. Cabral-Miranda, M.F. Bachmann, Major findings and recent advances in virus-like particle (VLP)-based vaccines, Semin. Immunol. 34 (2017) 123–132.
[65] Y. Qin, The preparation and characterization of chitosan wound dressings with different degrees of acetylation, Journal of Applied Polymer Science 107 (2008) 993–999.
[66] F. Mansoor, B. Earley, J.P. Cassidy, B. Markey, S. Doherty, M.D. Welsh, Comparing the immune response to a novel intranasal nanoparticle PLGA vaccine and a commercial BPI3V vaccine in dairy calves, BMC Veterinary Research 11 (2015) 220.

[67] C.P. Haycook, J.A. Balsamo, E.B. Glass, C.H. Williams, C.C. Hong, A.S. Major, et al., PEGylated PLGA nanoparticle delivery of eggmanone for T cell modulation: Applications in rheumatic autoimmunity, Int. J. Nanomed. 15 (2020) 1215–1228.

[68] E. Alphandéry, The potential of various nanotechnologies for coronavirus diagnosis/treatment highlighted through a literature analysis, Bioconjugate Chem. 31 (8) (2020) 1873–1882.

[69] P. Singh, D. Singh, P. Sa, P. Mohapatra, A. Khuntia, S. K Sahoo, Insights from nanotechnology in COVID-19: Prevention, detection, therapy and immunomodulation, Nanomedicine (London, England) 16 (14) (2021) 1219–1235.

[70] Y.T. Lee, E.J. Ko, Y. Lee, K.H. Kim, M.C. Kim, Y.N. Lee, et al., Intranasal vaccination with M2e5x virus-like particles induces humoral and cellular immune responses conferring cross-protection against heterosubtypic influenza viruses, PLoS One 13 (1) (2018) e0190868.

[71] G. Seo, G. Lee, M.J. Kim, S.H. Baek, M. Choi, K.B. Ku, et al., Rapid detection of COVID-19 Causative Virus (SARS-CoV-2) in Human Nasopharyngeal Swab Specimens Using Field-Effect Transistor-Based Biosensor, ACS nano 14 (4) (2020) 5135–5142.

[72] N. Sethuraman, S.S. Jeremiah, A. Ryo, Interpreting diagnostic tests for SARS-CoV-2, JAMA 323 (22) (2020) 2249–2251.

[73] L.L. Ren, Y.M. Wang, Z.Q. Wu, Z.C. Xiang, L. Guo, T. Xu, et al., Identification of a novel coronavirus causing severe pneumonia in human: A descriptive study, Chin. Med. J. (Engl.) 133 (9) (2020) 1015.

SECTION 4

Commercialization and Standardization Biosensing Technologies

CHAPTER

9

Nanobioengineering: A promising approach for early detection of COVID-19

Atal Gill[a], *Zondi Nate*[b], *Ruchika Chauhan*[a], *Mbuso Faya*[b], *Rajshekhar Karpoormath*[a], *and Calvin A. Omolo*[b,c]

[a]Department of Pharmaceutical Chemistry, College of Health Sciences, University of KwaZulu-Natal, Durban, South Africa, [b]Discipline of Pharmaceutical Sciences, College of Health Sciences, University of KwaZulu-Natal, Durban, South Africa, [c]United States International University-Africa, School of Pharmacy and Health Sciences, Department of Pharmaceutics, Nairobi, Kenya

9.1 Introduction

Unique pneumonia due to an unknown source emerged in December 2019 in the city of Wuhan, China [1,2]. Consequently, the World Health Organization (WHO) declared this condition as a new coronavirus disease-19 also known as COVID-19 on February 11, 2020, which on March 13, 2020 was declared as a pandemic [3]. The virus that causes COVID-19 was found to have a similar genome (80% similarity) with the previously known acute respiratory syndrome also known as SARS-CoV. The novel virus was later named severe acute respiratory syndrome coronavirus 2 (SARS-CoV-2) [4–6]. SARS-CoV-2 falls in the family of *Coronaviridae* which is further divided into *Nidovirales* and another subfamily called Orthocoronavirinae.

Sensing Tools and Techniques for COVID-19: Developments and Challenges in Analysis and Detection of Coronavirus
DOI: https://doi.org/10.1016/B978-0-323-90280-9.00004-3

The four generations of the coronaviruses belongs to the *Orthocoronavirinae* family that consists of alpha, beta, gamma and delta coronavirus, which are denoted as α-CoV, β-CoV, γ-CoV, δ-CoV respectively [7,8]. The α-CoV and β-CoVs are mainly known to infect mammals whereas γ-CoV and δ-CoV are generally found in birds. The β-CoVs also comprise of SARS-CoV and also include another virus that was found in the Middle East called the Middle East respiratory syndrome virus (MERS-CoV) and the cause of current pandemic SARS-CoV-2. These viruses initially cause the development of pneumonia in the patients and further development of a severe case of acute respiratory distress syndrome (ARDS) and other related symptoms that can be fatal leading to death [9].

Insights into the origin of SARS-CoV-2 and its transmission from one species to another are aiding in finding ways and means to prevent COVID-19 from further spreading. The natural host of SARS-CoV-2 has been suggested to be pangolins and bats [10]. The transmission takes place from animals to humans. Further, the virus also has the ability to spread from human to human with ease which creates a life-threatening disease [11]. The research to find the precise host for COVID-19 is still ongoing and novel diagnostic and strategies are paramount in early detection that can prevent the spread of the disease.

The genomic sequencing of the COVID-19 virus was performed and assisted in identifying the host and the evolutionary relationship between SARS-CoV and SARS-CoV-2. It has been found through phylogenetic analysis that SARS-CoV-2 and SARS-CoV share a highly conserved domain in nsp1 amino acid linkage [12]. Further studies confirmed that SAR-CoV-2 had 77.2% and 82.3% amino acid similarity profiles with coronaviruses in bat, i.e. SARS-CoV and SL-CoVZC45 respectively [13,14]. Other studies depicted that the RNA-binding domain of SARS-CoV-2 is an amino acid longer than SARS-CoV. These outputs provided insights into the ability of SARS-CoV-2 to use angiotensin converting enzyme 2 (ACE2) receptors to enter the cells, unlike SARS-CoV. Another study confirmed that SARS-CoV-2 possessed a spike(S) that had a furin-like cleavage site, which is generally not found in SARS-like CoVs [6]. The research for further understanding the genome pathology of SARS-CoV-2 is still going on all around a more effective treatment and sensitive diagnostic strategies are being reported as more research for disease progresses.

Following the outbreak and the spread of the COVID-19 pandemic has called for diagnostic systems that are; cost-effective, highly sensitive and have a short period between the testing and getting results tests for the masses worldwide. Several methods of molecular tests and immunoassays were rapidly developed. Nanoengineering has been at the forefront to bring to life sensitive nano-based diagnostic and biosensors systems [15]. This chapter covers the biomarkers of COVID-19 that have been employed to engineer diagnostic systems. Furthermore, we explore different nanosystems that have been developed for the diagnosis of COVID-19, and try to define future and emerging nanoengineered and the need for continuous development of accurate diagnostic testing is vital for quicker patient detection at the point of care in order to improved prevention and treatment, based on the information available to date .

9.2 Biomarkers

For effective management, prevention of and curbing the spread of COVID-19, accurate detection, and well-timed diagnosis are needed. For this to happen, biomarkers play a paramount role in the early detection of disease etiology, diagnosis, treatment, and prognosis. Several markers have been identified that have made the detection of the COVID-19 disease possible. This section of the book chapter will discuss those biomarkers and their application in the detection, treatment, and prevention of COVID-19.

9.2.1 C-reactive protein (CRP)

Inflammation has been a hallmark for COVID-19 and proinflammatory biomarkers have played a key in developing diagnostic systems for the disease. C-reactive protein (CRP) is a plasma protein produced in response to inflammatory mediators by the liver. CRP being a nonspecific protein is used as a biomarker for clinical evaluation of inflammatory conditions, and an increase in CRP level is directly proportional to the disease severity [16]. CPR was used as a biomarker in a study conducted in Wuhan, China, where the majority of nonsevere cohort patients demonstrated lower levels as compared to the severe cohort (33.2 mg/L vs. 57.9 mg/L $P < .001$) [17]. Another study found that patients with CPR level >41.8 mg/L could lead to fatal results due to COVID-19 infection [18]. These studies depicted that CPR can be used as an important marker to determine the severity of infection in the patients.

9.2.2 Interleukin-6

Interleukin 6 (IL-6) is an interleukin that is a pro-inflammatory cytokine and is considered one of the most important cytokines that are produced during infection. Moreover, IL-6 undergoes systemic upregulation during the acute phase of most viral infections [19]. Cytokine release syndrome (CRS) results in an extra release of the inflammatory mediators (like IL-6) and which also underlies the pathological process of acute respiratory distress syndrome (ARDS) [20]. ARDS has been associated with SARS, MERS, and infections from SARS-CoV-2 [21]. Therefore, this leads to the association of IL-6 to the severity of COVID-19 infection [22]. Several studies of Wuhan SARS-CoV-2 infected patients showed that there was an increase in the IL-6 concentration. Moreover, a comparison between severe COVID-19 patients and nonsevere patients there were up to 2.9 times higher in IL-6. Therefore, IL-6 has been concluded as a potential biomarker for the progression of the COVID-19 disease [23,24].

9.2.3 White cell count

White blood cells (WCC) are divided into two wide groups: agranulocytes and granulocytes. Granulocytes are subdivided into; basophils, neutrophils (NC), and eosinophils, whereas agranulocytes are subdivided into; monocytes and lymphocytes (LC). The ratio of these respective WBCs and the white cell count (WCC) sheds light on various infections inside the patient's body. A recent study illustrated that WCC had a difference between nonsevere COVID-19 and severe COVID-19 patients [17]. It was seen that the levels of monocytes basophils, eosinophils, and lymphocytes were lower which resulted in a higher neutrophil to lymphocyte ratio (NLR). NLR is considered as a biomarker that shows the inflammatory severity in the patient due to the disease. But there is a need to carry out an extensive study to check the effectiveness of NLR as a COVID-19 biomarker. It is seen that lymphoid tissue and disruption in IL-6 mechanisms by the virus can cause low LC count, which too should be studied [25].

9.2.4 Lactate dehydrogenase

Various biomarkers are being investigated for their role in prognosis COVID-19. Lactate dehydrogenase (LDH) is one such biomarker as its elevated levels have been associated progression of viral infections in patients before [26,27]. LDH is known to convert pyruvate into lactate during the glucose metabolism pathway. The secretion of the enzyme is kick started by the necrosis in the cell membrane, which shows either the damage in the lungs or viral infection like pneumonia caused by SARS-CoV-2 [28]. The evidence suggests that high levels of LDH in patients with COVID-19 are associated with a severe form of the disease [29]. The higher levels of LDH were associated with individuals in ICU as compared to non-ICU patients. Isozyme 3 of LDH is present in lung tissue and it has been reported, patients having COVID-19 infections release higher amounts of LDH in the circulation up to 6 fold, due to interstitial pneumonia, which causes acute respiratory distress syndrome (ARDS). ARDS is the hallmark of severe COVID-19 [30,31].

9.2.5 D-dimer

Fibrin Degradation Fragment (D-dimer) is one of the protein fragments produced due to the breakdown of blood clots. When cross-linked fibrin undergoes lysis D-dimers concentrations increase which indicates that fibrinolysis and coagulation have been activated [32]. The activation of the coagulation cascade has been associated with disseminated intravascular coagulation and adverse clinical outcomes in COVID-19 disease [33]. Initial observations in patients with hemostatic abnormalities suffering from COVID-19 had higher levels of D-dimer as compared to patients who survived [34]. A recent study with 191 patients concluded that D-dimer ($>1.0\ \mu g/mL$) levels and deaths were directly proportional. Further, the study concluded that patients with D-dimer levels equal to $2.0\ \mu g/mL$ or higher when admitted, predicted the mortality rate due to COVID-19 in hospitals [35]. Finally, studies reported that 90% of patients suffering from pneumonia had high coagulation, marking a rise in the D-dimer concentrations [36].

9.2.6 Platelet count

Previous studies revealed that COVID-19 led to hematological change causing thrombocytopenia. An analytical study consisting of 1799 individuals revealed a measurable decrease in the platelet counts of the patients [37]. It was observed that nonsurvivors had a reasonably lower amount of platelet count when mortality was considered as an endpoint. Even though the definition for the severity of the disease varies and thrombocytopenia has an impact on results, but platelet count still could be used as an indicator to clinically suggest the severity of the infection.

9.2.7 Cardiac troponin

Patients with cardiovascular disorders have a higher mortality rate when infected with COVID-19 as recently observed by various researchers [[38–40]]. Cardiac troponin (cTn) as a biomarker that may aid in classification and stratifying the risk for myocardial injury among patients with COVID-19. An increase in cardiac troponin is an indication of myocardial injury which is associated with COVID-19 prognosis [41]. A study was carried out in China with patients with positive COVID-19 infection, unveiled univariable odds with the death ratio at 80.1 for high-sensitive cardiac troponin (hs-TnI) (95% CI 10.3–620.4, $P < .0001$) [42]. The analysis of this marker is a risky task as compared to other biomarkers for COVID-19 and downstream analysis could put the life of the patient at a risk.

9.2.8 Renal markers

It has been also been seen that severe COVID-19 cases and chronic kidney infections have been related [43]. Several studies indicated a significant increase in the levels of renal biomarkers like creatinine and serum urea in severe COVID-19 cases [44]. High serum creatinine levels of patients co-related with the severity of COVID-19 was found in a study carried out by researchers involving 701 patients [45]. This study also concluded that these patients had more probability to need a mechanical ventilators. Additionally, another study depicted that urinalysis could also aid in the determination of disease severity [46]. Therefore, it suggested that urinalysis was more convenient as compared to kidney serum renal biomarkers. The effect of changes in biomarkers concentrations in patients is summarized in Table 9.1.

9.3 Detection techniques for COVID-19

Large-scale testing and contact tracing are vital in controlling Covid-19. Therefore rapid, reliable, and scalable detection assays are crucial for sensitive, specific, and large-scale surveillance of SARS-CoV2. Thus, rapid sensitive and effective are paramount in combating the COVID-19 pandemic. The current section discusses the existing molecular detection techniques for SARS-CoV2 as illustrated in Fig. 9.1.

9.3.1 PCR-based detection

The PCR program has three basic steps; the first step is denaturation, at a high temperature (around 90–95°C) the unwinding of the two strands of DNA takes place which is essential for the annealing of the oligonucleotide primers in presence of DNA polymerase enzyme start the replication of DNA (multiplication of copy number of the target sequence). The annealing usually takes place at a temperature ranging from 45 to 65°C depending upon the G-C content of the target sequence. Since Guanine (G) joins the Cytosine (C) with three hydrogen bonds (triple bond) compared to two hydrogen bonds (double bond) that forms between Adenine (A) and Thymine (T), therefore, the pair of GC rich oligonucleotide primers may require a relatively higher annealing temperature than an AT-rich oligonucleotide primer. The third step of a PCR reaction is an extension which prepares the amplified target sequence for the repetition of the PCR cycles. The extension takes place at 72°C. Each PCR reaction consists of a mixture of oligonucleotide primers, DNA template (target gene or sequence), dNTPs, PCR buffer, $MgCl_2$, and DNA polymerase enzyme. Agarose gel electrophoresis is used to detect the amplified product using a fluorescent dye (ethidium bromide) at the end of the various steps [47,48].

The PCR-based amplification provides more sensitive and targeted detection of the microbial pathogens compared to the serological or immunological assays. Several variations of the PCR assays have been in regular use for applications in forensics, biological research, archaeology,

TABLE 9.1 Summarizes changes observed of biomarkers during COVID-19.

Biomarker	Function	Presentation in COVID-19	References
CRP	Proinflammatory protein that triggers the classical innate immunological pathway by binding to microorganisms' polysaccharides such as phosphocholine (PCh)	Concentration increases in COVID-19 infections. Can be employed to classify severe from nonsevere presentation of the disease.	Qin et al. [17]; Sproston and Ashworth [231]
IL-6	A proinflammatory cytokine that is produced during an infection	Upregulation during acute viral infections. Underlies the pathological process of ARDS. Up to 3-fold higher in severe COVID-19 patients. Can be used to classify severity of the disease.	Coomes and Haghbayan [23]; Mahajan et al. [20]; Ulhaq and Soraya [24]; Velazquez-Salinas et al. [19]
LDH	Convert pyruvate into lactate in the glucose metabolism pathway. It reduces NAD+ to NADH and H+ through the oxidation of lactate to pyruvate. Employed to measure apoptotic and necrotic cell death.	Elevated hallmark for progression of viral infections in the body. Indicative of the damage in lungs or viral infection like pneumonia associated with COVID-19. Up to 6-fold higher in severe cases linked to interstitial pneumonia, that ARDS. Isoenzyme 3 found only in lungs used to detect ARDS and severe COVID-19	Ferrari et al. [29]; Henry et al. [30]; Mo et al. [31]
D-dimer	Protein fragments produced as a result of fibrinolysis.	Adverse clinical outcomes in COVID-19 disease are associated with disseminated intravascular coagulation. A predictor of COVID-19 mortality.	Milbrandt et al. [36]
Cardiac troponin	Found in skeletal and cardiac muscle fibers. They regulate muscular contraction. Elevated levels in blood are a measure of myocardial injury.	Sensitive biomarker assay in recognition of myocardial injury in acute illnesses. Used in testing and classification of pathogenesis in patients with myocardial injury. Aid in classification and stratifying the risk for myocardial injury among patients with COVID-19. Positive test COVID-19 patients my indicate prognosis of myocardial injury due to SARS-CoV-2 effects.	Chapman et al. [226]; Zhou et al. [42]

CRP = C-reactive protein; IL-6 = interleukin 6; LDH = lactate dehydrogenase.

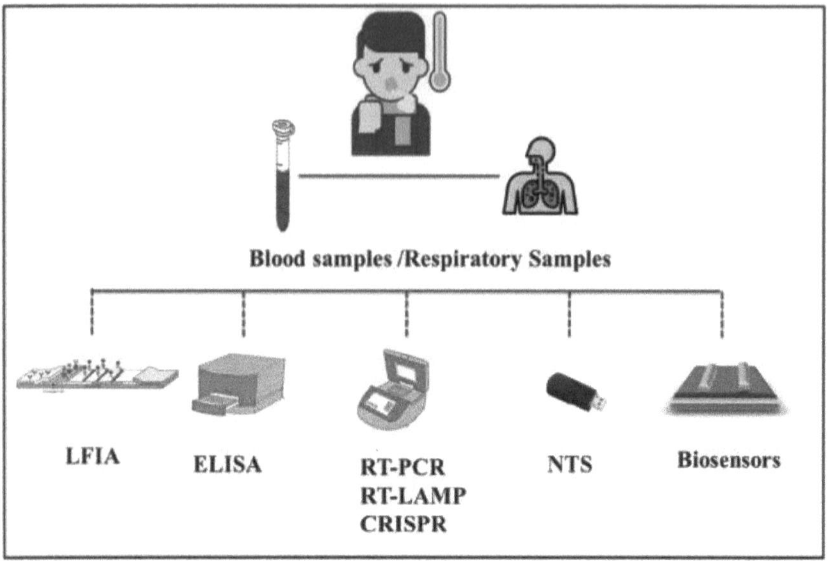

FIGURE 9.1 Diagnostic techniques for COVID-19 and new developments. Representation of different diagnostic methods for the detection of SARS-CoV-2 from respiratory and serum samples (permission granted) [226,227]. Talanta. 2020;220(1):121392. doi:10.1016/j.talanta.2020.121392.

anthropology, and, food technology [49,50]. There are several factors that may determine the sensitivity of a PCR assay including PCR protocol, target gene(s), oligonucleotide primer sequences, quality of nucleic acid template, and of course the handling. Since SARS-CoV2 has an RNA genome, therefore, the amplification of the target sequence for the detection of SARS-CoV2 requires an additional step of reverse transcription before the amplification step. The reverse transcriptase is an enzyme that converts RNA into DNA which can then be amplified to produce multiple copies of the target sequence required for the detection in the PCR assay.

The real time RT-PCR is based on fluorescent signaling amplification which is much faster and more specific due to the application of a TaqMan probe than the conventional RT-PCR assay. Nested RT-PCR is another version that is used for the detection of complicated target DNA to avoid false results due to mutation using two sets of primers [51]. For SARS-CoV-2 detection, several genes have been targeted such as N-gene (nucleocapsid), E gene, RdRP gene, ORF1ab gene, and S gene (spike protein) [52]. Research conducted by Chu et al. described two different one step real-time RT-PCR attempts to detect ORF1ab and N genes of the SARS-CoV-2 genome which showed a range of 0.0002–20 with 50% tissue culture infective dose (TCID50) per reaction and detection limit below 10 RNA copies per reaction [53].

The Charité University Hospital developed an RT-PCR based protocol for SARS-CoV-2 detection [54]. Additionally, the Centers for Disease Control and Prevention (CDC), Atlanta, GA, USA have developed a SARS-CoV-2 detection protocol along with nucleic acid extraction kits and the set of primers for SARS-CoV-2 detection [55]. Further, TaqMan chemistry has designed a single step real-time RT-PCR assay for SARS-CoV2 detection using N or ORF1b gene-based detection. Several other manufacturers have also developed SARS-CoV-2 detection assays targeting different genes (N gene, E gene, RdRp gene, ORF-1a gene, ORF-1b gene, etc.) in the SARS-CoV2 genome, with varying sensitivity [52]. The SYBR green-based real-time RT-PCR assay has also been introduced primarily to reduce the detection cost, with 91.7 % specificity [56].

9.3.2 ELISA-based detection

Enzyme-linked immunosorbent assays (ELISA) are based on antibody-antigen interaction to provide specific and reproducible results. This technique is cost-effective and easy to handle in comparison to RT-PCR assay. Several immunoassay kits have been developed for SARS-CoV-2 detection in recent months [57]. Currently numerous manual and automated immuno-test kits are available using fluorescence immunoassays, chemiluminescence enzyme immunoassays, lateral flow immunoassays for SARS-CoV2 detection. All these assays are based on SARS-CoV-2 nucleocapsid protein as an antigen and their respective antibodies (IgG, IgA, and IgM). The major challenge with ELISA kits is the lack of specificity as it has cross responsiveness of SARS-CoV-2 antibodies with other coronaviruses. The cross reactivity of S protein based antibodies of SARS-CoV-2 have been studied in 15 patients of SARS-CoV2 and observed the high frequency of cross reactivity with SARS-CoV [58]. Zhang and coworkers used Rp3 nucleocapsid protein of SARS-CoV-2 to detect the COVID-19 in patients, this protein resembles upto 80 % with SARS-CoV protein. Using IgG and IgM antibodies in sandwich ELISA, a study successfully detected SARS-CoV-2 [59]. The SARS-CoV2 consists of four proteins viz., S, M, E, and N, which have epitopes to bind with the paratope of antibody. ELISA exhibits high sensitivity with a relatively longer ($\geq$1.5 h) period of detection. The level of other protein biomarkers like C-protein, D-dimer, lymphocytes, leukocytes, and blood platelets also have been elevated in COVID-19 patients [38] but the level of these protein biomarkers may become abnormal in several other health conditions too, therefore, may not offer a reliable diagnosis.

There have been several approaches to detect SARS-CoV2 in blood, mucosal swabs, urine, etc. Several companies prepared immunoassays, for example, BioMedomics rapid test, Surescreen rapid test, Assay Genie POC kit, etc for reliable detection which is still unsure. The new diagnostic assays will have to find novelty related to the material, process, specificity, and sensitivity. The ELISA kits based on N, and S proteins- IgG interaction are available for SARS-CoV with 94.7% and 58.9% sensitivity, respectively [60]. The IgG and IgM antibodies were measured in the SARS-CoV-2 patients, the antibody concentration effected after 5–7 days of infection [61]. The multiplex digital detection of SARS-CoV-2 can be a fast and smart detection method using an optical reader [62]. Since a large number of SARS-CoV2 strains are in circulation (n = 104) therefore, maintaining the sensitivity of the immunoassays is quite challenging. There is a need to create an effective way to get an accurate diagnosis of COVID-19 that addresses sensitivity, specificity as well as cost implications.

9.3.3 CRISPR-based detection

The "Clustered regularly interspaced short palindromic repeats" (CRISPR) is the combination of two distinct characteristics of DNA, which has a repeated sequence of nucleotide and spacer. CRISPR is a genome-editing technique but has also been utilized in Covid-19 detection, for which the CRISPR-based SHERLOCK (Specific High Sensitivity Enzymatic Reporter UnLOCKing) technique has been developed. This uses a synthetic RNA fragment of SARS-CoV-2 which can detect 10–100 copies per microliter. It follows three steps and provides the results within an hour, following the 25 minutes of amplification of extracted nucleic acid then 30 minutes of incubation-detection of viral RNA sequence using Cas 13, and finally 2 minutes for reading the results with paper dipstick [63,61]. One advantage of the SHERLOCK technique is that this does not require skilled personnel and produces the result in less than one hour with the dipstick method.

Mammoth Biosciences have developed a CRISPR-DETECTOR (DNA endonuclease targeted CRISPR trans reporter) based lateral flow strip for the detection of SARS-CoV2. In this, Cas12 enzyme is used for E gene, N gene, and P gene RNA as a control for SARS-CoV-2, following the same protocol as in SHERLOCK, but the results can be visualized in lateral flow strip in 30 minutes with the range of 70–300 copies/μL [64,65]. Lucia and the group have used Cas12 enzyme with RdRp and ORF1ab genes for the detection of SARS-CoV-2 on the paper strips with labeled fluorescein ssDNA and WH-human1 sequence (as control). The detection limit of this assay is ~10 copies/μL [66].

9.3.4 Lateral flow-based detection

Lateral flow immunoassay (LFA) attracts interest due to its user-friendliness, portability, and cost-effectiveness. Most of the LFA assays are paper-based and immobilize with either nucleotides or antibodies. There have been several LFA kits available in the market for the detection of SARS-CoV2. The major drawback of LFA kits is that this assay detects the target only in the postsymptomatic patients, which means those individuals who are asymptomatic but infected may not get detected using LFA kits. However, immunoassays are rapid and cost-effective compared to the RT-PCR, but they have less specificity and low sensitivity in comparison to nucleic acid-based detection methods.

Blood and saliva tests based on LFA have been developed to detect anti-SARS-CoV2 antibodies [67–69]. LFA assays have the advantage of rapidity, no requirement of an experiment protocol, and no sample preparation. Many immunoglobulin antibodies (A, M, and G) based testing for SARS-CoV2 have been applied in clinical laboratories [67,68]. An elaborate study on LFA for SARS-CoV2 is required on different stages of the disease. These immunoassay kits can't be used at an early stage of the disease, which is one of the major disadvantages of these assays [11]. In the current COVID-19 pandemic, the LFA kits have been utilized for Sars-Cov-2 detection in patients with serious health conditions. One week after the appearance of symptoms, patients were tested with LFA kits which yielded 95–98% sensitivity [70]. Recently, a two-dimensional paper-based lateral immune assay has been developed [71]. The present LFA for detection of COVID-19 is based on IgG or IgM antibodies. The nucleotide sequence-based LFA assays are more sensitive than the immuno-based LFA assays. It has been reported that the labeling of capture probe with gold nanoparticles or other luminescent material may help to enhance the signal amplification in lateral flow assays during the detection of the target [72].

9.3.5 Biosensors: State-of-the-art biosensing devices

A biosensor is a device that consists of three parts: a bioreceptor, transducer, and detector. The bioreceptor loaded with biomolecules like antibodies, peptides, and DNA, which interact with their specific target(s). After the interaction, the signal is produced and read by the transducer, the signal is amplified and processed to the detector. Based on detectors, biosensors are classified into three classes: electrochemical, piezoelectric, and colorimetric. The bioreceptor plays an important role in a biosensor. Based on bioreceptor, the biosensors can be characterized into two categories: immunosensor and aptasensor. The immunosensors rely on antibody and antigen interaction while aptasensors rely on the aptamers (synthetic single-strand DNA complementary to the target). The following biosensing techniques have been utilized to detect SARS-CoV2:

9.3.5.1 Immunosensor for SARS-CoV-2 detection

In immunosensor the interaction of antibody and antigen is read by electrochemical, piezoelectrical, colorimetric, or impedimetric technique

[73–77]. It is observed that electrochemical immunosensors for influenza A virus subtype H5N1 have shown a higher sensitivity, selectivity toward the virus and appeared to be economically affordable than the conventional detection methods [78]. The signal amplification of immunosensor can determine sensitivity, based on two factors (a) electrochemical-chemical redox cycling, (b) enzyme reduction-electrochemical, which play a major role in signal amplification [79].

The major advantages offered by an immunosensor over other detection techniques are rapidity, portability, user-friendliness, and cost-effectiveness. Though immunosensor has some drawbacks like it requires the availability of a specific monoclonal antibody; shelf-life of assay and antigen variation makes it difficult to generate a similar type of antibody. During the ongoing COVID-19 pandemic the US- FDA has approved various immune kits for the detection of SARS-CoV2 [57]. Since the SARS-CoV2 has more than 100 different strains and each virus contains multiple proteins with unique epitopes over the surface, more research is required to develop a highly specific and sensitive immunosensor for SARS-CoV2 detection [80]. An ACE2 receptor based impedimetric biosensor has been developed for the detection of SARS-CoV2 [81] using human blood. To further improve the signal amplification, the redox active molecules (alkanethiol) and metal nanoparticles have been used [82].

9.3.5.2 *Aptasensor for SARS-CoV2 detection*

Aptamers are small peptides or oligonucleotides of RNA/DNA that act as the receptors for target DNA. The aptamers are synthesized using single-stranded RNA/DNA with 10–100 nucleotides in length. The nucleotides sequence for the aptamer is selected via the Systematic Evolution of Ligands by Exponential Enrichment (SELEX) process to get a specific RNA/DNA sequence toward a broad spectrum of the target. Aptamers are nanoscale in size and can detect up to nano- and picogram levels, therefore aptasensors are considered as the nanobiosensor [83,84].

Aptasensors are user-friendly, cost-effective, and stable analytical tools for rapid detection of viruses [85,86]. These nanosensors have great potential to detect the virus with various modifications at the electrode surface [87–90]. At present, there is no aptasensor for the detection of SARS-CoV2. In principle, oligonucleotide sequences of aptamer are specific for identifying the viral sequence via the hybridization process. The aptamers can be functionalized easily with fluorescence labels or any other redox material in order to improve the sensitivity and specificity of the aptasensor. The aptasensors have advantages over self-life and cost-effectiveness over immunosensors. Due to these properties, aptasensors are suitable for commercialization in the form of a device with the incorporation of other technologies for the rapid detection of SARS-CoV2.

Biosensors have the edge over other detection techniques like PCR, ELISA, and CRISPR due to portability, scalability, cost-effectiveness, and rapidity [91]. In recent days, several innovated hands-on biosensing devices have been developed using nanomaterials to overcome the drawbacks of lengthy protocols of other detection techniques. The use of nanomaterials enhances the surface to volume ratio and the tunneling property which helps in improving the sensitivity and signal amplification of the biosensing device [92,93].

9.3.6 Computed tomography (CT) scan

During the COVID-19 pandemic, the computed tomography (CT) of the chest of the patients has appeared to be one of the rapid and reliable diagnostic techniques with high sensitivity. It can give the results at the infection phase before the appearance of clinical symptoms. The CT scan measures cross-sectional images of the patient's chest from many angles using X-rays [94,95]. These images are then examined by a radiologist for establishing the diagnosis. The image features vary among the stages of infection. The thick lobular septa, multi-lobar ground glass opacity in the posterior and peripheral region have been observed in COVID-19 patients in chest CT images [96,97]. A study reported CT images (56%) at the initial two days of disease progression with a follow up till the appearance of the clinical symptoms (~10 days) which found hazy opacity in the peripheral region of the lungs [98].

A few studies reported that the imaging feature of CT shows a higher sensitivity up to 86-98% in comparison to other detection techniques [38,99–101]. Unfortunately, the CT imaging features have a major limitation of low specificity (25%), which may collapse with the other viral infections (pneumonia, etc.) or in an event of lung disease [100]. Another limitation of the CT scanning technique is that it requires an expensive instrument and expert technician, and the cost is higher than the other detection techniques. The CT scans are primarily conducted to eliminate the false-negative results of the assay kits. Due to the overwhelming numbers of infected people and limited numbers of available test kits, the CT scans have been useful for the screening of the serious as well as the mild COVID-19 cases with the objective to streamline the life-threatening cases as earliest as possible.

9.4 Role of nanomaterials in biosensing of COVID-19

The COVID-19 pandemic has ravaged the globe, unlike previous pandemics the outbreak was met with the advancement in technology. Nano-based technologies have been at the forefront in fighting the pandemic.

Different studies are being carried out to find an effective drug, sensitive detection methods, and vaccines against SARS-Cov2 infection. In both scenarios, this is still a dream in the pipeline. For effective treatment of SARS-Cov2 infection prevention and contract tracing, it is essential to have diagnostic tools that are rapid, cost-effective, sensitive, selective, reproducible, multiplexing, disposable, and easy to use. Different diagnostic tools such as reverse transcription polymerase chain reaction (RT-PCR), serological testing, and chest computed tomography (CT) scan are used. However, some of these methods contain drawbacks; the high cost of CT scan diagnosis and its inability to differentiate between SARS-Cov2 and other viruses' infections limit its application. RT-PCR is time-consuming and requires sophisticated equipment. Also, during the earlier stages of the outbreak, many false negative and false positive results were reported from the RT-PCR method [102,103].

Nanomaterials, such as silver or gold, polymeric, and silica nanoparticles (NPs), carbon nanotubes, and quantum dots, are being employed in the detection of viruses [104]. In these types of nanosystems, their surfaces were modified with biomolecules (such as DNA, RNA, antigen, antibody, peptide) obtained from the virus [105]. It is also important to note that the high surface and volume ratios of nanomaterials improve the interactions between the sensor and analyte, thus increasing the detection limit and decreasing the detection time [106]. Nano-based probes are widely used for the assembly of biosensors and they improve the sensor's response, either by obtaining electrical, optical or catalytical properties, thus offering superior analytical sensitivity for diagnosis [107]. Amongst the nanosystems utilized as detection tools, gold nanoparticles have stood out thanks to their photonic, electrical, and catalytic properties [108]. The gold NPs have been functionalized with probes modified with thiol on the surface, that hybridize with the target, inhibiting the aggregation of the NPs by salts and this results in a color change; therefore this type of system can be easily modified for COVID-19 diagnosis [109]. Other diagnostic systems have incorporated gold nanoparticles fused with antibodies against SARS-CoV-2 IgG/IgM. These systems have showed the potential for application as a rapid symptomatic and asymptomatic screening for COVID-19 [110]. Field-effect transistors (FET) supported graphene have previously reported to detect the viral load of SARS-CoV-2 from nasopharyngeal swab specimens using specific antibodies from the virus [111]. Selected antibody against SARS-CoV-2 spike protein coated graphene sheets of a FET were employed to fabricate the biosensor. SARS- CoV-2 spike antibody was then immobilized onto the fabricated device through 1-pyrenebutanoic acid succinimidyl ester, an efficient interface coupling agent used as a probe linker. The sensor was found to sensitive enough to differentiate the SARS-CoV-2 and MERS-CoV proteins. Moitra et al. also reported a gold NPs capped with thiol-modified antisense oligonucleotides (ASO) biosensor

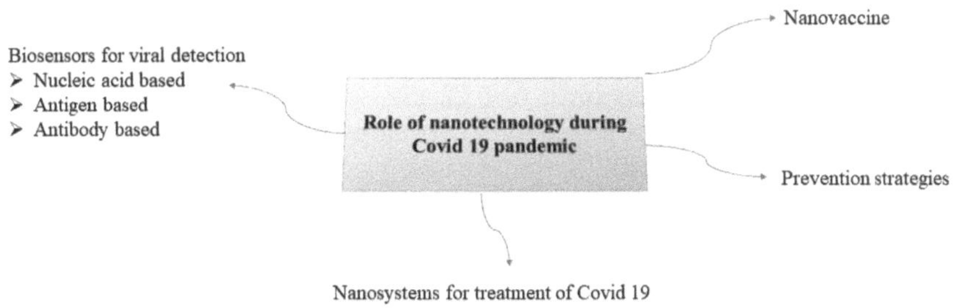

FIGURE 9.2 Role of nanotechnology during the COVID 19 pandemic.

system with specificity for the N-gene (nucleocapsid phosphoprotein) of SARS-CoV-2. The reported system could provide diagnostic results for COVID-19 in a couple of minutes [112]. Furthermore, Baker et al., have also reported gold nanoparticles that had been polymer-stabilized multivalent bearing sialic acid derivatives and that interacted with SARS-CoV-2 spike glycoprotein [113]. From the study α, N-acetylneuraminic acid binds strongly with the spike glycoprotein and functioned as a prototype for the lateral flow diagnostic device detection unit.

As the world continues to fight the SARS-Cov2 pandemic and the number of cases increases at an alarming rate, testing kits' availability has not been sufficient; this has resulted in under-reporting of the actual active SARS-Cov2 cases worldwide. Different studies are being conducted to fabricate diagnostic tools from nanomaterials such as gold, magnetic, and polymeric nanoparticles to solve this challenge. It is important to note that nanomaterials' morphology and size play a crucial role in their application. Nanomaterial-based diagnostic tools are known to have several advantages compared to conventional methods. Fig. 9.2 shows the role of nanotechnology during the COVID 19 pandemic. This section will discuss various nanomaterial-based diagnostic tools that have been reported so far in the literature.

9.4.1 Gold nanostructure-based biosensors

In recent years, the application of nanotechnology in the fight against different diseases has increased. Gold nanoparticles are among the frequently used nanomaterials to fabricate biosensors for SARS-CoV2 infection. This is mainly due to their stability, high surface chemistry, biocompatibility, and surface molecular biochemical conjugation. A biosensor is a device made up of a biological recognition component (anti-bodies, enzymes, nucleic acids, organic and biological receptors) and a transducer. A transducer's function is to transform the interaction between the analytes and biological components into a measurable signal. Qiu et al [114] developed a dual functional plasmonic photothermal biosensor for detection of SARS-CoV2. Gold nanoislands factionalized with complementary DNA receptors were used to detect SARS-CoV2 through nucleic acid hybridization. A detection limit of 0.22 pM was obtained. This method is highly sensitive, fast, and reliable. Moitra et al. [112] reported the development of a colorimetric biosensor using gold NPs capped with thiol modified Antisense oligonucleotides specific for N-gene of SARS-CoV2. The presence of SARS-CoV2 was detected within 10min. The biosensor showed selectivity toward SARS-CoV2 in the presence of MERS-Cov viral RNA with a low limit of detection of 0.18 ng/uL. This method is cost-effective and eliminates the need for sophisticated instruments. Also, detecting the presence of SARS-CoV2 in 10 min from the total RNA isolated from infected biosamples is an added advantage.

Ventura et al [115] developed a colorimetric biosensor for fast detection of SAR-CoV2 in nasal and throat swabs. Colloidal gold nanoparticles (20 nm) functionalized with antibodies targeting three surface proteins of SARS-CoV2 (spike, envelope, and membrane) were used. A low limit of detection, which was close to that of real-time PCR was obtained. This method can be used for mass screening of SARS-CoV2 since only the interaction between the functionalized gold nanoparticles and virions is required. Therefore, this eliminates the pretreatment step, such as RNA extraction and amplification required in other methods. Mahari et al [116] reported the fabrication of two biosensors for the detection of SARS-Cov2. The first sensor was fabricated using a fluorine-doped tin oxide electrode with gold nanoparticles (FTO/AuNPs). This electrode was immobilized with nCovid 19 monoclonal antibody. The second sensor used a screen-printed carbon electrode (SPCE) immobilized with nCovid 19 monoclonal antibodies. These two biosensors were used to detect nCovid antigen in spiked saliva samples. The detection limit was found to be 90 fM and 120 fM for the first and second sensors, respectively. A colorimetric sensor that is based on gold nanoparticles was reported by Kumar et al [117]. In this study, the detection of RNA polymerase (RdRp) gene of SARS-CoV2 was achieved in less than 30 min with a detection limit of 0.5 ng. Recent studies on the application of gold nanoparticles to fabricate biosensors for SARS-Cov2 detection are shown in Table 9.2.

9.4.2 Graphene-based biosensors

Carbon-based materials such as graphene and its derivatives (graphene oxide and reduced graphene oxide) are frequently used in sensing and drug delivery systems. These materials have a high surface area, high conductivity, and excellent mechanical, thermal, and electrical properties [118,119]. G3raphene is used in optical sensors because of its ability to quench photoluminescence. Its high electrocatalytic activity makes it useful in electrochemical sensors as illustrated in Fig. 9.3. While graphene's

TABLE 9.2 Recent gold nanoparticles based biosensors for detection of SARS-CoV2.

Biosensor description	Nanoparticles	Detection limit	Time	Benefits	Ref.
Plasmonic fiber-optic absorbance biosensor	Gold nanoparticles	–	15 min	Rapid, cost-effective, sensitive, and selective	Kumar et al. [232]
Rapid IgM-IgG combined antibody test	Gold nanoparticles (40 nm in diameter)	–	15 min	Minimal sample preprocessing, fast and sensitive. No additional equipment is required.	Whitman et al. [69]
Opto-microfluidic sensing platform with gold nanospikes	Gold nanospikes	0.08 ng/mL	30 min	Cost-effective, easy to use, rapid, and feasible for mass production.	Murugan et al. [233]
Colloidal gold nanoparticles based lateral-flow assay	Gold nanoparticles (30 nm in diameter)	–	15 min	Small sample used (10–20 uL), excellent specificity, stability, low cost, easy operation, and less time-consuming.	Funari, Chu, and Shen [234]
Nanomaterial-based optical sensing platform	Gold nanoparticles (10 nm)	0.5 ng	30 min	Fast sensitive and selective, No sophisticated equipment	Mahari et al. [117]
FTO/AuNPs/nCovid-19 Ab SPCE/nCovid-19 Ab	Gold nanoparticles	120 fM 90 fM	10–30 s	Onsite application, cheap, no sophisticated laboratory set up, the device is potable	Ventura et al. [116]
Colorimetric biosensor based on gold nanoparticles	Gold nanoparticles (20 nm)	–	–	Pretreatment is not required, relies on its sensitivity to the virion rather than to its content (RNA)	Qiu et al. [115]
A colorimetric assay based gold nanoparticles capped with thiol modified antisense oligonucleotides	Gold nanoparticles	0.18 ng/uL	10 min	Rapid, selective, and visual 'naked eye' detection, no sophisticated laboratory equipment.	Seo et al. [112]
Dual-functional plasmonic photothermal biosensor	Gold nanoislands	0.22 pM	–	Highly sensitive, fast, and reliable.	Baker et al. [114]

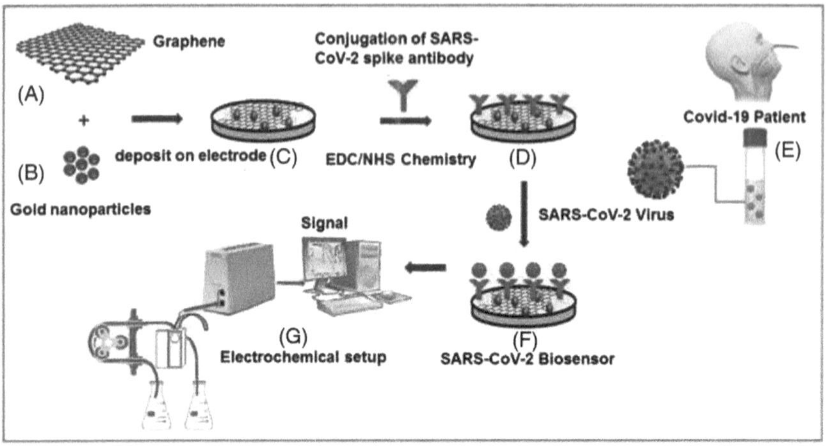

FIGURE 9.3 Potential of graphene-based materials to combat COVID-19: properties, perspectives, and prospects. Illustrations of how electrochemical graphene-based biosensors can be employed for the detection of the SARS-CoV-2 virus. Adapted with permissions from [119,228]. Materials Today Chemistry. 2020;18:100385. doi:10.1016/j.mtchem.2020.100385.

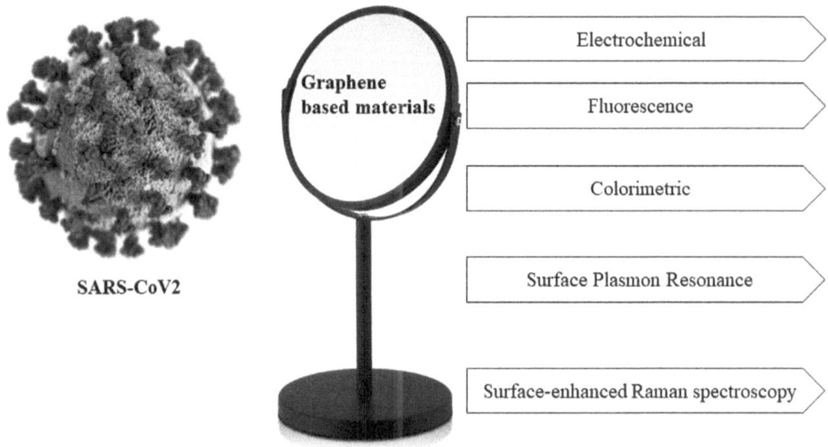

FIGURE 9.4 Typical graphene-based biosensors.

ability to adsorb guest molecules on its surface is advantageous for biosensors, as shown in Fig. 9.4.

However, graphene's application is limited by its hydrophobic nature and lack of chemically reactive functional groups that could be used to immobilize different biological molecules. To overcome these drawbacks, covalent and noncovalent functionalization, synthesis of graphene-based nanomaterials, doping with metal and metal oxide nanoparticles, and the introduction of defects are among the commonly used strategies. Graphene oxide (GO) and reduced graphene oxide (rGO) are the most used graphene-based materials in biosensors. Graphene oxide has many oxygen-containing functional groups such as hydroxyl, epoxy, and carbonyl groups. These functional groups make the attachments of biological recognition elements and surface functionalization easy. However, the presence of these oxygenated groups decreases its electrical conductivity. Therefore, reduced graphene oxide is synthesized as an alternative using various methods (hydrothermal, chemical, electrochemical, and thermal). Torrente-Rodríguez et al [120] developed an electrochemical graphene-based sensor for simultaneous detection of different SARS-CoV2 biomarkers. Viral antigen nucleocapsid proteins, IgM and IgG antibodies, and inflammatory biomarkers were used. The sensor showed high sensitivity and selectivity due to graphene's unique properties and the specificity and sensitivity of the immune-sensing strategies.

A field-effect transistor biosensor for SARS-CoV2 detection in clinical samples was reported by Seo et al [111]. The antibody specific for SARS-CoV2 was coated on the graphene sheet. A limit of detection of 1 fg/mL and 100 fg/mL was obtained for SARS-CoV2 in spike phosphate-buffered saline and clinical transport medium, respectively. This method is selective and sensitive; also, no sample pretreatment and labeling were required. A graphene oxide decorated Au/fiber Bragg grating probe was used to detect COVID-19 in human saliva by Samavati et al [121]. An ultrasensitive super sandwich-type electrochemical sensor for SARS-CoV2 was developed using calixarene functionalized graphene oxide Zhao et al [122]. This sensor is based on the RNA measurement of SARS-CoV2.a limit of detection of 200 copies/mL was obtained. In this method, a small sample of 10 uL per assay was required. The sensor works as a portable electrochemical smartphone; thus, nucleic acid amplification and reverse transcription are not needed.

9.4.3 Magnetic nanomaterial-based biosensors

With the increasing demand for SAR-CoV2 testing worldwide, several studies have reported the potential use of magnetic nanoparticles (iron oxide, ferrites of manganese, nickel, cobalt, and magnesium) based biosensors such as magnetic resonance, electrochemical, fluorescence, and rolling circle amplification [119]. Iron oxide nanoparticles are commonly used because of their biocompatibility and superparamagnetic properties. Magnetic biosensors are designed by modifying the magnetic tags with magnetic nanoparticles that are later functionalized with various biological components that are selective to the target analytes. These biosensors' main advantages include low sample matrix and low background noise since most of the biological environments are nonmagnetic.

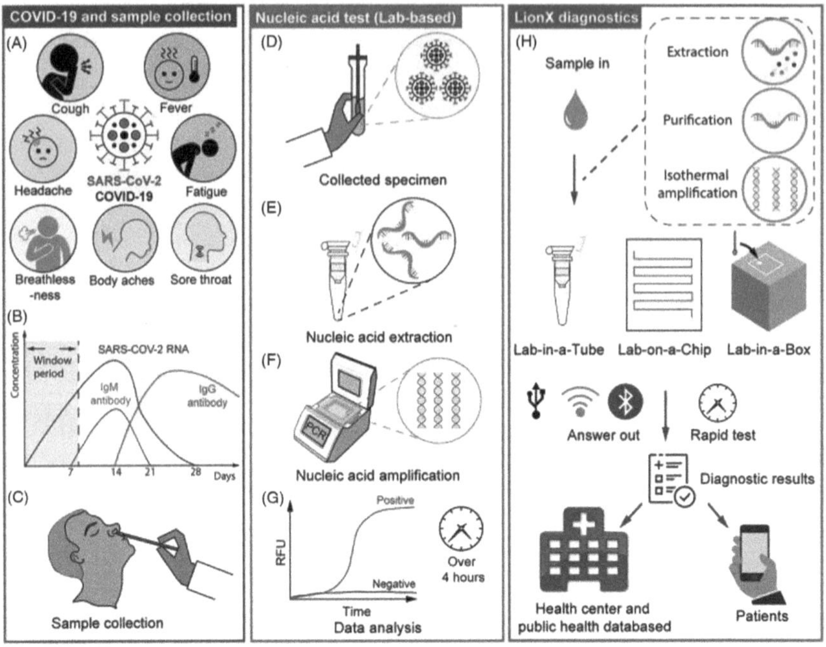

FIGURE 9.5 COVID-19 diagnosis with laboratory-based and POC tests (with permission) [229,177]. Fighting COVID-19: Integrated Micro- and nanosystems for viral infection diagnostics. Matter. 2020;3(3):628-651. doi:10.1016/j.matt.2020.06.015.

9.5 Emerging diagnostic techniques for COVID-19

This pandemic has caused researchers to begin to revamp conventional and novel systems in order to curb its global reach. Coronavirus treatment requires both restricting viral proliferation and the limitation of inappropriate immune reaction [123]. At present, numerous diagnostic kits and methods to test for COVID-19 are accessible, and repurposing strategies for COVID-19 have demonstrated to be clinically viable (Fig. 9.5; [124,125]. As the worldwide interest for diagnostics and therapeutics keeps on rising, it is important to quickly create different procedures to effectively distinguish and contain the infection. With the infection rates of COVID-19 rising exponentially, precise and quick diagnostic methods are needed to detect and control the virus. Confirmed identification and detection of the virus followed by nonpharmaceutical interventions are the primary measures to curb the spread of SARS-CoV-2 [126,127]. In spite of the fact that there are a number of strategies available for virus diagnosis, these conventional strategies and techniques have their own drawbacks. Right now, the World Health Organization (WHO) has recommended different analytic instruments and recommended the collection of upper respiratory specimen utilizing NP swabs [128,129]. The CDC and WHO also suggest the collection of Oropharyngeal (OP) specimens, sputum, endotracheal aspirate, bronchoalveolar lavage, blood, stool, and postmortem material and tissue from the lungs [130]. As of late, the WHO has affirmed the utilization of qSARS-CoV-2 IgG/IgM quick serological tests, in which venepuncture blood gathered by clinical experts can be utilized [131,132]. Be that as it may, the utilization of this test is restricted to approved research facilities. This test distinguishes the IgM and IgG antibodies created by the patient in light of SARS-CoV-2 contamination. Furthermore, the tests accessible check for the presence of viral nucleic acid through RT-PCR. For diagnosis confirmation, at least two targets must be picked on the SARS-CoV-2 genome, with one being specific for SARS-CoV-2. Another downside is that the development of antibodies only occurs a few weeks after infection which inevitably increases the possibility of false negatives [133]. Until further notice, the utilization of RT-PCR as a diagnostic tool for covid-19 is recommended. As the worldwide interest for real-time diagnostic tools keeps on rising, it is of utmost importance to hasten the development of novel diagnostic systems at the point of care. COVID-19 diagnosis is a critical step in tracing the virus to understand its epidemiology [134]. The usual tests for pathology are gathered from upper and lower respiratory tracts (throat, oropharyngeal, nasopharyngeal, bronchoalveolar liquid, and sputum) through swabs for RT-PCR test [135,136]. Urine and blood samples are not considered useful specimens due to the fact that that the virus is still absent at the time of detection in the upper and lower respiratory tracts [137]. Reports of irregularity in RT-PCR test results for CoV-SARS-2 in different tissues indicates a huge gap in research in the relationship between the viral load and its bio-distribution in different tissues [138]. Therefore, novel methods are necessary to galvanize the detection and diagnosis of SARS-CoV-2 viral infection in order to halt their propagation. To date, numerous diagnostic tools and assays have been commercialized and approved for COVID-19 diagnosis and others are currently at their respective phases of development. In this section, we will discuss some novel methods used for Covid-19 detection.

9.5.1 Nano-based automated systems

Nanotechnology innovation has gotten a lot of attention in biomedical applications and diagnostics [104,139]. One of the most significant applications of advanced nanomaterials is the probability of employing multiple probes concurrently with biological and nonbiological labels in virus detection [140,141]. The advantage of using nanobiosensors in detecting SARS-CoV-2 is due to their small size, fast response, high sensitivity as well as portability which is possible to use in point-of-care settings [142,143]. Since

SARS-CoV-2 mutation is expected to potentially produce more virulent strains globally [144,145], it is, therefore, essential to provide diagnostic nano-biosensors especially at the point of care for rapid detection in order to halt the possibility of mutations and further outbreaks.

9.5.2 Point of care diagnostics for personalized health

Rapid diagnosis is an important control measure for effective management of Covid-19 propagation [146]. Currently, RT-PCR is the gold standard for diagnosis [147], however central lab facilities are needed. Point of Care (POC) testing provides results within minutes of the test being conducted, thereby providing on-time decision making regarding patient care [148]. POC testing is also applicable in community settings and regions where rapid diagnosis is not readily available [149]. POC testing can be used for SARS-CoV-2 diagnosis in settings such as doctors' rooms, clinics, pharmacies, schools, nursing homes, as well as in drive-through sites manages by different organizations [150,151]. A point-of-care test for acute coronavirus 2 in hospitalized adults is possible, accurate, and improves the duration of outcomes compared with laboratory PCR [152]. This type of testing is associated with advances in the use of infection control methods, patient flow, and patient enrolment in clinical trials. At this juncture, it is important to direct efforts toward personal testing as well as medical care center testing in preparation for the second wave of COVID-19 [153]. WHO has emphasized on rapid point-of-care diagnostics for detection of SARS-CoV-2 as vital strategy in tackling the pandemic [154,155]. The Foundation for Innovative Diagnostics has identified over 90 point-of-care near-patient or mobile diagnostic tests for SARS-CoV-2 detection [156]. However, most of these tests require the handling of samples, which limits their use at the point-of-care. Point-of-care tests decentralizes testing and this strategy helps to reduce the work load of the centralized laboratories, thereby increasing overall testing capacity (Fig. 9.5; [157]). Another important factor in point-of-care diagnostics is their ability to accelerate clinical decision making, thus enabling effective and timely infection and control measures [158]. Conversely, point-of-care diagnostics may still need some extend of sample handling and processing, which impacts their efficiency [159]. The SARS-CoV-2 symptoms are nonspecific and during this scenario of viral detection, a point-of-care (POC) device can be used onsite and within the field, and an added advantage is that these devices do not necessarily require a trained technician to work [2].

In order to circumvent the present time-consuming detection procedures which usesRT-qPCR, an alternate molecular amplification technique should be deployed. The Loop-mediated isothermal amplification (LAMP) reaction may be a novel technique that amplifies DNA with high specificity and speed under isothermal conditions [160]. Specially designed primers are used in this method and a DNA polymerase is also employed with strand displacement activity to synthesize target DNA up to 109 copies in an hour and at a 65C [161]. The ultimate products are stem-loop DNAs with multiple inverted repeats of the target, bearing structures with a cauliflower-like appearance. LAMP is specific, sensitive, and is straightforward as consequently, it became very popular after its initial development as a pathogen detection tool [162]. The US Food and Drug Administration has also issued an emergency authorization of use for the American company, Cepheid's POC COVID-19 diagnostic device named Xpert Xpress SARS-CoV-2 [163]. The tests detect the virus within 45 minutes, using specimens collected from nasopharyngeal swabs and nasal wash/aspirate. Another POC diagnostic tool, Xpert Xpress SARS-CoV-2 test cartridge is devoted to detecting the virus using RT-qPCR without the use of reagents [164].

With the amount of COVID-19 positive cases and infections rising exponentially, mass public testing is vital for speed detection, identification, and treatment of infected persons. At present, the varied RT-PCR tests and immunoassays present limitations in terms of accessibility and in rapid diagnosis [125,165]. After the patient's samples are collected, transport logistics derail the delivery of samples for analysis. Current turnaround times for standardized tests can take up to 2 weeks [166]. This lack of rapid testing for a large number of infected individuals has been a great limiting factor in halting the virus propagation. POC diagnostic devices can help control the spread of SARS-CoV-2 because they supply a rapid and easy way for broad-spectrum testing in community settings [167]. The design of POC diagnostic tests is in such a way that they can be user friendly without the necessity for a skilled professional. They also do not require complicated facilities and machinery and therefore may be ideal for usage in at-home settings by consumers [168].

Several POC tests have been developed and received Emergency Use Authorization (EUAs) to be used under the patient-care settings (Table 9.3). Among these tests is Cepheid's Xpert Xpress SARS-CoV-2 test mentioned earlier, except for use on Cepheid's GeneXpert Xpress System compact and simplified system utilized in physician offices and clinics [169]. The other tests are the Abbott Diagnostic's ID NOW COVID-19 Test, Mesa Biotech's Accula SARS-CoV-2 Test, and the Cue Health's Cue COVID-19 Test [170,171]. The ID NOW COVID-19 test relies on isothermal macromolecule amplification, targeting a singular region of the RNA-dependent RNA polymerase (RdRP) gene of SARS-CoV-2 [172] which has a shorter diagnostic time (19 tests in 13 minutes).The system also has high analytical sensitivity of 125 copies/mL [4,173]. The Mesa Biotech's Accula SARS-CoV-2 Test employs a combinatorial approach of RT-PCR and lateral flow immunoassay and targets the N gene of SARS-CoV-2 from nasal and throat samples [174]. The test is performed on the Accula Dock or Silaris Dock

TABLE 9.3 Emergency Use Authorization (EUA) for use at the point of care (POC).

Product	Manufacturer	Sample type	FDA approval	Type	Target	Reference
Xpert SARSCoV-2	Cepheid (USA)	Nasopharyngeal swab, nasal aspirate	Yes	RT-PCR	SARS-CoV-2 RNA	Qin et al. [235]
VitaPCR COVID-19 assay	Credo (Singapore)	Nasopharyngeal or oropharyngeal swabs	Pending	RT-PCR	SARS-CoV-2 RNA	Lauxmann, Santucci, and A.M.J.I.b.j.u., [236]
RapiPrep COVID-19	Microsens Dx (London)	Sputum or swabs	Yes	LAMP amplification technology	SARS-CoV-2 RNA	Fournier et al. [237]
ePlex SARSCoV-2	GenMark Diagnostics (United States)	Nasopharyngeal swab	Yes	RT-PCR	SARS-CoV-2 RNA	Green et al. [238]
Accula SARSCoV-2	Mesa Biotech (United States)	Throat and nasal swabs (in same collection tube)	Yes	RT-PCR + lateral flow	SARS-CoV-2 RNA	Younes et al. [239]
ID NOW COVID-19	Abbott Diagnostics	Throat, nasal, nasopharyngeal and oropharyngeal swabs	Yes	Isothermal nucleic acid amplification	SARS-CoV-2 RNA	Sidiq et al. [240]
GT-100 SARS- CoV-2 IgG/IgM kit	Goldsite Diagnostic Inc. (China)	Human serum and plasma (20 uL)	Yes	Time- resolved fluorescence immunoassay	IgG/IgM	Fournier et al. [237]
Rapid POC kit	Assay Genie (Acro Biotech, Inc) (Ireland)	Blood, serum and plasma	Yes	Colloidal gold immuno-chromatography	IgG/IgM	Green et al. [238]

(*continued on next page*)

COVID- 19 IgM- IgG Rapid Test	BioMedomics, BD (United States)	Finger prick/venous blood	Yes	Lateral flow immunoassay	IgG/IgM	Fournier et al. [237]
COVID- 19 Rapid Test Cassette	SureScreen Diagnostics (England)	Finger prick	Yes	Lateral flow immunoassay	IgG/IgM	Harrington et al. [241]
VivaDiag COVID- 19 IgG - IgM test	VivaChek (China)	10uL volume - finger prick / venous blood, plasma or serum	Yes	Colloidal gold immune-chromatography	IgG/IgM	Leightley et al. [242]
BioFire Respiratory Panel 2.1-EZ (RP2.1-EZ)	BioFire Diagnostics, LLC	Nasopharyngeal swab	Yes	RT, Nested multiplex PCR, Multianalyte	SARS-CoV-2 RNA	Cassaniti et al. [243]
Xpert Xpress SARS-CoV-2 test	Cepheid	Nasopharyngeal swab, nasal swab or nasal wash/aspirate	Yes	Real-time RT-PCR, multianalyte	SARS-CoV-2 RNA	Paret et al. [244]
Cobas SARS-CoV-2 & Influenza A/B Nucleic Acid Test for use on the Cobas Liat System	Roche Molecular Systems, Inc.	Nasopharyngeal and nasal swabs and self-collected nasal swabs	Yes	Real-time RT-PCR, Multi-analyte	SARS-CoV-2 RNA	McCormick-Baw et al. [245]
ID NOW COVID-19	Abbott Diagnostics Scarborough, Inc.	Direct nasal, nasopharyngeal or throat swabs	Yes	RT, Isothermal amplification	SARS-CoV-2 RNA	Sidiq et al. [240]
Accula SARS-CoV-2 Test	Mesa Biotech Inc.	n throat swab and nasal swabs	Yes	RT and amplification	SARS-CoV-2 RNA	Hansen et al. [246]
Cue COVID-19 Test	Cue Health Inc.	Nasal swabs	Yes	RT, Isothermal amplification	SARS-CoV-2 RNA	Wong et al. [163]

and is comparatively straightforward to use. The swab from the sample is dipped into a buffer vial and transferred into a test cassette which has all of the reaction reagents [163]. The test cassette sits within the dock for about half-hour, and afterwards results are optically interpreted. The test comprise of the interior positive process control line, the SARS-CoV-2 test line, and therefore the internal negative process control line. The observation of any shade of blue at the SARS-CoV-2 test line indicates a positive result. However, any appearance of blue at the negative process control line indicates an invalid test, and therefore the test must be performed again. Cue Health's Cue COVID-19 Test also uses isothermal amplification on nasal swabs and also targets the N gene of SARS-CoV-2 [163]. Moreover, Cue's disposable POC test cartridge forms a connected diagnostic platform with a mobile device that permits a patient to possess their health status. All 4 tests are quite sensitive and specific however, Abbott and Cue Health's tests both use isothermal amplification and are therefore easier to use, have shorter turnaround times, and consume less power compared to Mesa Biotech and Cepheid's tests that use RT-PCR. From this we can deduce that isothermal amplification may be the preferred detection technique for POC SARS-CoV-2 detection compared to RT-PCR [175]. An advantage of using the RT-PCR based Cepheid's Xpert Xpress Test is that it is the sole diagnostic POC tool that has been approved by the FDA-EUA for usage in patient care settings and can detect both the N2 and E gene of SARS-CoV-2, thus offering a further assurance of accuracy [176]. In the case of Cue Health's test for POC application, it is portable, simple to use, and connect to mobile devices to supply patients' health status at their fingertips [177].

9.5.3 AI supported POC diagnostic systems

The management of the COVID-19 pandemic requires an amalgamation of approaches from various fields to ensure rapid, selective, and sensitive diagnosis at its early stages of infection. The application of artificial intelligence (AI) is important for the efficient and smart diagnosis of SARS-CoV-2. There is a huge scope in research for the development of smart sensors employing AI for rapid detection of SARS-CoV-2 proteins at the picomolar level. Due to the nature of COVID-19 infection and transmission, its diagnosis requires point-of-care techniques to eliminate the burden of specialized labs and expert personnel. Researchers have indicated that big data, data banks and vast bioinformatics collection associated with the COVID-19 pandemic are required for effective control of the spread of the pandemic [178,179]. A number of AI-based diagnostic systems have been designed and developed to predict the probability of which population groups or patients can become critically ill [180,181]. The consequences of SARS-CoV-2 are population dynamics related and the usage of smart technology is required to characterize those aspects for accurate prediction, surveillance, and diagnosis. AI-supported deep learning, machine learning algorithms, and internet of things (IoT) techniques have recently emerged as tool to halt the propagation of SARS-CoV-2 [182]. Wang et al. [105] reported that Taiwan employs such technology at the battlefront to explore big data analysis, new technologies, and proactive testing [183]. This research reported on the identification of pandemic zones, optimization of resources, and understanding of emergency and timely diagnosis decisions. Bai et al. [184] reported that IoT-based methods were very useful to medical personnel to establish the full spectrum of COVID-19 and allowed for intelligent processing and accurate diagnosis [185].

BlueDot, a Canadian company specializing in communicable disease forecasting revealed the importance of AI application toward COVID-19 [184]. Using an AI engine that continuously gathers disease data from various sources globally, the company was ready to predict the COVID-19 outbreak and alert its users even before the WHO did. In Singapore, an AI-powered Chabot (SGDormBot) has been used for the mass screening of COVID-19 symptomatic migrant workers [186]. Another fully automated 3D deep-learning framework developed for the detection of COVID-19 (COVNet) has been designed and developed to extract relevant patient information from 2D and 3D images obtained from a CT scan images in order to differentiate patients with COVID-19 from patients with non–COVID-19 community-acquired pneumonia [187]. The AI-system also identifies COVID-19 features in CT scans of patients with false-negative RT-PCR results [188]. Even with the great prospects of AI to be implemented within the fight against COVID-19, such systems are still at the early stages of development. A drawback to AI algorithms usage is the access needed to massive datasets and this raises ethical considerations regarding the privacy of patients [189]. Efficacy for COVID-19 detection also can be increased by usage of the smart city data network which employs a terminal tracking system and data sharing for better urban management and therefore the location of sporadic occurrences might be predicted [190]. The usage of smartphones for self-screening and surveys can help identify population groups under quarantine [191].

Machine learning algorithms are also important in the processing of COVID-19 patient symptoms [192]. The symptoms are evaluated by asking basic questions from infected patients. Using that information and that from emergency care admission exams, the algorithm is then employed to diagnose covid-19 [193]. An ensemble of machine learning algorithms has been utilized in a variety of works to diagnose the disease. These approaches include logistic regression, Support Vector Machine, Decision Tree, and Random Forest and call all process patient data and diagnose covid-19 cases [194]. Other approaches include K-Nearest Neighbour (KNN), artificial Neural Networks (ANN) and Naive Bayes algorithms and can also be used to diagnose the disease [195]. In the case of smart devices, they can provide a good platform for developing AI methods for diagnosis. They're widely available, they will collect an excellent deal of knowledge from people from symptoms to behavior and travel patterns, and that they can inform people of any risk they'll face [178]. An AI-based algorithm linked to a mobile App that monitors people's cough has been proposed to spot potential COVID-19 cases [28]. In these devices, an end-to-end portable system supported machine learning records cough data from patients and uses them to coach a classifier for diagnosing the disease.

The application of Deep Learning algorithms is also of great importance in order to accelerate the method of COVID-19 diagnosis. Some of these approaches include Generative Adversarial Networks (GANs), Extreme Learning Machine (ELM), and Long /Short Term Memory (LSTM) [196]. These approaches have the ability to put together a continuum of structured and unstructured data sources. A framework called CovidDeep has also been proposed and employs the usage of DNN with wearable medical sensors for pervasive testing of the virus and therefore the disease [197]. The algorithm functions on the info collected from wearable devices and a few easy-to-answer questions during a questionnaire.

Another machine learning-based method has been proposed to research blood exams as input and find the suspected cases of COVID-19 [198]. Another approach proposed uses white blood cell counts, and therefore the platelets and plasma levels as features for machine learning algorithms to detect COVID-19 infection [199]. In another study, it was found that COVID-19 patients tend to possess higher plasma fibrinogen levels, low platelet counts, and around 25% of patients showing outright thrombocytopenia [200]. This information was then fed to a neural network-powered extraction system for the analysis and diagnosis of COVID-19 infection. Text analytics and processing also allow for the extraction of tons of knowledge around the disease and more text-based algorithms should be developed [201]. In this method, a web questionnaire is developed to gather data about COVID-19 patients [202]. The information is then provided to a predictive machine learning algorithm such as SVM, Logistic Regression, and MLP for the accurate prediction of possible Covid-19 infected persons based on specific descriptors, indications, and symptoms identified.

9.5.4 Nano-based therapies for COVID-19

Recently, broad-based antiviral drugs and conventional dosage forms have been seen to be prone to resistance, and the rise of newer strains poses an even greater danger [203]. Nanotechnology has been recently explored as therapeutic use toward COVID-19 [204]. Within the medical

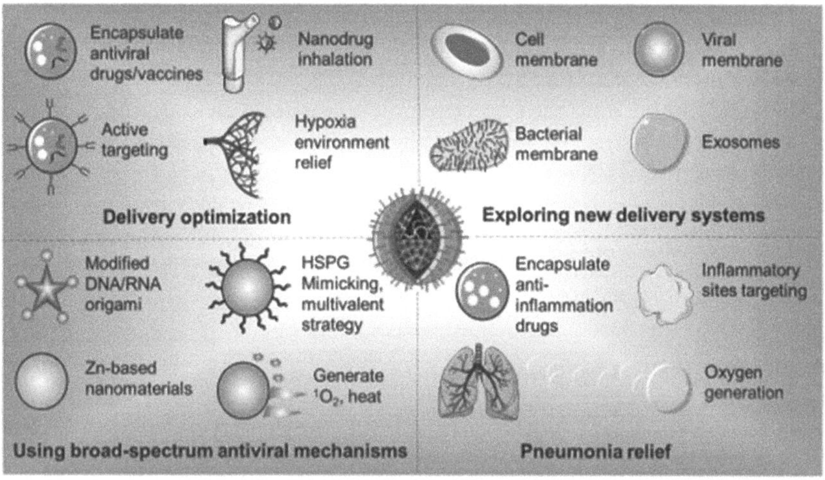

FIGURE 9.6 Application of nanotechnology against COVID-19 (permission granted) [230,212]. Insights from nanotechnology in COVID-19 treatment. Nanotoday. 2020;36:101019. doi:10.1016/j.nantod.2020.101019.

field, it incorporates the utilization of nanomaterials for diagnosis, treatment, control, and prevention of diseases [205]. This technology holds great potential as a diagnostic and therapeutic tool for the prevention of COVID-19. Different nano-based approaches could be taken to curb the spread of COVID-19. The design and development of highly specific and sensitive nano-based sensors for quick identification of infection and/or immunological response [206], nano-based vaccination to enhance the immune response [207] and nano-based formulations with targeted delivery antivirals [208,209,[210]] could be the trinity that could assist in the elimination of the disease; the development of nano-based vaccines using differing [211–213]. The application of nanotechnology in combating COVID-19 has marked potential, ranging from delivery optimization, and exploration of novel delivery systems to using broad-spectrum antiviral mechanisms (Fig. 9.6). Two nano-based vaccines have successfully passed phase III clinical trials and one of them has been approved in the United Kingdom, Canada United States of America and the manufacturers are seeking approval in the European Union. Therefore, much research is still required for the optimization of antiviral drugs and vaccines in nanosystems for the prevention of COVID-19 infections.

Similar to other viruses, SARS-CoV-2 requires rapid response testing and operational simplicity. It is quite a difficult task to sanitize surfaces and other inanimate objects all the time, and there's no guarantee that the surface won't be contaminated again. Coating surfaces with virus active molecules is also a strategy. In this regard, nanotechnology can function as a useful tool for the design and prevention of contamination of the equipment. A study by Jorge et al. (2020) [247] reported that metal-loaded and metal embedded copper nanoparticles in polymer matrices are very effective against viral pathogens and consequently can be employed toward COVID-19 [214]. Furthermore, it has also been reported that the mixture of copper nanoparticles with quaternary ammonium shell that have exhibited potential antiviral activity [215]. Other studies have reported that various metal and metal oxide nanoparticles such as zinc nanoparticles (ZnONPs), cuprous oxide nanoparticles (CuONPs), silver nanoparticles (AgNPs), nanosized copper (I) iodide particles (CuINPs), gold nanoparticles (AuNPs), silica nanoparticles (SiONPs)and quaternary ammonium cations commonly called QUATs are capable of deactivating SARS-CoV-2 [216–218]. Bhattacharjee et al., study showed that graphene oxide grafted metal nanoparticles can be employed to surface coat private protective equipment (PPE) [219] for killing and prevention of viral transmission. Balagna et al., evaluated also studied the antiviral effect of silver nanocluster/silica composite by depositing the on the filtering facepiece-3 (FFP3) facial mask SARS-CoV-2 transmission prevention [220]. The co-sputtering process was employed in depositing silver nanocluster/silica on the PEE thus adding an antiviral surface coat on the face mask due to the antiviral effect of the silver NPs [221]. The study demonstrated that the silver nanocluster/silica coating deposited on the facial mask possessed virucidal activity [222]. Surface-coated copper nanoparticles also can also effectively block the viral infection. Polymer-based copper nanoparticles and other metal nanoparticles are often used as an antiviral coating which may be applied or sprayed on surfaces [222,223]. These nanosystems release metal ions on the coated surface and these ions act as antiviral agents [224]. This coating technique can very well limit the spread of COVID-19 and the metallic ions contained in them have varied properties which makes them ideal candidates to destroy SARS-CoV-2 [225].

It is envisaged that these emerging diagnostic approaches to tackle COVID-19 will bring about new perspectives and overhaul to current dosage forms and treatment regimens. Moreover, more studies are needed to elucidate the SARS-CoV-2 inhibition by nanosystems in order to provide a rational approach in target-based therapy. Undoubtedly, the COVID-19 would need to involve a myriad of novel approaches from different scientific fields in order to probe further on its pathogenesis. The application of Nanotechnology should be the hegemonic strategy of researchers coupled with biosensors and POC systems in order to ease the burden on medical personnel and healthcare facilities. With the combinatorial approaches of biosensors, AI can very well provide rapid communication between hospitals, patients, and the wider community in order to timeously respond to future possibilities of outbreaks.

9.6 Future perspectives

COVID-19 a pandemic has affected the global healthcare and economy. Overcoming the pandemic has become a matter of urgency. The design, development, validation, verification, and implementation of diagnostic tests have been rapidly developed by manufacturers and approved by regulatory authorities while several are waiting for clinical approval. This chapter has summarized the important biomarkers for the detection and classification of the COVID-19 disease. Standard systems being employed for detection of COVID-19 have successfully assisted in early detection control of the disease have also been discussed. However, their disadvantages such as longer time taken for detection vs. the rapid spread of the disease, lack of sensitivity and universal tests for different strains of the virus have resulted to the need of better alternatives. Due to the dire need for faster, reliable, and sensitive diagnostic systems which are being developed with the aid of nanotechnology. Systems such as point of care diagnostics for personalized health and artificial intelligence-based systems are revolutionizing the detection of COVID-19. The designs of nanotherapeutics that are target specific, effective, and safe to treat COVID-19 are also required. The development of novel nanomedicines that target the virus are showing potential to be potent, effective and at the same time they could overcome possible SARS-CoV-2 resistance. Nanotechnology has provided the two effective vaccines against the virus that have been approved in some markets. This indicates the coming of age of nanomedicine. However, as the advancement of nanotechnology continues there is also a need for continuous global improvement in diagnostic tests and nanotherapeutics that provide rapid detection and treatment of patients, possibly at the point of care. Moreover, these diagnostic tests and treatment options need to be developed so that they can be sensitive and cost-effective and can be applied in both industrialized and resource-limited countries.

Acknowledgments

The authors acknowledge the University of KwaZulu-Natal (UKZN), United States International University Africa, and National Research Foundation South Africa (NRF-SA) grant number 121804 for financial support.

Conflict of interest

Authors declare no conflict of interest.

References

[1] E. Morales-Narváez, C. Dincer, The impact of biosensing in a pandemic outbreak: COVID-19, Biosens. Bioelectron. 163 (2020) 112274.

[2] B. Udugama, et al., Diagnosing COVID-19: The disease and tools for detection, ACS Nano 14 (4) (2020) 3822–3835.

[3] D. Hassan, et al., Novel chitosan-based pH-responsive lipid-polymer hybrid nanovesicles (OLA-LPHVs) for delivery of vancomycin against methicillin-resistant Staphylococcus aureus infections, Int. J. Biol. Macromol. 147 (2020) 385–398.

[4] C. Callahan et al. Nasal-Swab Testing Misses Patients with Low SARS-CoV-2 Viral Loads, (2020).

[5] R. Lu, et al., Genomic characterisation and epidemiology of 2019 novel coronavirus: Implications for virus origins and receptor binding, Lancet 395 (10224) (2020) 565–574.

[6] P. Zhou, et al., A pneumonia outbreak associated with a new coronavirus of probable bat origin, Nature 579 (7798) (2020) 270–273.

[7] A. Assiri, et al., Epidemiological, demographic, and clinical characteristics of 47 cases of Middle East respiratory syndrome coronavirus disease from Saudi Arabia: A descriptive study, Lancet Infect. Dis. 13 (9) (2013) 752–761.

[8] A. Banerjee, et al., Bats and coronaviruses, Viruses 11 (1) (2019).

[9] C. Liu, et al., Research and development on therapeutic agents and vaccines for COVID-19 and related human coronavirus diseases, ACS Cent Sci 6 (3) (2020) 315–331.

[10] T.T.-Y. Lam, et al., Identifying SARS-CoV-2-related coronaviruses in Malayan pangolins, Nature 583 (7815) (2020) 282–285.

[11] M. Chen et al. Titanium incorporation into Zr-porphyrinic metal–organic frameworks with enhanced antibacterial activity against multidrug-resistant pathogens. 16 (7) (2020) 1906240.

[12] M.I. Khan et al., Comparative genome analysis of novel coronavirus (SARS-CoV-2) from different geographical locations and the effect of mutations on major target proteins: An in silico insight. 15 (9) (2020b) e0238344.

[13] C.A. Omolo, et al., Update on therapeutic approaches and emerging therapies for SARS-CoV-2 virus, Eur. J. Pharmacol. 883 (2020) 173348.

[14] F. Wu, et al., A new coronavirus associated with human respiratory disease in China, Nature 579 (7798) (2020) 265–269.

[15] S. Patra et al. Emerging molecular prospective of SARS-CoV-2: Feasible nanotechnology based detection and inhibition. 11 (2020) 2098.

[16] J. Gong, et al., Correlation analysis between disease severity and inflammation-related parameters in patients with COVID-19 pneumonia, medRxiv (2020) 2020.02.25.20025643.

[17] C. Qin, et al., Dysregulation of immune response in patients with coronavirus 2019 (COVID-19) in Wuhan, China, Clin. Infect. Dis. 71 (15) (2020) 762–768.

[18] F. Liu, et al., Prognostic value of interleukin-6, C-reactive protein, and procalcitonin in patients with COVID-19, J. Clin. Virol. 127 (2020) 104370.

[19] L. Velazquez-Salinas et al. The role of interleukin 6 during viral infections. 10 (1057) (2019).

[20] S. Mahajan, et al., Plcγ2/Tmem178 dependent pathway in myeloid cells modulates the pathogenesis of cytokine storm syndrome, J. Autoimmun. 100 (2019) 62–74.

[21] W.H. Mahallawi, et al., MERS-CoV infection in humans is associated with a pro-inflammatory Th1 and Th17 cytokine profile, Cytokine 104 (2018) 8–13.

[22] N. Chen, et al., Epidemiological and clinical characteristics of 99 cases of 2019 novel coronavirus pneumonia in Wuhan, China: A descriptive study, Lancet 395 (10223) (2020) 507–513.

[23] E.A. Coomes, H. Haghbayan, Interleukin-6 in Covid-19: A systematic review and meta-analysis, Rev. Med. Virol. 30 (6) (2020) 1–9.

[24] Z.S. Ulhaq, G.V. Soraya, Interleukin-6 as a potential biomarker of COVID-19 progression, Med. Mal. Infect. 50 (4) (2020) 382–383.

[25] L. Tan, et al., Lymphopenia predicts disease severity of COVID-19: A descriptive and predictive study, Signal Transduct Target Ther 5 (1) (2020) 33.
[26] C.Y. Chen, et al., Clinical features and outcomes of severe acute respiratory syndrome and predictive factors for acute respiratory distress syndrome, J. Chin. Med. Assoc. 68 (1) (2005) 4–10.
[27] R.-J. Tao, et al., Viral infection in community acquired pneumonia patients with fever: A prospective observational study, J. Thorac. Dis. 10 (7) (2018) 4387.
[28] A. Imran et al. AI4COVID-19: AI enabled preliminary diagnosis for COVID-19 from cough samples via an app, (2020).
[29] D. Ferrari, et al., Routine blood tests as a potential diagnostic tool for COVID-19, Clin. Chem. Lab. Med. 58 (7) (2020) 1095–1099.
[30] B.M. Henry, et al., Lactate dehydrogenase levels predict coronavirus disease 2019 (COVID-19) severity and mortality: A pooled analysis, Am. J. Emerg. Med. 38 (9) (2020) 1722–1726.
[31] P. Mo, et al., Clinical characteristics of refractory COVID-19 pneumonia in Wuhan, China, Clin. Infect. Dis. (2020).
[32] L. Zhang, et al., Use of D-dimer in oral anticoagulation therapy, Int J Lab Hematol (2018).
[33] P. Paliogiannis, et al., D-dimer concentrations and COVID-19 severity: A systematic review and meta-analysis, Frontiers in Public Health 8 (432) (2020) 1–7.
[34] N. Tang, et al., Abnormal coagulation parameters are associated with poor prognosis in patients with novel coronavirus pneumonia, J. Thromb. Haemost. 18 (4) (2020) 844–847.
[35] L. Zhang, et al., D-dimer levels on admission to predict in-hospital mortality in patients with Covid-19, J. Thromb. Haemost. 18 (6) (2020) 1324–1329.
[36] E.B. Milbrandt, et al., Prevalence and significance of coagulation abnormalities in community-acquired pneumonia, Mol. Med. 15 (11-12) (2009) 438–445.
[37] G. Lippi, M. Plebani, B.M. Henry, Thrombocytopenia is associated with severe coronavirus disease 2019 (COVID-19) infections: A meta-analysis, Clin. Chim. Acta 506 (2020) 145–148.
[38] W.J. Guan, et al., Clinical characteristics of coronavirus disease 2019 in China, N. Engl. J. Med. 382 (18) (2020) 1708–1720.
[39] I.H. Khan, et al., At the heart of COVID-19, J. Card. Surg. 35 (6) (2020) 1287–1294.
[40] T.R. Khashkhusha, J.S.K. Chan, A. Harky, ACE inhibitors and COVID-19: We don't know yet, J. Card. Surg. 35 (6) (2020) 1172–1173.
[41] P.A. Kavsak, et al., Cardiac troponin testing in patients with COVID-19: A strategy for testing and reporting results, Clin. Chem. (2020).
[42] F. Zhou, et al., Clinical course and risk factors for mortality of adult inpatients with COVID-19 in Wuhan, China: A retrospective cohort study, Lancet 395 (10229) (2020) 1054–1062.
[43] B.M. Henry, G. Lippi, Chronic kidney disease is associated with severe coronavirus disease 2019 (COVID-19) infection, Int. Urol. Nephrol. 52 (6) (2020) 1193–1194.
[44] J. Xiang et al. Potential biochemical markers to identify severe cases among COVID-19 patients. medRxiv, (2020).
[45] Y. Cheng, et al., Kidney disease is associated with in-hospital death of patients with COVID-19, Kidney Int. 97 (5) (2020) 829–838.
[46] H. Zhou et al. Urinalysis, but not blood biochemistry, detects the early renal-impairment in patients with COVID-19, (2020).
[47] R. Higuchi, et al., Simultaneous amplification and detection of specific DNA sequences, Bio/Technology 10 (4) (1992) 413–417.
[48] R. Higuchi, et al., Kinetic PCR analysis: real-time monitoring of DNA amplification reactions, Bio/Technology 11 (9) (1993) 1026–1030.

[49] E. Pilli, et al., Monitoring DNA contamination in handled vs. directly excavated ancient human skeletal remains, PLoS One 8 (1) (2013) e52524.
[50] Jagtar Singh, N.B., S. Sinha, A. Goswami, A critical review on PCR, its types and applications, International Journal of Advance Research in Biological Science 1 (7) (2014) 65–80.
[51] H. Chen, et al., Clinical characteristics and intrauterine vertical transmission potential of COVID-19 infection in nine pregnant women: A retrospective review of medical records, Lancet 395 (10226) (2020) 809–815.
[52] P. Pokhrel, C. Hu, H. Mao, Detecting the Coronavirus (COVID-19), ACS Sensors 5 (8) (2020) 2283–2296.
[53] D.K.W. Chu, et al., Molecular diagnosis of a novel coronavirus (2019-nCoV) causing an outbreak of pneumonia, Clin. Chem. 66 (4) (2020) 549–555.
[54] I. Smyrlaki, et al., Massive and rapid COVID-19 testing is feasible by extraction-free SARS-CoV-2 RT-PCR, Nat. Commun. 11 (1) (2020) 4812.
[55] Centers for Disease Control and Prevention, C., CDC 2019-Novel Coronavirus (2019-nCoV), (2022).
[56] Real-Time RT-PCR Diagnostic Panel (2019).
[57] K.K. To, et al., Consistent detection of 2019 novel coronavirus in saliva, Clin. Infect. Dis. 71 (15) (2020) 841–843.
[58] X. Yuan, et al., Current and perspective diagnostic techniques for COVID-19, ACS Infectious Diseases 6 (8) (2020) 1998–2016.
[59] H. Lv, et al., Cross-reactive antibody response between SARS-CoV-2 and SARS-CoV infections, bioRxiv (2020) 2020.03.15.993097.
[60] J. Xiang et al. Evaluation of enzyme-linked immunoassay and colloidal gold-immunochromatographic assay kit for detection of novel coronavirus (SARS-Cov-2) causing an outbreak of pneumonia (COVID-19). medRxiv, (2020) P. 2020.02.27. 20028787.
[61] C.B.F. Vogels, et al., Analytical sensitivity and efficiency comparisons of SARS-COV-2 qRT-PCR primer-probe sets, medRxiv (2020) 2020.03.30.20048108.
[62] J. Zhang, et al., Serological detection of 2019-nCoV respond to the epidemic: A useful complement to nucleic acid testing, medRxiv (2020) 2020.03.04.20030916.
[63] M. Natesan, et al., A smartphone-based rapid telemonitoring system for Ebola and Marburg disease surveillance, ACS Sensors 4 (1) (2019) 61–68.
[64] M. Biosciences, J.P.B. Deng, F. Wayne, L. Clare, J. Singh, Chiu, Y. Charles, Chen, S. Janice. A protocol for rapid detection of the 2019 novel coronavirus SARS-CoV-2 using CRISPR diagnostics: SARS-CoV-2 DETECTR V.3 (2020). https://www.protocols.io/workspaces/coronavirus-method-development-community.
[65] M.J. Kellner, et al., SHERLOCK: Nucleic acid detection with CRISPR nucleases, Nat. Protoc. 14 (10) (2019) 2986–3012.
[66] J.P. Broughton, et al., Rapid detection of 2019 novel Coronavirus SARS-CoV-2 Using a CRISPR-based DETECTR Lateral Flow Assay, medRxiv (2020) 2020.03.06.20032334.
[67] R. Lassaunière, et al., Evaluation of nine commercial SARS-CoV-2 immunoassays, medRxiv (2020) 2020.04.09.20056325.
[68] C. Lucia, P.-B. Federico, G.C. Alejandra, An ultrasensitive, rapid, and portable coronavirus SARS-CoV-2 sequence detection method based on CRISPR-Cas12, bioRxiv (2020) 2020.02.29.971127.
[69] J.D. Whitman, et al., Test performance evaluation of SARS-CoV-2 serological assays, medRxiv (2020) 2020.04.25.20074856.
[70] B. Ragnesola, et al., COVID19 antibody detection using lateral flow assay tests in a cohort of convalescent plasma donors, BMC Research Notes 13 (1) (2020) 372.
[71] F. Rudolf et al. Clinical characterisation of eleven lateral flow assays for detection of COVID-19 antibodies in a population. medRxiv, (2020) P. 2020.08.18.20177204.

[72] K.M. Byers, et al., Fully dried two-dimensional paper network for enzymatically enhanced detection of nucleic acid amplicons, ACS Omega 5 (9) (2020) 4673–4681.
[73] D. Lee, et al., Carbon nanotube electric immunoassay for the detection of swine influenza virus H1N1, Biosens. Bioelectron. 26 (8) (2011) 3482–3487.
[74] D. Nidzworski, et al., Universal biosensor for detection of influenza virus, Biosens. Bioelectron. 59 (2014) 239–242.
[75] H. Shafiee, et al., Nanostructured optical photonic crystal biosensor for HIV viral load measurement, Sci. Rep. 4 (2014) 4116.
[76] R. Tanaka, et al., A novel enhancement assay for immunochromatographic test strips using gold nanoparticles, Analytical and Bioanalytical Chemistry 385 (8) (2006) 1414–1420 In this issue.
[77] A.A. Yanik, et al., An optofluidic nanoplasmonic biosensor for direct detection of live viruses from biological media, Nano Lett. 10 (12) (2010) 4962–4969.
[78] H. Ashiba, et al., Detection of norovirus virus-like particles using a surface plasmon resonance-assisted fluoroimmunosensor optimized for quantum dot fluorescent labels, Biosens. Bioelectron. 93 (2017) 260–266.
[79] J.M. Yang, K.R. Kim, C.S. Kim, Biosensor for rapid and sensitive detection of influenza virus, Biotechnol. Bioprocess Eng. 23 (4) (2018) 371–382.
[80] P. Nandhakumar, et al., Carboxyl esterase-like activity of DT-diaphorase and its use for signal amplification, ACS Sensors 4 (11) (2019) 2966–2973.
[81] S. Jiang, L. Du, Z. Shi, An emerging coronavirus causing pneumonia outbreak in Wuhan, China: Calling for developing therapeutic and prophylactic strategies, Emerg Microbes Infect 9 (1) (2020) 275–277.
[82] R.M. Mayall, et al., Enhanced signal amplification in a toll-like receptor-4 biosensor utilizing ferrocene-Terminated mixed monolayers, ACS Sensors 4 (1) (2019) 143–151.
[83] L. Cheng, et al., An electrochemical molecular recognition-based aptasensor for multiple protein detection, Anal. Biochem. 491 (2015) 31–36.
[84] W. Li, et al., High-activity Fe(3)O(4) nanozyme as signal amplifier: A simple, low-cost but efficient strategy for ultrasensitive photoelectrochemical immunoassay, Biosens. Bioelectron. 127 (2019) 64–71.
[85] M. Negahdary, et al., Electrochemical aptasensing of human cardiac troponin I based on an array of gold nanodumbbells-Applied to early detection of myocardial infarction, Sens. Actuators B 252 (2017) 62–71.
[86] N. Sattarahmady, A. Rahi, H. Heli, A signal-on built in-marker electrochemical aptasensor for human prostate-specific antigen based on a hairbrush-like gold nanostructure, Sci. Rep. 7 (1) (2017) 11238.
[87] H.A.M. Faria, V. Zucolotto, Label-free electrochemical DNA biosensor for zika virus identification, Biosens. Bioelectron. 131 (2019) 149–155.
[88] H. Ilkhani, S. Farhad, A novel electrochemical DNA biosensor for Ebola virus detection, Anal. Biochem. 557 (2018) 151–155.
[89] M. Negahdary, et al., An aptamer-based biosensor for troponin I detection in diagnosis of myocardial infarction, Journal of Biomedical Physics and Engineering 8 (2) (2018) 167–178.
[90] C. Singhal, et al., Paper based DNA biosensor for detection of chikungunya virus using gold shells coated magnetic nanocubes, Process Biochem. 74 (2018) 35–42.
[91] M. Steinmetz, et al., A sensitive label-free impedimetric DNA biosensor based on silsesquioxane-functionalized gold nanoparticles for Zika Virus detection, Biosens. Bioelectron. 141 (2019) 111351.
[92] S.M. Azab, A.M. Fekry, Electrochemical design of a new nanosensor based on cobalt nanoparticles, chitosan and MWCNT for the determination of daclatasvir: A hepatitis C antiviral drug, RSC Advances, 7 (2) (2017) 1118–1126.

[93] A. Mokhtarzadeh, et al., Nanomaterial-based biosensors for detection of pathogenic virus, Trends in Analytical Chemistry: TRAC 97 (2017) 445–457.
[94] M. Liebel, J.T. Hugall, N.F. van Hulst, Ultrasensitive label-free nanosensing and high-speed tracking of single proteins, Nano Lett. 17 (2) (2017) 1277–1281.
[95] S.P. Power, et al., Computed tomography and patient risk: Facts, perceptions and uncertainties, World Journal of Radiology 8 (12) (2016) 902–915.
[96] E.Y.P. Lee, M.Y. Ng, P.L. Khong, COVID-19 pneumonia: What has CT taught us? Lancet Infect. Dis. 20 (4) (2020) 384–385.
[97] Y. Pan, et al., Initial CT findings and temporal changes in patients with the novel coronavirus pneumonia (2019-nCoV): A study of 63 patients in Wuhan, China, Eur. Radiol. 30 (6) (2020) 3306–3309.
[98] H. Shi, X. Han, C. Zheng, Evolution of CT manifestations in a patient recovered from 2019 novel coronavirus (2019-nCoV) pneumonia in Wuhan, China. Radiology 295 (1) (2020) 20.
[99] T. Ai, et al., Correlation of chest CT and RT-PCR testing for coronavirus disease 2019 (COVID-19) in China: A report of 1014 cases, Radiology 296 (2) (2020) E32–e40.
[100] A. Bernheim, et al., Chest CT findings in Coronavirus Disease-19 (COVID-19): Relationship to duration of infection, Radiology 295 (3) (2020) 200463.
[101] Y. Fang, et al., Sensitivity of Chest CT for COVID-19: Comparison to RT-PCR, Radiology 296 (2) (2020) E115–e117.
[102] D.J. Bacich, et al., False negative results from using common PCR reagents, BMC Research Notes 4 (1) (2011) 457.
[103] X. Xie, et al., Chest CT for typical coronavirus disease 2019 (COVID-19) pneumonia: Relationship to negative RT-PCR testing, Radiology 296 (2) (2020) E41–e45.
[104] D. Li, et al., False-negative results of real-time reverse-transcriptase polymerase chain reaction for severe acute respiratory syndrome coronavirus 2: Role of deep-learning-based CT diagnosis and insights from two cases, Korean J. Radiol. 21 (4) (2020) 505–508.
[105] X. Wang, F. Li, Y.J.F.i.C. Guo. Recent trends in nanomaterial-based biosensors for point-of-care testing. 8 (2020).
[106] M. Nasrollahzadeh et al. Nanomaterials and nanotechnology-associated innovations against viral infections with a focus on coronaviruses. 10 (6) (2020) P. 1072.
[107] G. Konvalina, H.J.A.o.c.r. Haick, & Sensors for breath testing: From nanomaterials to comprehensive disease detection. 47 (1) (2014) P. 66–76.
[108] R. Batool et al. A review of the construction of nano-hybrids for electrochemical biosensing of glucose. 9 (1) (2019) P. 46.
[109] S.K. Das et al. Synthesis, characterization and catalytic activity of gold nanoparticles biosynthesized with Rhizopus oryzae protein extract. 14 (5) (2012) 1322–1334.
[110] M.S. Draz, H. Shafiee, Applications of gold nanoparticles in virus detection, Theranostics 8 (7) (2018) 1985–2017.
[111] I.P. Pavlova, et al., The rapid coronavirus antibody test: Can we improve accuracy? Frontiers in Medicine 7 (2020) 569.
[112] G. Seo, et al., Rapid detection of COVID-19 causative virus (SARS-CoV-2) in human nasopharyngeal swab specimens using field-effect transistor-based biosensor, ACS Nano 14 (4) (2020) 5135–5142.
[113] P. Moitra, et al., Selective naked-eye detection of SARS-CoV-2 mediated by N gene targeted antisense oligonucleotide capped plasmonic nanoparticles, ACS Nano 14 (6) (2020) 7617–7627.
[114] A.N. Baker, et al., The SARS-COV-2 spike protein binds sialic acids and enables rapid detection in a lateral flow point of care diagnostic device, ACS Central Science (2020) acscentsci.0c00855.

[115] G. Qiu, et al., Dual-functional plasmonic photothermal biosensors for highly accurate severe acute respiratory syndrome coronavirus 2 detection, ACS Nano 14 (5) (2020) 5268–5277.

[116] B.D. Ventura, et al., Colorimetric test for fast detection of SARS-CoV-2 in nasal and throat swabs, ACS Sensors 5 (10) (2020) 3043–3048.

[117] S. Mahari, et al., eCovSens-Ultrasensitive novel in-house built printed circuit board based electrochemical device for rapid detection of nCovid-19 antigen, a spike protein domain 1 of SARS-CoV-2, bioRxiv (2020) 2020.04.24.059204.

[118] C. Huang, et al., Rapid detection of IgM antibodies against the SARS-CoV-2 virus via colloidal gold nanoparticle-based lateral-flow assay, ACS Omega 5 (21) (2020) 12550–12556.

[119] J. Peña-Bahamonde, et al., Recent advances in graphene-based biosensor technology with applications in life sciences, Journal of Nanobiotechnology 16 (1) (2018) 75.

[120] A.K. Srivastava, et al., Potential of graphene-based materials to combat COVID-19: Properties, perspectives, and prospects, Materials Today Chemistry 18 (2020) 100385.

[121] R.M. Torrente-Rodríguez, et al., SARS-CoV-2 RapidPlex: A graphene-based multiplexed telemedicine platform for rapid and low-cost COVID-19 diagnosis and monitoring, Matter 3 (6) (2020) 1981–1998, doi:10.1016/j.matt.2020.09.027.

[122] A. Samavati, et al., Sustainable and fast saliva-based COVID-19 virus diagnosis kit using a novel GO-decorated Au/FBG sensor, Chem. Eng. J. 420 (Part 2, 15) (2020) 127655.

[123] H. Zhao, et al., Ultrasensitive supersandwich-type electrochemical sensor for SARS-CoV-2 from the infected COVID-19 patients using a smartphone, Sens. Actuators B 327 (2021) 128899.

[124] W. Feng et al. Molecular diagnosis of COVID-19: Challenges and research needs. 92 (15) (2020) 10196–10209.

[125] P.P. Liu et al. The science underlying COVID-19: Implications for the cardiovascular system, (2020c).

[126] C.H. Chau et al. COVID-19 clinical diagnostics and testing technology. 40 (8) (2020) 857–868.

[127] S. Flaxman et al. Estimating the effects of non-pharmaceutical interventions on COVID-19 in Europe. 584 (7820) (2020) 257–261.

[128] U. Ghoshal, S. Vasanth, N.J.I.J.o.G. Tejan. A guide to laboratory diagnosis of Corona Virus Disease-19 for the gastroenterologists. (2020), pp. 1–7.

[129] C. Mondal et al. Mitigating the transmission of infection and death due to SARS-CoV-2 through non-pharmaceutical interventions and repurposing drugs, (2020).

[130] B. Fung et al. Direct comparison of SARS-CoV-2 analytical limits of detection across seven molecular assays. 58 (9) (2020).

[131] M.J.T.B.J.o.O. Al-Muharraqi, M. Surgery. Testing recommendation for COVID-19 (SARS-CoV-2) in patients planned for surgery-continuing the service and 'suppressing' the pandemic, (2020).

[132] F. Calabrese, et al., Pulmonary pathology and COVID-19: Lessons from autopsy. The experience of European Pulmonary Pathologists, Virchows Arch. 477 (3) (2020) 359–372.

[133] Z. Li, et al., Development and clinical application of a rapid IgM-IgG combined antibody test for SARS-CoV-2 infection diagnosis, J. Med. Virol. (2020).

[134] A. La Marca, et al., Testing for SARS-CoV-2 (COVID-19): A systematic review and clinical guide to molecular and serological in-vitro diagnostic assays, Reprod. Biomed. Online 41 (3) (2020) 483–499.

[135] S.P. Adhikari et al. Epidemiology, causes, clinical manifestation and diagnosis, prevention and control of coronavirus disease (COVID-19) during the early outbreak period: A scoping review. 9 (1) (2020) 1–12.

[136] S. Cheuk et al. Posterior oropharyngeal saliva for the detection of SARS-CoV-2 (2020).
[137] T. Gulholm et al. Laboratory diagnosis of severe acute respiratory syndrome coronavirus 2 (2020).
[138] K.A. Walsh, et al., SARS-CoV-2 detection, viral load and infectivity over the course of an infection, J. Infect. 81 (3) (2020) 357–371.
[139] M. Abd Elkodous et al., Therapeutic and diagnostic potential of nanomaterials for enhanced biomedical applications. 180 (2019) 411–428.
[140] A. El-Sayed, M. Kamel, Advances in nanomedical applications: Diagnostic, therapeutic, immunization, and vaccine production, Environmental Science and Pollution Research 27 (16) (2020) 19200–19213.
[141] M. Sharifi, et al., Rapid diagnostics of coronavirus disease 2019 in early stages using nanobiosensors: Challenges and opportunities, Talanta 223 (2020) 121704.
[142] Y. Orooji, et al., An overview on SARS-CoV-2 (COVID-19) and other human coronaviruses and their detection capability via amplification assay, chemical sensing, biosensing, immunosensing, and clinical assays, Nano-Micro Letters 13 (1) (2020) 18.
[143] M. Sharifi, et al., Rapid diagnostics of coronavirus disease 2019 in early stages using nanobiosensors: Challenges and opportunities, Talanta 223 (2021) 121704.
[144] M.R. Islam, et al., Genome-wide analysis of SARS-CoV-2 virus strains circulating worldwide implicates heterogeneity, Sci. Rep. 10 (1) (2020) 14004.
[145] R. Samson, G.R. Navale, M.S. Dharne, Biosensors: Frontiers in rapid detection of COVID-19, 3 Biotech 10 (9) (2020) 385.
[146] X. Tang, Xiaolu, Changcheng Wu, Xiang Li, Yuhe Song, Xinmin Yao, Xinkai Wu, Yuange Duan, et al., On the origin and continuing evolution of SARS-CoV-2, Natl. Sci. Rev. 7 (6) (2020) 1012–1023.
[147] L. Li et al. Propagation analysis and prediction of the COVID-19. 5 (2020b) 282–292.
[148] M. Dramé et al. Should RT-PCR be considered a gold standard in the diagnosis of Covid-19? (2020).
[149] J. Rodriguez-Manzano et al. A handheld point-of-care system for rapid detection of SARS-CoV-2 in under 20 minutes (2020).
[150] P.K. Drain, et al., Diagnostic point-of-care tests in resource-limited settings, The Lancet. Infectious Diseases 14 (3) (2014) 239–249.
[151] G. Earl, et al., Pharmacists' role in infectious pandemics: Illustration with COVID-19, Elsevier, Remington, 2020, pp. 849–876.
[152] R. Hulshizer et al. Horizon Scanning COVID-19 Supplement High Impact Report, Volume 1, Issue (2020).
[153] N.J Brendish et al. Clinical impact of molecular point-of-care testing for suspected COVID-19 in hospital (COV-19POC): A prospective, interventional, non-randomised, controlled study (2020).
[154] V. Adebowale et al. Covid-19: Call for a rapid forward looking review of the UK's preparedness for a second wave—an open letter to the leaders of all UK political parties. 369 (2020).
[155] J. Jacobs et al. Implementing COVID-19 (SARS-CoV-2) rapid diagnostic tests in Sub-Saharan Africa: A review. 7 (684) (2020).
[156] Z. Luo, et al., Combating the coronavirus pandemic: Early detection, medical treatment, and a concerted effort by the global community, Research (Washington, D.C.) 2020 (2020) 6925296.
[157] N. Ravi, et al., Diagnostics for SARS-CoV-2 detection: A comprehensive review of the FDA-EUA COVID-19 testing landscape, Biosens. Bioelectron. 165 (2020) 112454.
[158] R. Junker, H. Schlebusch, P.B. Luppa, Point-of-care testing in hospitals and primary care, Deutsches Arzteblatt International 107 (33) (2010) 561–567.
[159] A. St John, The evidence to support point-of-care testing, The Clinical Biochemist. Reviews 31 (3) (2010) 111–119.

[160] A. St John, C.P. Price, Existing and emerging technologies for point-of-care testing, The Clinical Biochemist. Reviews 35 (3) (2014) 155–167.

[161] R. Augustine et al. Loop-mediated isothermal amplification (Lamp): A rapid, sensitive, specific, and cost-effective point-of-care test for coronaviruses in the context of covid-19 pandemic. 9 (8) (2020) 182.

[162] X. Zhu et al. Multiplex reverse transcription loop-mediated isothermal amplification combined with nanoparticle-based lateral flow biosensor for the diagnosis of COVID-19. 166 (2020) 112437.

[163] Y.P. Wong, et al., Loop-mediated isothermal amplification (LAMP): A versatile technique for detection of micro-organisms, J. Appl. Microbiol. 124 (3) (2018) 626–643.

[164] N. Ravi et al. Diagnostics for SARS-CoV-2 detection: A comprehensive review of the FDA-EUA COVID-19 testing landscape. 165 (2020) 112454.

[165] M.C. Smithgall et al. Types of assays for SARS-CoV-2 testing: A review. 51 (5) (2020) e59–e65.

[166] S. Ward, et al., Clinical testing for COVID-19, J. Allergy Clin. Immunol. 146 (1) (2020) 23–34.

[167] N.J. Beeching, T.E. Fletcher, M.B. Beadsworth, Covid-19: Testing times, British Medical Journal Publishing Group (2020).

[168] J. Dinnes et al. Rapid, point-of-care antigen and molecular-based tests for diagnosis of SARS-CoV-2 infection (8) (2020).

[169] B. Heidt, et al., Point of care diagnostics in resource-limited settings, A Review of the Present and Future of PoC in Its Most Needed Environment 10 (10) (2020) 133.

[170] N. Ravi, et al., Diagnostics for SARS-CoV-2 detection: A comprehensive review of the FDA-EUA COVID-19 testing landscape, Biosens. Bioelectron. 165 (2020) 112454.

[171] S. Zhang et al. Nucleic acid testing for coronavirus disease 2019: Demand, research progression, and perspective. (2020) 1–12.

[172] J.-V.R. Goode et al. The pharmacist's role in SARS-CoV-2 diagnostic testing (2020).

[173] L.J. Carter, et al., Assay techniques and test development for COVID-19 diagnosis, ACS central science 6 (5) (2020) 591–605.

[174] B.A. Rabe, C.J.P.o.t.N.A.o.S. Cepko. SARS-CoV-2 detection using isothermal amplification and a rapid, inexpensive protocol for sample inactivation and purification. 117 (39) (2020) 24450–24458.

[175] C.A. Hogan, et al., Comparison of the accula SARS-CoV-2 test with a laboratory-developed assay for detection of SARS-CoV-2 RNA in clinical nasopharyngeal specimens, J. Clin. Microbiol. 58 (8) (2020) e01072–e01120.

[176] M. Osterdahl et al. Detecting SARS-CoV-2 at point of care: Preliminary data comparing Loop-mediated isothermal amplification (LAMP) to PCR (2020).

[177] M.P. Cheng et al. Diagnostic testing for severe acute respiratory syndrome–related coronavirus-2: A narrative review (2020).

[178] C.A. Hogan et al. Comparison of the Accula SARS-CoV-2 test with a laboratory-developed assay for detection of SARS-CoV-2 RNA in clinical nasopharyngeal specimens (2020).

[179] P. Radanliev et al. COVID-19 what have we learned? The rise of social machines and connected devices in pandemic management following the concepts of predictive, preventive and personalized medicine. (2020) pp. 1–22.

[180] Q.-V. Pham et al. Artificial Intelligence (AI) and Big Data for Coronavirus (COVID-19) Pandemic: A Survey on the State-of-the-Arts (2020).

[181] L. Wynants et al. Prediction models for diagnosis and prognosis of covid-19: Systematic review and critical appraisal. 369 (2020).

[182] K. Zhang et al. Clinically applicable AI system for accurate diagnosis, quantitative measurements, and prognosis of covid-19 pneumonia using computed tomography (2020).

[183] J. Chen et al. A survey on applications of artificial intelligence in fighting against covid-19 (2020).

[184] L. Bai et al. Chinese experts' consensus on the Internet of Things-aided diagnosis and treatment of coronavirus disease 2019 (COVID-19). 3 (2020) 7–15.

[185] C.-A. Hu et al. Using a machine learning approach to predict mortality in critically ill influenza patients: A cross-sectional retrospective multicentre study in Taiwan. 10 (2) (2020) e033898.

[186] T. Nguyen, R.G. Gosine, P. Warrian, Digitalization of the oil and gas industry: Practical lessons learned from digital responses during the first stage of the COVID-19 outbreak, in: Proceedings of the Future Technologies Conference, Springer, 2020.

[187] M.H. Chew, et al., Clinical assessment of COVID-19 outbreak among migrant workers residing in a large dormitory in Singapore, J. Hosp. Infect. 106 (1) (2020) 202–203.

[188] L. Li et al. Artificial intelligence distinguishes COVID-19 from community acquired pneumonia on chest CT (2020b).

[189] A. Oulefki et al. Automatic COVID-19 Lung Infected Region Segmentation and Measurement Using CT-Scans Images. (2020) 107747.

[190] D.S. Char, M.D. Abràmoff, C.J.T.A.J.o.B. Feudtner. Identifying ethical considerations for machine learning healthcare applications. 20 (11) (2020) 7–17.

[191] F. Patruno, F. Vitali. Identification of intensive care unit planning criteria during a COVID-19 emergency: ICU reorganizational protocol (2020).

[192] S. Ruth Weissman, Stephanie Bauer, J. Thomas Jennifer, Access to Evidence-Based Care for Eating Disorders during the COVID-19 Crisis 53 (5) (2020) 639–646.

[193] M. Otoom et al. An IoT-based framework for early identification and monitoring of COVID-19 cases. 62 (2020) 102149.

[194] S. Debnath et al. Machine learning to assist clinical decision-making during the COVID-19 pandemic. 6 (1) (2020) 1–8.

[195] A.K. Das, S. Mishra, S.S.J.P. Gopalan. Predicting CoVID-19 community mortality risk using machine learning and development of an online prognostic tool. 8 (2020) e10083.

[196] R. Devika, S.V. Avilala, V. Subramaniyaswamy, Comparative Study of Classifier for Chronic Kidney Disease prediction using Naive Bayes, KNN and Random Forest, in: 2019 3rd International Conference on Computing Methodologies and Communication (ICCMC), IEEE, 2019.

[197] M. Jamshidi et al. Artificial intelligence and COVID-19: Deep learning approaches for diagnosis and treatment. 8 (2020) 109581–109595.

[198] S. Hassantabar et al. Coviddeep: Sars-cov-2/covid-19 test based on wearable medical sensors and efficient neural networks (2020).

[199] F. Soares, A novel specific artificial intelligence-based method to identify COVID-19 cases using simple blood exams, medRxiv (2020) 2020.04.10.20061036.

[200] D. Brinati et al. Detection of COVID-19 infection from routine blood exams with machine learning: A feasibility study (2020).

[201] C. Pawlowski et al. Longitudinal laboratory testing tied to PCR diagnostics in COVID-19 patients reveals temporal evolution of distinctive coagulopathy signatures (2020).

[202] J.S. Obeid et al. An artificial intelligence approach to COVID-19 infection risk assessment in virtual visits: A case report. 27 (8) (2020) 1321–1325.

[203] E. Sirotich et al. Capturing patient-reported outcomes during the COVID-19 pandemic: Development of the COVID-19 global rheumatology alliance patient experience survey (2020).

[204] K.A. Kormuth, S.S.J.N.M. Lakdawala. Emerging antiviral resistance. 5 (1) (2020) 4–5.

[205] G. Chauhan et al. Nanotechnology for COVID-19: Therapeutics and vaccine research. 14 (7) (2020) 7760–7782.

[206] E.J.B.c. Alphandéry. The potential of various nanotechnologies for Coronavirus diagnosis/treatment highlighted through a literature analysis. 31 (8) (2020) 1873–1882.

[207] S. Talebian et al. Nanotechnology-based disinfectants and sensors for SARS-CoV-2. 15 (8) (2020) 618–621.

[208] M.A. Ansari et al. Recent nano-based therapeutic intervention of bioactive sesquiterpenes: Prospects in cancer therapeutics. 26 (11) (2020) 1138–1144.

[209] Y. Li et al. Nano-based approaches in the development of antiviral agents and vaccines. (2020d), p. 118761.

[210] S. Mukherjee et al. Biomedical application, drug delivery and metabolic pathway of antiviral nanotherapeutics for combating viral pandemic: A review. 191 (2020) 110119.

[211] F.A. Khan, Major nano-based products: Nanomedicine, nanosensors, and nanodiagnostics, applications of nanomaterials in human health, Springer, 2020, pp. 211–228.

[212] S. Parveen, N. Yadav, M. Banerjee, Nano-based drug delivery tools for personalized nanomedicine, Nanomaterials and Environmental Biotechnology, Springer, California, USA, 2020, pp. 189–199.

[213] A.P. Singh et al. Targeted therapy in chronic diseases using nanomaterial-based drug delivery vehicles. 4 (1) (2019) 1–21.

[214] Z. Tang, et al., Insights from nanotechnology in COVID-19 treatment, Nano Today 36 (2021) 101019.

[215] J.M. Zuniga, A. Cortes, The role of additive manufacturing and antimicrobial polymers in the COVID-19 pandemic, Expert Review of Medical Devices 17 (6) (2020) 477–481.

[216] S. Iravani et al. Synthesis of silver nanoparticles: Chemical, physical and biological methods. 9 (6) (2014) 385.

[217] Y. Jiao, et al., Quaternary ammonium-based biomedical materials: State-of-the-art, toxicological aspects and antimicrobial resistance, Prog. Polym. Sci. 71 (2017) 53–90.

[218] V. Kandi, S.J.E. Kandi and Health, Antimicrobial properties of nanomolecules: Potential candidates as antibiotics in the era of multi-drug resistance. 37 (2015).

[219] Y. Wang, Y.J.N.l. Xia. Bottom-up and top-down approaches to the synthesis of monodispersed spherical colloids of low melting-point metals. 4 (10) (2004) 2047–2050.

[220] S. Bhattacharjee, et al., Graphene modified multifunctional personal protective clothing, Advanced Materials Interfaces 6 (21) (2019) 1900622.

[221] C. Balagna, et al., Virucidal effect against coronavirus SARS-CoV-2 of a silver nanocluster/silica composite sputtered coating, Open Ceramics 1 (2020) 100006.

[222] J. Zhou, et al., Progress and perspective of antiviral protective material, Advanced Fiber Materials 2 (3) (2020) 123–139.

[223] M. Rai, et al., Nanotechnology-based promising strategies for the management of COVID-19: Current development and constraints, Expert Rev. Anti Infect. Ther. (2020) 1–10.

[224] S.A. Meguid, A. Elzaabalawy, Potential of combating transmission of COVID-19 using novel self-cleaning superhydrophobic surfaces: Part I—protection strategies against fomites, Int. J. Mech. Mater. Des. 16 (3) (2020) 423–431.

[225] S. Galdiero, et al., Silver nanoparticles as potential antiviral agents, Molecules 16 (10) (2011) 8894–8918.

[226] A.R. Chapman et al. High-sensitivity cardiac troponin and the universal definition of myocardial infarction. 141 (3) (2020) 161–171.

[227] E. Sheikhzadeh, et al., Diagnostic techniques for COVID-19 and new developments, Talanta 220 (2020) 121392.

[228] M. Srivastava, et al., Prospects of nanomaterials-enabled biosensors for COVID-19 detection, Sci. Total Environ. 754 (2021) 142363.

[229] M.J.I.J.o.C.M. Bhattacharya, P. Health. India and COVID-19: The Task Ahead. 7 (10) (2020) 4199.

[230] P. Navya et al. Current trends and challenges in cancer management and therapy using designer nanomaterials. 6 (1) (2019) 23.
[231] N.R. Sproston, J.J. Ashworth, Role of C-reactive protein at sites of inflammation and infection, Front. Immunol. 9 (754) (2018).
[232] V. Kumar et al. Development of RNA-based assay for rapid detection of SARS-CoV-2 in clinical samples. bioRxiv, (2020) P. 2020.06.30.172833.
[233] D. Murugan, et al., P-FAB: A fiber-optic biosensor device for rapid detection of COVID-19, Transactions of the Indian National Academy of Engineering 5 (2) (2020) 211–215.
[234] R. Funari, K.-Y. Chu, A.Q. Shen, Detection of antibodies against SARS-CoV-2 spike protein by gold nanospikes in an opto-microfluidic chip, Biosens. Bioelectron. 169 (2020) 112578.
[235] Z. Qin, et al., Fighting COVID-19: Integrated micro- and nanosystems for viral infection diagnostics, Matter 3 (3) (2020) 628–651.
[236] M.A. Lauxmann, N.E. Santucci, A.M.J.I.b.j.u., Autrán-Gómez. The SARS-CoV-2 coronavirus and the COVID-19 outbreak. 46 (2020) 6–18.
[237] P.-E. Fournier et al. Contribution of VitaPCR SARS-CoV-2 to the emergency diagnosis of COVID-19. (2020) P. 104682.
[238] K. Green, et al., Molecular and antibody point-of-care tests to support the screening, diagnosis and monitoring of COVID-19. 2020, The Centre for Evidence-Based Medicine Develops, Promotes and Disseminates (2022).
[239] N. Younes et al. Challenges in laboratory diagnosis of the novel coronavirus SARS-CoV-2. 12 (6) (2020) 582.
[240] Z. Sidiq et al. Laboratory diagnosis of Novel corona virus (2019-nCoV)-present and the future (2020).
[241] A. Harrington et al. Comparison of Abbott ID Now and Abbott m2000 methods for the detection of SARS-CoV-2 from nasopharyngeal and nasal swabs from symptomatic patients (2020).
[242] D. Leightley et al. The King's College London Coronavirus Health and Experiences of Colleagues at King's Study: SARS-CoV-2 antibody response in an occupational sample (2020).
[243] I. Cassaniti et al. Performance of VivaDiag COVID-19 IgM/IgG Rapid Test is inadequate for diagnosis of COVID-19 in acute patients referring to emergency room department (2020).
[244] M. Paret et al. SARS-CoV-2 infection (COVID-19) in febrile infants without respiratory distress (2020).
[245] C. McCormick-Baw et al. Saliva as an alternate specimen source for detection of SARS-CoV-2 in symptomatic patients using Cepheid Xpert Xpress SARS-CoV-2 (2020).
[246] G. Hansen et al. Clinical performance of the point-of-care cobas Liat for detection of SARS-CoV-2 in 20 minutes: A multicenter study (2020).
[247] A. Cortes Aaron, Jorge M.Zuñiga, The use of copper to help prevent transmission of SARS-coronavirus and influenza viruses. A general review, Diagnostic Microbiology and Infectious Disease 98 (4) (2020) 115176. doi:https://doi.org/10.1016/j.diagmicrobio.2020.115176. In this issue.

CHAPTER

10

Electrochemical biosensors for detection of SARS-CoV-2

Yang Liu[a,d], *and Blake N. Johnson*[a,b,c,d]

[a] Department of Industrial and Systems Engineering, Virginia Tech, Blacksburg, VA, USA, [b] Department of Materials Science and Engineering, Virginia Tech, Blacksburg, VA, USA, [c] Department of Chemical Engineering, Virginia Tech, Blacksburg, VA, USA, [d] Macromolecules Innovation Institute, Virginia Tech, Blacksburg, VA, USA

10.1 Introduction

Pathogen detection is an essential application of electrochemical biosensors [1]. Through the integration of selective biorecognition elements with sensitive transducers, electrochemical biosensors have enabled the rapid, sensitive, and selective detection of viruses. While various studies have achieved impressive detection limits, in some cases a single virus or tens to hundreds of viral RNA molecules, the developed approaches for electrochemical detection of virus particles significantly vary in regard to device and measurement approach, such as the electrode, biorecognition element, electrochemical method utilized for transduction of target binding, and measurement format utilized (e.g., sample collection, preparation, and handling protocols). Thus, the reagents, materials, and measurement approach must be carefully considered to accurately assess the utility and time-to-results (TTR) for a given electrochemical biosensor-based assay for pathogen detection in a pandemic setting.

Since the beginning of the ongoing COVID-19 pandemic, several studies have examined the detection of severe acute respiratory syndrome coronavirus 2 (SARS-CoV-2) using electrochemical biosensors. While

Sensing Tools and Techniques for COVID-19: Developments and Challenges in Analysis and Detection of Coronavirus
DOI: https://doi.org/10.1016/B978-0-323-90280-9.00011-0

polymerase chain reaction (PCR)-based assays are the gold standard for SARS-CoV-2 detection (i.e., SARS-CoV-2 antigen and antibody testing), such assays require trained analysts, PCR analyzers, various reagents, and sample preparation and handling steps. Thus, PCR-based assays typically exhibit TTR near 2-4 hours because of a combination of sample preparation and detection time (i.e., the time to prepare the sample vs. the time dedicated to binding of the target analyte to the sensor and the associated electrochemical transduction process used for detection). The clinical and public demand for rapid assays for SARS-CoV-2 antigen and antibody testing as well as mobile real-time screening platforms has led to the investigation of various biosensors for SARS-CoV-2 detection. Among these, electrochemical biosensors have received considerable attention given their synergy with low-cost functional materials for transducer fabrication, fabrication processes, and readout systems, such as miniature impedance analyzers. While all assays for pathogen detection should exhibit high selectivity and low probabilities of false negative and positive results, the demand for safe, user-friendly, and rapid biosensor-based assays for pandemic management significantly constrains the design and measurement format associated with typical biosensors. In particular, it places significant weight on the cost, reliability, simplicity, and safety of the device and measurement approach.

The development of low-cost robust electrochemical biosensors for pandemic management will require investment in research and development. An effective biosensor for use in pandemic management should exhibit a highly stable and selective biorecognition element, safe and user-friendly measurement formats (e.g., sample preparation-free formats), and mobile data acquisition and readout platforms, such as those based on smartphones and miniature analyzers. While it may be possible for experienced analysts and researchers to establish the proof of concept for pathogen detection using electrochemical biosensors in controlled research settings, such as the various molecular targets associated with SARS-CoV-2 antigen and antibody testing, there are various challenges associated with creating robust, low-cost commercial biosensors for pandemic management.

Given their potential for mass production, commercialization, and implementation in mobile, low-cost measurement formats, here, we discuss recent developments in the application of electrochemical biosensors for detection of SARS-CoV-2 (i.e., electrochemical biosensor-based assays for SARS-CoV-2 antigen and antibody testing). In addition to highlighting various electrochemical biosensors that have enabled the detection of SARS-CoV-2, we highlight advances in biosensor design and measurement formats for use in point-of-care and field-based settings. We also highlight emerging areas in the field of electrochemical biosensors for pandemic management and future challenges and directions in applications to SARS-CoV-2 rapid antigen and antibody testing.

COVID-19 disease is caused by SARS-CoV-2 infection. SARS-CoV-2 is a positive-sense single-stranded coronavirus that exhibits structural and molecular characteristics similar to SARS-CoV and Middle East respiratory syndrome coronavirus (MERS-CoV). Thus, the target species associated with SARS-CoV-2 antigen detection (testing) includes the active or inactivated virus, protein-containing viral fragments, or viral RNA. In addition, the detection of SARS-CoV-2 antibody serves as a critical target of interest for antibody testing applications.

Electrochemical biosensors can be classified as biocatalytic or biocomplexing in nature, depending on the type of biorecognition element utilized. A comprehensive review of pathogen detection using electrochemical biosensors can be found elsewhere [1]. As shown in Table 10.1, electrochemical biosensors for SARS-CoV-2 antigen and antibody detection have been primarily based on biocomplexing reactions with immobilized antibodies or single-stranded DNA probes, which are highly selective biorecognition elements. Thus, the majority of electrochemical biosensor-based assays for SARS-CoV-2 detection can be broadly classified as antibody- or DNA-based assays.

As shown in Fig. 10.1 and Table 10.1, SARS-CoV-2 (antigens) and SARS-CoV-2 antibodies have been detected through the recognition of several target species using a variety of transducers, biorecognition elements, and measurement formats [2–27]. For example, the SARS-CoV-2 spike (S) protein, nucleocapsid (N) protein, and glycoproteins have served as the target species for several antibody-based electrochemical biosensing applications to SARS-CoV-2 antigen testing. Monoclonal and polyclonal antibodies against the aforementioned protein targets have been the most commonly used biorecognition element. Monoclonal antibodies exhibit several advantages, including high reproducibility and specificity but may be vulnerable to change of the epitope, such as via S protein mutation. Thus, tracking changes associated with the genome of SARS-CoV-2 is also a critical aspect of developing rapid and selective assays for SARS-CoV-2 detection in addition to vaccine development. For example, the spike protein, a common target of electrochemical biosensor-based assays for SARS-CoV-2 detection has mutated since the onset of the COVID-19 pandemic [9]. Alternatively, polyclonal antibodies are relatively less expensive, exhibit relatively shorter production time and higher stability, and can identify multiple epitopes of a target. However, polyclonal antibodies may exhibit relatively increased batch-to-batch variability. Thus, monoclonal and polyclonal antibodies exhibit advantages and disadvantages as biorecognition elements for use in electrochemical biosensor-based SARS-CoV-2 screening technologies for rapid antigen and antibody testing.

Given their use as targets for PCR-based SARS-CoV-2 antigen testing assays, which remain the gold standard for COVID-19 diagnostics, the S

TABLE 10.1 Classification of electrochemical biosensors for detection of SARS-CoV-2 in terms of target species, sample type, working electrode, biorecognition element, electrochemical method, and limit of detection.

Target species	Sample type	Working electrode	Biorecognition element	Electrochemical method	Limit of detection	Ref.
SARS-CoV-2 RNA	Cell lysate	Au electrode	CRISPR-Cas9	SWV	N/A	[5]
SARS-CoV-2	Transport medium and human cells	Perfluorocarbon SAM-modified Au electrode	Angiotensin converting enzyme 2	EIS	37.8 dC/mL	[23]
SARS-CoV-2	Saliva	Screen printed carbon electrode	SARS-CoV-2 monoclonal antibody	DPV, CV	90 fM	[14]
SARS-CoV-2 N gene	Nasal swab; saliva	Graphene-based Au electrode	Antisense oligonucleotides	N/A	6.9 copies/μL	[2]
SARS-CoV-2	Nasal swab	Graphene-based Au/Cr electrode	SARS-CoV-2 S protein antibody	FET	2.42×10^2 copies/mL (clinical sample)	[19]
N protein, immunoglobulins against SARS-CoV-2 S protein (S1) (S1-IgM and S1-IgG); C-reactive protein (CRP)	Blood; saliva	Graphene electrode	N protein monoclonal antibody; CRP monoclonal antibody; CRP polyclonal antibody; S protein-RBD monoclonal antibody	AMP	N/A	[21]
SARS-CoV-2 S protein	Saliva	Au electrode	Anti-S protein antibody	CA	N/A	[25]
SARS-CoV-2 S protein	Saliva	Shrinky-Dink wrinkled Au electrodes	Aptamer	N/A	1 ag/mL (S1 protein)	[26]
SARS-CoV-2	Nasal swab	Carbon nanofiber-modified screen-printed carbon electrodes	Anti-N protein antibody	SWV	0.8 pg/mL	[6]

(continued on next page)

SARS-CoV-2 N protein	Nasal swab	Poly-m-phenylenediamine (PmPD) modified Au-based thin-film electrodes	N protein imprinted PmPD	DPV	15 fM	[18]
SARS-CoV-2 S protein	Nasal swab; saliva	Cu_2O nanocubes modified screen printed carbon electrode	Anti-S protein monoclonal antibody	CV, EIS	0.04 fg/mL	[16]
SARS-CoV-2 antibody	Serum	ZnO nanowire functionalized paper-based carbon electrode	SARS-CoV-2 S protein receptor-binding domain	EIS	N/A	[12]
SARS-CoV-2 S and N proteins	Saliva	Carbon black-based screen-printed electrode	Monoclonal anti-N protein antibody; polyclonal anti-N protein antibody; Monoclonal anti-S protein antibody; polyclonal anti-S protein antibody	DPV	19 ng/mL (S protein); 8 ng/mL (N protein)	[7]
SARS-CoV-2 N protein	Serum	Screen-printed Au electrode	Anti-SARS-CoV-2 monoclonal N protein antibody	CA	50 pg/mL	[11]
SARS-CoV-2 antibody	Serum	Graphene oxide modified graphene electrode	S protein receptor-binding domain	SWV	1 ng/mL	[24]
SARS-CoV-2 antibody	Serum	Au electrode	S protein receptor-binding domain	EIS	N/A	[17]

(*continued on next page*)

TABLE 10.1 Classification of electrochemical biosensors for detection of SARS-CoV-2 in terms of target species, sample type, working electrode, biorecognition element, electrochemical method, and limit of detection—cont'd

Target species	Sample type	Working electrode	Biorecognition element	Electrochemical method	Limit of detection	Ref.
SARS-CoV-2 S protein; SARS CoV-2	Solution	Graphene electrode	Anti-S protein antibody	SWV	20 μg/mL (S protein), 5.5×10^5 PFU/mL (SARS-CoV-2)	[15]
SARS-CoV-2 S S1 antibody; SARS-CoV-2 S RBD antibody	Solution	Reduced graphene oxide-coated Au micropillar array	SARS-CoV-2 S protein RBD-His protein; SARS-CoV-2 S protein S1-His protein	EIS	2.8×10^{-15} M (spike protein), 16.9×10^{-15} M (spike protein RBD)	[3]
SARS-CoV-2 RNA	N/A	Screen printed carbon electrode	ssDNA capture probe to ORF1ab	DPV	200 copies/mL	[27]
SARS-CoV-2 glycoprotein	Nasal swab; saliva; blood	Glassy carbon electrode	Graphene oxide with sensitive chemical compounds along with Au nanostars	DPV	1.68×10^{-22} μg/mL	[8]
SARS-CoV-2 S protein	Saliva	MXene–graphene	S protein monoclonal antibody	FET	1 fg/mL	[13]
SARS-CoV-2 S and N proteins	Nasal swab	Single-walled carbon nanotube	S protein polyclonal antibody; N protein polyclonal antibody	FET	0.55 fg/mL (spike antigen), 0.016 fg/mL (nucleocapsid antigen)	[20]
MERS-CoV	N/A	AuNPs on carbon electrode	MERS-CoV antigen-antibody complex	SWV; $Fe(CN)_6{}^{3-/4-}$; MERS CoV-antibody complex	400 fg/mL	[10]
SARS-CoV-2 S and N genes	Nasal swab	Screen-printed carbon electrodes	S/N gene specific ssDNA probe	DPV; RCA	1 copy/μL	[4]

Abbreviations: EIS, electrochemical impedance spectroscopy; CV, cyclic voltammetry; SWV, square wave voltammetry; DPV, differential pulse voltammetry; CA, chronoamperometry; FET, field-effect transistor; RCA, rolling circle amplification; Au, gold; AMP, amperometry.

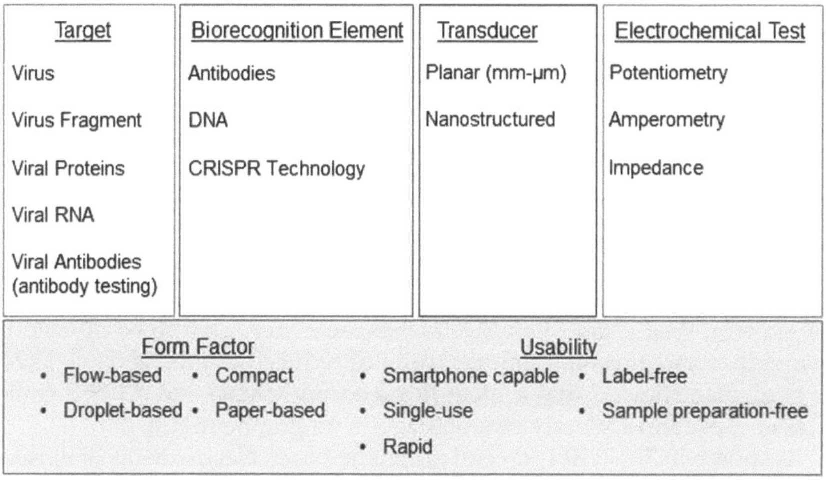

FIGURE 10.1 Components and measurement formats associated with electrochemical biosensors for SARS-CoV-2 antigen and antibody detection.

and N genes of SARS-CoV-2 have been utilized to develop selective single-stranded DNA (ssDNA) probes for electrochemical biosensor-based detection of SARS-CoV-2. A discussion of ssDNA probe design is beyond the scope of this chapter. Commercially available software now exists for probe design and optimization. In addition to antibodies and ssDNA, Clustered Regularly Interspaced Short Palindromic Repeats (CRISPR) technology has recently received attention in the biosensing field as a novel biorecognition element, particularly for nucleic acid sensing applications. We direct the interested reader to various recently published reviews of CRISPR-based biosensors. As shown in Table 10.1, Dai et al. recently designed a CRISPR-based heterogeneous biochemical circuit, which enabled the detection of SARS-CoV-2 genome fragments via electrochemistry [5]. In that study, the target gene fragments were first specifically recognized and transformed by a pair of CRISPR/Cas9 D10A nucleases. The obtained strand structure was then translated into an arbitrary output and amplified into a concatemer via a primer exchange reaction mediated circuit wiring. The output of the heterogeneous biochemical circuit was examined by an electrochemical biosensing platform. The integrated platform was applied for SARS-CoV-2 genome analysis in human cell lysate.

As shown in Table 10.1, various working electrodes and electrode formats have been utilized for the electrochemical detection of SARS-CoV-2. Similar to recent trends in pathogen detection using electrochemical biosensors [1], the majority of electrochemical biosensor-based assays for SARS-CoV-2 antigen and antibody testing have examined planar and nano-structured and -functionalized (Au) electrodes. In particular, graphene-functionalized electrodes have been used in several studies (see Table 10.1). One advantage of graphene is its ability to be integrated with flexible substrates, such as paper and polyimide films [21,24]. Another advantage is the availability of facile bioconjugation techniques for immobilization of biorecognition elements. Several methods exist for protein immobilization on graphene, which could be a SARS-CoV-2 antibody or antigen depending on the application. One method uses 1-pyrenebutyric acid N-hydroxysuccinimide ester or 1-pyrenebutyric acid as linker [15,19,21]. The pyrene group, which contains π-electrons, adsorbs to graphene allowing the carboxylic/ester group to react with available functional groups on the protein. Well-established EDC/NHS chemistry can also be utilized to immobilize proteins on graphene oxide or reduced graphene oxide [3,24] Li et al. also showed that a method based on MXene and APTES could achieve S protein antibody immobilization on graphene [13].

As shown in Table 10.1, several graphene-based electrochemical biosensors have achieved sensitive detection of SARS-CoV-2 using various biorecognition elements and electrochemical methods. For example, Seo et al. immobilized S protein antibody on a graphene-based field effect transistor (FET), which enabled the detection of SARS-CoV-2 viral RNA in clinical samples with a limit-of-detection (LOD) of 2.42 × 10 copies/mL [19]. In another study, Ali et al. fabricated graphene oxide-functionalized aerosol jet nano-printed 3D electrodes for SARS-CoV-2 detection [3]. The sensor-enabled SARS-CoV-2 S1 protein detection at a LOD of 2.8 fM. In addition to the S protein antibody and antigen, specific antisense oligonucleotides targeting the viral N gene were also used for detection of SARS-CoV-2 viral RNA of 6.9 copies/μL using a graphene-based electrochemical biosensor chip.

10.2 Future directions

10.2.1 Rapid and sample preparation-free assays

While Table 10.1 shows that various studies have examined the detection of SARS-CoV-2 antigens and antibodies, the target species often required sample preparation, such as extraction and amplification prior to, and sometimes during, detection. Sample preparation steps were most commonly reported in nucleic acid-based SARS-CoV-2 biosensing applications. While such approaches may provide sensitive and robust assays when performed in controlled laboratory settings by experienced analysts, they present a number of challenges for field and public use. For example, sample handling should be minimized to prevent cross-contamination. Further, the reagents associated with amplification reactions exhibit stability concerns and may impose challenging handling and storage requirements on end users. In contrast, as shown in Table 10.2, several recently developed antibody-based electrochemical biosensor assays for SARS-CoV-2 antigen and antibody detection exhibit sample preparation-free formats. Ultimately, it is desirable to establish low-cost electrochemical biosensors for SARS-CoV-2 antigen and antibody testing that exhibit sample preparation-free measurement formats. It is desirable to avoid sample preparation steps within assays as sample preparation can increase biosafety hazards, TTR, the potential for false results, and assay cost. A list of 'rapid' electrochemical biosensor-based assays for SARS-CoV-2 antigen and antibody testing is provided in Table 10.2. Although the assays in Table 10.2 were classified as 'rapid' given the reported detection time was less than two hours (i.e., the time for target binding to be transduced to a point that the concentration can be identified or quantified), their actual TTR may be significantly increased based on sample preparation requirements, which must be well understood for each assay. In the assessment of electrochemical biosensors for SARS-CoV-2 antigen and antibody testing, we recommend that one should also consider the sample matrix (e.g., type of body fluid), the sample collection method, and the required sample volume, all of which also impact the environmental safety hazards associated with the assay. We recommend that these aspects of biosensor-based assays be described in future studies related to pathogen detection.

10.2.2 Mobile- and smartphone-based measurement platforms

In addition to creating safe, reliable, and user-friendly electrochemical biosensor-based assays for SARS-CoV-2 antigen and antibody detection, there remains a demand for mobile screening platforms. Zhao et al. recently established a smartphone-based electrochemical sensor for SARS-CoV-2 antigen testing [27]. In that study, a SARS-CoV-2 S protein RBD His protein-functionalized reduced-graphene-oxide-coated Au micropillar array electrode was interfaced with a smartphone. The smartphone-based platform enabled detection of SARS-CoV-2 antibody via electrochemical impedance spectroscopy [27].

10.2.3 Mass production of biosensors – Considerations in biosensor design and packaging

While mobile electrochemical biosensing platforms are now emerging for SARS-CoV-2 antigen and antibody testing [27], additional research is required to understand the stability of biorecognition elements used in field-based mobile biosensing applications. Creating highly stable biosensors for field, home, or clinical use is a multi-faceted challenge that will require innovative solutions in electrode and biorecognition layer design, biorecognition element engineering, packaging, and perhaps even

TABLE 10.2 Classification of electrochemical biosensors employing rapid measurement formats.

Target species	Sample collection method and type	Working electrode	Biorecognition element	Electrochemical method	Detection time	Limit of detection	Ref.
S-RBD protein	Protein solution	Cobalt -functionalized TiO2 nanotubes	Cobalt-functionalized TiO2 nanotubes	AMP	~ 30 s	0.7 nM	[22]
SARS-CoV-2 virus	Saliva spiked with Covid-19	Screen printed carbon electrode	SARS-CoV-2 monoclonal antibody	DPV, CV	10 - 30 s	90 fM	[14]
SARS-CoV-2 N-Gene	Nasal swab; saliva	Graphene-based Au electrode	Antisense oligonucleotides	N/A	< 5 m	6.9 copies/μL	[2]
SARS-CoV-2	Nasal swab	Graphene-based Au/Cr electrode	SARS-CoV-2 S protein antibody	FET	< 10 m	2.42×10^2 copies/mL (clinical sample)	[19]
N protein, IGs against SARS-CoV-2 S protein (S1) (S1-IgM and S1-IgG); C-reactive protein (CRP)	Blood; saliva	Graphene electrode	N protein monoclonal antibody; CRP monoclonal antibody; CRP polyclonal antibody; anti-S protein RBD monoclonal antibody	AMP	1 min	N/A	[21]
SARS-CoV-2 S protein	Saliva	Au coating	Anti-S protein antibody	CA	5 min	N/A	[25]
SARS-CoV-2 S protein	Nasal swab, saliva	Cu_2O nanocubes modified screen printed carbon electrode	Anti-S protein monoclonal antibody	CV, EIS	20 min	0.04 fg/mL	[16]

(continued on next page)

SARS-CoV-2 antibody	Serum	ZnO nanowire functionalized paper-based carbon electrode	SARS-CoV-2 S protein receptor-binding domain	EIS	30 min	N/A	[12]
SARS-CoV-2 N protein	Serum	Screen-printed Au electrode	Anti-SARS-CoV-2 monoclonal N protein antibody	CA	< 1 h	50 pg/mL	[11]
S protein; SARS CoV-2	Solution	Graphene electrode	Anti-S protein antibody	SWV	45 min	20 μg/mL (S protein); 5.5×10^5 PFU/mL (SARS-CoV-2)	[15]
SARS-CoV-2 S1 antibody; SARS-CoV-2 Spike RBD antibody	Solution	Reduced graphene oxide coated Au micropillar array	SARS-CoV-2 S protein RBD-His protein; SARS-CoV-2 S protein S1-His protein	EIS	Seconds	2.8×10^{-15} M (spike protein); 16.9×10^{-15} M (spike protein RBD)	[3]
SARS-CoV-2 S protein	Saliva	MXene– graphene	S protein monoclonal antibody	FET	~ 50 ms	1 fg/mL	[13]
SARS-CoV-2 S protein; N protein	Nasal swab	Single-walled carbon nanotube	S protein polyclonal antibody, N protein polyclonal antibody	FET	2 min	0.55 fg/mL (spike antigen); 0.016 fg/mL (nucleocapsid antigen)	[20]
SARS-CoV-2 S gene; SARS-CoV-2 N gene	Nasal swab	Screen-printed carbon electrodes	S/N gene specific ssDNA probe	DPV	< 2 h	1 copy/μL	[4]

Abbreviations: EIS, electrochemical impedance spectroscopy; CV, cyclic voltammetry; Au, gold; AMP, amperometry.

machine learning and artificial intelligence. For example, while ssDNA has enabled selective detection of SARS-CoV-2 via viral RNA, it may be advantageous to consider probes based on alternative oligonucleotide chemistry, such as probes that employ locked nucleic acids or peptide nucleic acids. In addition to considering alternative biorecognition elements, it may be useful to consider materials-based biorecognition technology, such as molecularly-imprinted polymers, as opposed to molecular biorecognition elements (e.g., antibodies, ssDNA, and enzymes). Raziq et al. recently used a nucleocapsid-imprinted poly-m-phenylenediamine (PmPD) biorecognition layer on Au thin-film electrodes for detection of SARS-CoV-2 N protein with a detection limit of 15 fM [18].

10.3 Conclusions

Here, we summarize recent progress in SARS-CoV-2 antigen and antibody detection using electrochemical biosensors. A comprehensive analysis of studies reported since the beginning of the COVID-19 pandemic was provided in terms of transducer design, biorecognition element, electrochemical method, and measurement format. Critical aspects of biosensor and assay design and performance characteristics for pandemic management applications are highlighted including rapid, sample preparation-free, and mobile measurement formats.

References

[1] E. Cesewski, B.N. Johnson, Electrochemical biosensors for pathogen detection, Biosens. Bioelectron. 159 (2020) 112214. http://doi.org/10.1016/j.bios.2020.112214.

[2] M. Alafeef, K. Dighe, P. Moitra, D. Pan, Rapid, ultrasensitive, and quantitative detection of SARS-CoV-2 using antisense oligonucleotides directed electrochemical biosensor chip, ACS Nano 14 (12) (2020) 17028–17045. http://doi.org/10.1021/acsnano.0c06392.

[3] M.A. Ali, C. Hu, S. Jahan, B. Yuan, M.S. Saleh, E. Ju, et al., Sensing of COVID-19 antibodies in seconds via aerosol jet nanoprinted reduced-graphene-oxide-coated 3D electrodes, Adv. Mater. 33 (7) (2021). http://doi.org/10.1002/adma.202006647.

[4] T. Chaibun, J. Puenpa, T. Ngamdee, N. Boonapatcharoen, P. Athamanolap, A.P. O'Mullane, et al., Rapid electrochemical detection of coronavirus SARS-CoV-2, Nat. Commun. 12 (1) (2021). http://doi.org/10.1038/s41467-021-21121-7.

[5] Y. Dai, W. Xu, R.A. Somoza, J.F. Welter, A.I. Caplan, C.C. Liu, An Integrated multi-function heterogeneous biochemical circuit for high-resolution electrochemistry-based genetic analysis, Angewandte Chemie - International Edition 59 (46) (2020) 20545–20551. http://doi.org/10.1002/anie.202010648.

[6] S. Eissa, M. Zourob, Development of a low-cost cotton-tipped electrochemical immunosensor for the detection of SARS-CoV-2, Anal. Chem. 93 (3) (2021) 1826–1833. http://doi.org/10.1021/acs.analchem.0c04719.

[7] L. Fabiani, M. Saroglia, G. Galatà, et al., Magnetic beads combined with carbon black-based screen-printed electrodes for COVID-19: A reliable and miniaturized electrochemical immunosensor for SARS-CoV-2 detection in saliva, Biosens. Bioelectron. 171 (2021). http://doi.org/10.1016/j.bios.2020.112686.

[8] S.A. Hashemi, N.G. Golab Behbahan, S. Bahrani, et al., Ultra-sensitive viral glycoprotein detection NanoSystem toward accurate tracing SARS-CoV-2 in biological/non-biological media, Biosens. Bioelectron. 171 (2021) 112731. http://doi.org/10.1016/j.bios.2020.112731.

[9] Korber, B., Fischer, W. M., & Gnanakaran (2020). et al. Tracking changes in SARS-CoV.;*182*:812–827.

[10] L.A. Layqah, S. Eissa, An electrochemical immunosensor for the corona virus associated with the Middle East respiratory syndrome using an array of gold nanoparticle-modified carbon electrodes, Microchim. Acta 186 (4) (2019) 224. http://doi.org/10.1007/s00604-019-3345-5.

[11] J. Li, P.B. Lillehoj, Microfluidic magneto immunosensor for rapid, high sensitivity measurements of SARS-CoV-2 nucleocapsid protein in serum, ACS Sensors 6 (2021) 1270–1278. http://doi.org/10.1021/acssensors.0c02561.

[12] X. Li, Z. Qin, H. Fu, R. Peng, Z. Li, J.M. Rini, et al., Enhancing the performance of paper-based electrochemical impedance spectroscopy nanobiosensors: An experimental approach, Biosens. Bioelectron. 177 (2021) 112672. http://doi.org/10.1016/j.bios.2020.112672.

[13] Y. Li, Z. Peng, N.J. Holl, et al., MXene-Graphene Field-Effect Transistor Sensing of Influenza Virus and SARS-CoV-2, ACS Omega (2021). http://doi.org/10.1021/acsomega.0c05421.

[14] Mahari, S., Roberts, A., Shahdeo, D., & Gandhi, S. (2004)eCovSens-ultrasensitive novel in-house built printed circuit board based electrochemical device for rapid detection of nCovid-19 antigen, a spike protein domain 1 of SARS-CoV-2. bioRxiv.;.

[15] B. Mojsoska, S. Larsen, D.A. Olsen, J.S. Madsen, I. Brandslund, F.A. Alatraktchi, Rapid SARS-CoV-2 detection using electrochemical immunosensor, Sensors (Switzerland) 21 (2) (2021) 1–11. http://doi.org/10.3390/s21020390.

[16] Z. Rahmati, M. Roushani, H. Hosseini, H. Choobin, Electrochemical immunosensor with Cu2O nanocube coating for detection of SARS-CoV-2 spike protein, Microchim. Acta 188 (3) (2021) 105. http://doi.org/10.1007/s00604-021-04762-9.

[17] M.Z. Rashed, J.A. Kopechek, M.C. Priddy, K.T. Hamorsky, K.E. Palmer, N. Mittal, et al., Rapid detection of SARS-CoV-2 antibodies using electrochemical impedance-based detector, Biosens. Bioelectron. 171 (2021) 112709. http://doi.org/10.1016/j.bios.2020.112709.

[18] A. Raziq, A. Kidakova, R. Boroznjak, J. Reut, A. Öpik, V. Syritski, Development of a portable MIP-based electrochemical sensor for detection of SARS-CoV-2 antigen, Biosens. Bioelectron. 178 (2021) 113029. http://doi.org/10.1016/j.bios.2021.113029.

[19] G. Seo, G. Lee, M.J. Kim, S.-H. Baek, M. Choi, K.B. Ku, et al., Rapid Detection of COVID-19 Causative Virus (SARS-CoV-2) in Human Nasopharyngeal Swab Specimens Using Field-Effect Transistor-Based Biosensor, ACS Nano 14 (4) (2020) 5135–5142. http://doi.org/10.1021/acsnano.0c02823.

[20] W. Shao, M.R. Shurin, S.E. Wheeler, X. He, A. Star, Rapid detection of SARS-CoV-2 antigens using high-purity semiconducting single-walled carbon nanotube-based field-effect transistors, ACS Appl. Mater. Interfaces 13 (2021) 10321–10327. http://doi.org/10.1021/acsami.0c22589.

[21] R.M. Torrente-Rodríguez, H. Lukas, J. Tu, J. Min, Y. Yang, C. Xu, et al., SARS-CoV-2 RapidPlex: A graphene-based multiplexed telemedicine platform for rapid and low-cost COVID-19, Diagnosis and Monitoring Matter 3 (2020) 1981–1998.

[22] B.S. Vadlamani, T. Uppal, S.C. Verma, M. Misra, Functionalized tio2 nanotube-based electrochemical biosensor for rapid detection of SARS-CoV-2, Sensors (Switzerland) 20 (20) (2020) 1–10. http://doi.org/10.3390/s20205871.

[23] Vezza, V. J., Butterworth, A., & Lasserre, P. et al. (2020).An uncomplicated electrochemical sensor combining a perfluorocarbon SAM and ACE2 as the bio-recognition element

to sensitively and specifically detect SARS-CoV-2 in complex samples. ChemRxiv. doi: http://doi.org/10.26434/chemrxiv.13416272.v1

[24] A. Yakoh, U. Pimpitak, S. Rengpipat, N. Hirankarn, O. Chailapakul, S. Chaiyo, Paper-based electrochemical biosensor for diagnosing COVID-19: Detection of SARS-CoV-2 antibodies and antigen, Biosens. Bioelectron. 176 (2021) 112912. http://doi.org/10.1016/j.bios.2020.112912.

[25] H. Yousefi, A. Mahmud, D. Chang, J. Das, S. Gomis, J.B. Chen, et al., Detection of SARS-CoV-2 viral particles using direct, reagent-free electrochemical sensing, J. Am. Chem. Soc. 143 (4) (2021) 1722–1727. http://doi.org/10.1021/jacs.0c10810.

[26] J.A. Zakashansky, A.H. Imamura, D.F. Salgado, H.C.R. Mercieca, R.F.L. Aguas, A.M. Lao, et al., Detection of the SARS-CoV-2 spike protein in saliva with Shrinky-Dink© electrodes, Anal. Methods 13 (7) (2021) 874–883. http://doi.org/10.1039/d1ay00041a.

[27] H. Zhao, F. Liu, W. Xie, T.-C. Zhou, J. OuYang, L. Jin, et al., Ultrasensitive supersandwich-type electrochemical sensor for SARS-CoV-2 from the infected COVID-19 patients using a smartphone, Sens. Actuators B 327 (2021). http://doi.org/10.1016/j.snb.2020.128899.

SECTION 5

Application of Sensing Tools and Techniques

CHAPTER

11

Spatial analysis tools to address the geographic dimension of COVID-19

Ivan Franch-Pardo[a], *Isabel Barea-Navarro*[b], *and Emily Sturdivant*[c]

[a]Universidad Nacional Autónoma de México, Escuela Nacional de Estudios Superiores, Morelia, Michoacan, Mexico, [b]Soil Erosion and Degradation Research Group, Department of Geography, Valencia University, Valencia, Spain, [c]Universidad Nacional Autónoma de México, Centro de Investigaciones en Geografía Ambiental, Morelia, Michoacan, Mexico

11.1 Introduction

Discussions of spatial analysis generally refer to the study of any information that includes a geographic dimension. In other words, it studies information that is spatially identifiable and, thus able to be mapped. This is independent of whether a given analysis is qualitative or quantitative.

The spatial analysis of geographic data with geographic information systems (GIS) seeks to describe, evaluate, and understand processes occurring in geographic space and possibly using that understanding to inform decision-making [1]. The GIS analyst incorporates co-incident variables and other subject matter in spatiotemporal models. GIS is an information system to capture, manage, manipulate, analyze, model, and represent geographic, spatial, and georeferenced data. By working with varied data layers, we can identify correlations between spatially distributed variables. In this work we highlight the diverse use of these tools for studying the COVID-19 pandemic.

Studies of the pandemic in the spheres of medicine, mathematics, social sciences, as well as interdisciplinary synergies include geographic

Sensing Tools and Techniques for COVID-19: Developments and Challenges in Analysis and Detection of Coronavirus
DOI: https://doi.org/10.1016/B978-0-323-90280-9.00014-6

components. The various geographic facets of these studies are addressed by the spatial analysis tools presented in this chapter.

The fields of medicine and health are not new to GIS [2–4]. In the midst of the COVID-19 global health crisis, health systems depend to a large extent on contact tracing to prevent COVID-19 propagation. Within this framework, GIS, and spatial analysis tools have been especially designed for the study and execution of nonpharmaceutical intervention strategies [5].

GIS has numerous uses specific to the field of public health [6–8]. It is used to monitor the epidemiological evolution of the disease and create explanatory maps for distribution and communication. It is involved in the spatial surveillance of the disease and rapid detection, understanding the dynamics of the pandemic, studying and controlling outbreaks, and for identifying problems of a spatial nature. It is used to improve public safety, such as by responding effectively to health emergencies, anticipating logistical needs and distributing resources, and strategizing policy measures. Furthermore, researchers use spatial analysis to contextualize and understand coincident variables and make predictive models.

Specific to this review, we highlight the spatial analysis tools (GIS, spatial statistics, remote sensing, GPS) that have seen the most use in addressing the COVID-19 pandemic. We focus on papers published in the second half of 2020. This allows us to compare with previous work including reviews or reflections on the utility of this technology for pandemic research [5–7,9–14]. Only months ago, due to logistics, novelty and lack of research, we had far less information. We have seen an enormous increase in publications related to the COVID-19 pandemic. In just a few months the number of studies using spatial analysis has greatly multiplied. To conduct this review, we consulted ScienceDirect, Web of Science, and Google Scholar, identifying 114 articles on the topics of COVID-19, GIS, spatial statistics and their specific tools, as are presented in the following pages. The order of appearance was determined based on the respective volume of work addressing COVID-19.

11.2 Spatial analysis tools

11.2.1 Spatial statistics in GIS for COVID-19

One of the most commonly used spatial analysis methods is ordinary least squares (OLS). OLS is a form of global multiple linear regression that minimizes the sum of squares of the vertical distances between the data values and those of the estimated model [15]. It indicates the strength and the average meaning of the independent variables [16] and, as such, is essentially one-size-fits-all [17]. It is has been used for analyses of socioeconomic variables [18–20], urban environments [17,21–23], and social vulnerability [24,25], among others.

Certain cases require refining a geographically weighted regression (GWR) model, using the variables previously included in the OLS regression. In the case of Karaye and Horney [24] , for example, this is caused by the nonstationary vulnerability to COVID-19 among U.S. counties.

GWR regression creates a local model and calculates the parameters for all points of the sample considering the spatial variation in the relationships [26,27]. It considers nonstationary variables (such as climate, demographic factors, and characters of the physical context) and models the local relationships between those predictors and the patterns under study. Although GWR is a useful exploratory technique, its utility as a predictive tool is controversial (Statistics solutions). It facilitates analysis of spatial variation in a phenomenon in a given place [28] , following Tobler's first law of geography (1970) that "everything is related to everything else, but near things are more related than distant things."

Murgante et al. [28] look for geographic parallels between affected areas in the valley region of Po and the metropolitan region of Wuhan. They find that both pollution and land use play important roles in explaining the distribution of COVID-19 in the two regions. Wu et al. [17] use GWR for each study site, basing the local regression on data from adjacent neighborhoods, thus estimating local regression coefficients for each of the predictor variables [16]. Mansour, Al Kindi, Al-Said, Al-Said, and Atkinson [29], in Oman, correlate sociodemographic variables with COVID-19.

Iyanda et al. [18] note that GWR has encountered recent critiques given its inability to address spatial multicolinearity, its inability to mitigate for spatial and atypical autocorrelation, and its bandwidth with a single core that fails to consider the size of geographic units. This last major weakness of GWR, which assumes that all the processes being modeled operate at the same spatial scale, lead to the development of multiscale geographically weighted regression (MGWR) [27,30]. The latter model allows different processes to operate at different spatial scales. Iyanda et al. [18] use MGWR with socio-demographic variables and out-of-pocket expenditure at the global level. Mollalo, Vahedi, and Rivera [31], on the other hand, use it for a local-level examination of the spatial nonstationarity between 35 environmental, socioeconomic, topographic, and demographic variables by U.S. counties. Likewise for European countries, Sannigrahi et al. [19] estimate local spatial correlation coefficients between socio-demographic variables and COVID-19 data.

Among generalized linear models, logistic regression has had special prominence in geographic studies of COVID-19. Poisson distribution describes the expected frequency of a set of probabilities for a discrete variable [32]. Each point of the distribution represents the probability that a given number of events occurs during a time period in a specific place or population. This distribution is typically used to study events with small probabilities or anomalous eventsDiaz Quijano [33].

To explore infection dynamics at the neighborhood level, Harris [34] analyzes differences in the number of COVID-19 deaths across borders of neighboring locations and looks for relationships with other differences in the make-up of neighbors. The assumption is that, all else being equal, nearby locations should have similar mortality levels because they share the same geographic context and because the infection is transmitted through close contact between people. Although based on a simple principle, when applied it induces a clustering effect in the data that is addressed through a multilevel Poisson model.

Yip et al. [23] use the Poisson distribution, among other regression, to evaluate the association between notified confirmed cases and the constructed environment in Hong Kong. Other significant uses of Poisson include studies by Andersen, Harden, Sugg, Runkle, and Lundquist [35], Desjardins et al. [36], and Hohl, Delmelle, Desjardins, and Lan [37]. Their surveillance studies assume that COVID-19 cases have a Poisson distribution and use that to infer high-risk geographic hotspots. Das et al. [38] use Poisson as well as other models to evaluate the impact of living environment in hotspots in the city of Kolkata (India), similar to Harris [34] in London. Sugg, Spaulding, and Lane [39] attempt to determine the driving factors in accumulated COVID-19 cases in nursing homes at local scales.

As can be seen, it is crucial to be able to detect hotspots. Hotspot identification allows a focused approach to assessment and response by local, state, and federal agencies. One of the most common tools to measure local clustering is the Getis-Ord Gi statistic [40] (e.g. [34,39,41–47]). The statistic is calculated by analyzing every entity within the context of neighboring entities. An entity with a high value is relevant, but it may not be a statistically significant hotspot. Following Tobler's first law (1970) [48], the Gi statistic uses a global index to measure the level of spatial autocorrelation, that is, the degree to which objects or activities in a geographic unit are similar to other objects or activities in nearby geographic units [49].

For measuring spatial autocorrelation, Global Univariate Moran's is one of the most used methods (e. g. [17,18,20,28,43,46,50–54]. The global version, Moran's I evaluates the pattern and general tendency of the data. As a global indicator, it overlooks the instability of local spatial processes, which leads to the development of the local version of Moran [55] which identifies both the spatial clustering of entities with similar values and the occurrence of divergent values. This latter version is known as local indicators of spatial association (LISA).

Xie et al. [53] in China, Murgante et al. [28] in the United States, and Santana Juárez et al. [52] in Mexico create LISA cluster maps to analyze the characteristics of local spatial correlation of the COVID-19 epidemic. Wu et al. [17] use LISA to test whether the stationarity of the incidence of COVID-19 in China. Sun, Di, Sprigg, Tong, and Casal [56] use LISA to map COVID-19 risk in the U.S. as well as to identify hotspots related to demographic variables. Mollalo et al. [46] examine the spatial pattern of mortality associated with lower respiratory tract infections and adjusted for age with global and local indices of spatial autocorrelation. They use the Moran's I and the Getis-Ord General G to investigate the extent to which nearby counties have similar rates of lower respiratory tract infections. Yao et al. (2020), analyzing 49 Chinese cities, investigate the associations between concentrations of particulate matter and the COVID-19 fatality rate. Sun et al. [56] use Moran to spatially relate the COVID-19 mortality rates with non-COVID-19 mortality in the United States.

11.2.2 Multicriteria analysis

Our review of correlation methods for multicriteria analysis highlights the importance of the Pearson correlation coefficient. The test statistic measures the strength of association between two continuous variables. It is considered the best method to measure the association between variable of interest because it is based on the covariance method. The statistic indicates the magnitude of correlation as well as the direction of the relationship (Statistics solutions).

Chatterjee et al. [57] measured variables relating social behavior with age groups, socio-political, exposure, and comorbidities. Mollalo et al. [31] use the Pearson correlation coefficient in a U.S. case study to investigate the correlations between socioeconomic, behavioral, environmental, topographic, and demographic variables. Wu et al. [17] use the same to evaluate the relationship between environmental variables and COVID-19. Bherwani, Anjum, and Kumar [58] study population, density, and area by state in India. Tao, Downs, Beckie, Chen, and McNelley [59] analyzed the Pearson coefficient for the correlation between population density and accessibility of COVID-19 testing sites. Using atmospheric data, Coccia [60] and Marquès, Rovira, Nadal, and Domingo [61] correlated atmospheric stability and stability of pollution levels.

While the Pearson coefficient evaluates the linear relationship between two continuous variables (a relationship is linear when a change in one variable is associated with a proportional change in the other variable), the Spearman and Kendall correlation evaluates the monotonic relationship between two continuous or ordinal variables, without considering the linearity of the relationship. Prunet, Lezeaux, Camy-Peyret, and Thevenon [62] use Spearman to extract the correlation between remotely sensed NO_2 data (S5P) and in situ NO_2 data. Pani, Lin, and RavindraBabu [63] correlate COVID-19 with meteorological parameters. Sun et al. [56], in the U.S., design a probability model to estimate the community risk of exposure to pandemics, which uses Spearman in different time intervals, as well as to study the relationship in estimation of risk between neighboring counties. Ran, Zhao, and Han [64] use Spearman to evaluate the relationship between the ozone and the spatial distribution of COVID-19 [62].

Analytic Hierarchy Process (AHP) is a multicriteria decision-making model [65]. It is an additive and compensatory technique of pair-based comparison, based on three principles: decomposition, comparative evaluation, and establishment of priorities. It is a process for identifying, understanding, and evaluating the interactions of a system in a holistic way by providing a scale to measure intangible factors and a method to establish priorities [66]. It is widely used for issues related to environment and health. Requia et al. [66], working in Brazil, establish a hierarchical network for issues of land use, socioeconomic, population, health conditions, and the healthcare system. Mishra et al. [67] use AHP to generate a COVID vulnerability index for urban environments in India and Fang et al. [45] perform a similar process for the island of Xiamén, China. Yao et al. [68] use multicriteria analysis to model the potential distribution of soya crops and how the soya trade has changed since the COVID-19 pandemic.

Following this account of interesting works, Wei, Liu, and Zhu [69] look at the intersection of meteorologic and transportation variables, factors that may play an important role in COVID-19 transmission in continental China. Their study shows that counties crossed by railroads or major highways or that have airports have significantly greater risk of COVID-19, and at the same time, they find that the greatest COVID-19 attack rates are significantly associated with lower average temperature, moderate cumulative precipitation, and greater wind speeds.

According to Karaye and Horney [24], it is essential to address social determinants of health, such as housing, education, and environmental and economic justice in order to reduce the inequities associated with health disasters in the U.S. As such, they use socio-economic, climatic, and PM2.5 data to show that the socio-economic conditions of the infected population profoundly affect the health of socially vulnerable communities. The study finds that minority status and language, household composition and disability, and housing and transportation can be used to predict COVID-19 case counts in U.S. counties. This is consistent with findings by Harris [34] for London neighborhoods. Karaye and Horney [24] position their analysis as further evidence that age, relative wealth, and ethnicity are key risk factors associated with the highest COVID-19 mortality rates. In a study also addressing socio-economic disparities, Cuadros, Xiao, and Mukandavire [70] develop a mathematic model to classify the population by Susceptible-Infected-Hospitalized-Recuperated-Dead (SIHRD).

11.2.3 Geostatistics

Geostatistics refers to a collection of spatial statistical tools and techniques to analyze and predict values of a continuous variable [71]. Here we focus on interpolation methods used for the analysis of COVID-19 spatiotemporal patterns.

Inverse distance weight (IDW) interpolation: Saha, Barman, and Chouhan [72] analyze the impact of the COVID-19 lockdown on community mobility by changing spatiotemporal series in different Indian states. They use IDW to show movement trends before and after the lockdown. They use data from COVID-19 community mobility reports from Google, 2020 to address the spatiotemporal dynamics centered on the social isolation policies, with the use of human mobility variables. Park, Jo, and Cho [73] use IDW to model the concentration of PM 2.5 in the city of Guro-gu (Seoul). Prunet et al. [62] use IDW to map S5P L2 NO2 data in a customized grid that allows them to calculate temporal averages while assuring that the results are equivalent for cities at different latitudes (Paris, Milan, Athens, and Madrid). Their use of IDW minimizes interpolation error by weighting measurements based on their proximity to interpolation points.

Another form of interpolation is kriging. The method is most commonly used for climatic variables [54,6869,74] and common atmospheric contaminants [50,64]. In the latter case, research has used kriging to identify associations between air pollution and COVID-19.

A related technique is cokriging. Kerimray, Baimatova, and Ibragimova [75] use cokriging to map distributions of PM 2.5 and benzene in Almaty in 2018–2019 and 2020, respectively. In particular, they are interested in the effect of government-mandated lockdowns on concentrations of these materials.

Finally, spline interpolations are another technique to use discrete data points to model a continuous variable. Sui, Zhang, and Shang [76] use cubic spline interpolation (CSI) to estimate the second-by-second speed of buses and taxis, using data from vehicle GPS devices.

11.2.4 Other models

Geographically Weighted Principal Component Analysis (GWPCA). GWPCA is an extension of the classic principal components analysis (PCA) that adapts the approach for use with geographic data by considering the spatial heterogeneity of the data [77]. Covariances are weighted based on the distance between the feature object and neighboring features. In essence, GWPCA performs a local PCA using the neighborhood surrounding each spatial feature [38]. Basu et al. [42] note that GWPCA does not require additional data to weight the variables. It is successfully applied in a wide range of fields such as economic development modeling, deprivation, pollution sources, water resources, environmental health, soil characteristics, and landslides. Das et al. [38] use GWPCA in developing their improved Index of Multiple Deprivations (IMD) in districts of Calcutta, using housing condition, household amenities, water, sanitation, and hygiene (WaSH), asset possession, and gender disparity.

Voronoi: Bherwani et al. [58] use Bayesian probabilistic modeling to understand the relationship between COVID-19 cases and population density in a given region together with GIS-based Voronoi diagram to identify high-risk areas. Subsequently, Thiessen polygons delineate the risk zone boundaries.

Spatial autocorrelation from Clifford, Richardson, and Hemon [78]: Nomura, Yoneoka, and Shi [79], in the prefecture of Fukuoka, Japan, execute this spatial autocorrelation analysis to relate the number of PCR-confirmed COVID-19 cases to a social network application that provides real-time monitoring of self-reported COVID-19 symptoms. The highly significant correlation they report indicates the utility for these methods to structure epidemiological evaluation and assist in policy evaluations such as emergency declarations.

Spatial error model (SEM) and spatial lag model (SLM): Iyanda et al. [18]; Maiti et al. [27]; Mollalo et al. [31]; Sannigrahi et al. [19]; Sun et al. [20].

Spatial autoregressive combined (SAC): Sun et al. [20]; Zulkarnain and Ramadani [80]

Self-organizing maps (SOM), also referred to as Kohonen [81]

Spatial Markov [82]

Birthday paradox [56]

Topological Weighted Centroid (TWC) by Buscema, Della Torre, Breda, Massini, and Grossi [83], a new algorithm used in Italy

Space-time scan statistic from Kulldorff [84]: Andersen et al. [35]

Spatiotemporal refined risk model [85].

Geodetector: Wu et al., [17], Xie et al. [53].

Random forest: Kerimray et al. [75]; Mollalo et al. [46]; Pourghasemi, Pouyan, and Heidari [86].

11.2.5 Remote sensing (RS) and unmanned aerial vehicle (UAV)

By Earth's spheres (integrated biosphere):

The atmosphere:

The subject of relevant atmospheric analyses tends to fall into one of two categories: climatic conditions or pollution. Regarding the spatiotemporal dynamics of climatic conditions, these are analyzed at all scales

[17,54,63,87–92]. About spatiotemporal analyses of pollution levels during the pandemic, these tend to focus on the consequence of reductions in mobility and economic production [17,91,93,94]. There is special attention paid to countries that adopted lockdowns to control the pandemic, as well as to issues specifically affecting urban areas.

In the lithosphere:

Land use change, with significant examples such as the case of Yao et al. [68] where analyze the potential distribution of soya crops and how COVID-19 has affected the soya market. Brancalion et al. [95] look at a possibly faster pace of tropical deforestation during 2020. Wang, Peng, and Yu [96] conduct a land use analysis of crops in China to evaluate whether the pandemic led to increased cultivation and land exploitation; their findings are relevant for the development of agricultural policy to guarantee food security. RS and UAV also have applications for the real estate market as they are used for property inspection [97]

Analysis in urban environments associated with the level of pollutants they emit [38,62]. Lighting is considered an indicator of economic recovery or recession, as in the case of several Chinese cities, where lighting levels are compared during peak closure with the same variable a year earlier [98].

Research on the dynamics of human movement during the pandemic demonstrates important applications of RS and UAV. Minetto, Segundo, Rotich, and Sarkar [99] employ a deep learning technique to automatically detect objects, such as cars and aircraft from satellite imagery. They suggest that the ability to automatically identify these objects in image time series will allow for temporal analysis of societal indicators [99]. Similarly, Wu et al. [17] use deep learning to identify vehicles within Wuhan, China from RS imagery and thus evaluate the effect of a transportation ban on the city. In a very different type of analysis, Okyere, Chuku, and Ekumah [100] use UAV to monitor fishing boats and assess adherence the effects of physical distancing mandates and risks of exposure in the fishing sector.

There are also studies in areas with armed conflict as in the case of Syria [101]. Using RS and spatial models, the authors design the "Risk of Vulnerability to COVID-19 in War Zones Index" (Id_Covid19_WZ) to identify areas that are vulnerable to the pandemic and thus help decision-makers to limit risk and avoid and/or manage widespread infection.

The hydrosphere, with special attention to lockdowns:

Addressing the concepts of rivers and pollution: changes in water quality as a result of a major lockdown are evaluated in both the Ganges River [102] and the Sabarmati River [103] in India using Lansat 8 and Sentinel-2 imagery, respectively. Similar methods are applied to measure lake pollution, such as by Yunus, Masago, and Hijioka [104] who use Landsat 8 to measure water turbidity in Vembanad Lake in India, finding an improvement in water quality as a result of lockdown.

In another hemisphere, Sentinel-2 and Sentinel-3 imagery was applied to detect a harmful algal bloom (HAB) in salmonid aquaculture in Chile. The analysis technique combined with rapid delivery of the high-resolution satellite imagery allowed for near real-time monitoring and decision-making at a time when in situ sampling was restricted by a mandated lockdown [105].

The tools are also highly relevant to topics related to ocean economic activities. For example, UAV was used in Ghana to monitoring water-based activity during COVID-19, in order to provide solid scientific evidence as a basis for decision-making in the artisanal fishing sector [100].

11.2.6 Participative GIS, mapping, voluntary geographic information (VGI), public participation GIS (PPGIS)

Green areas in the cities, nature, public spaces play an important role in times of crisis [106,107]. Public participation GIS (PPGIS) and Voluntary geographic information (VGI) are important tools, in addition to existing information sources, for gathering data from the population to fight COVID-19. Gorayeb, de Oliveira, and da Cunha [108] describes information gathering through citizen surveys in Fortaleza, Brazil. The information provided by the population through the surveys exhibits similarities with data provided by official maps, suggesting that these are promising tools for rapid data collection. In Israel, an online questionnaire was carried out to identify possible symptoms and, with this, to follow up with infected persons over time [109]. In the interpretation of the data, differences in the proportion of reported symptoms in participants from different cities and different neighborhoods that are geographically close to each other are revealed, which could suggest the ability to detect changes at a high geographical resolution.

The implementation of PPGIS in Greece [110] and Turkey [111] during the spring of 2020 was motivated by the need to rapidly acquire data based on location. These studies find that crowdsourcing applications are important tools for real-time mapping and monitoring to allow health authorities to make decisions and design effective management approaches [110,112].

11.2.7 GPS networks

Throughout this overview, we have detailed numerous works employing GPS. In the following section, we focus on data mining and analysis of communication and transportation networks.

Studies using data mining to study human mobility tend to focus on areas where lockdowns were enacted. We differentiate these studies by their inputs: a) cellular telephone data [113–117, 74,118]; b) data from

Google, Facebook, Apple, and Baidu, and geotagged tweets from Twitter [119–125]; c) data from bicycle share systems in urban contexts [126] ; d) geo-tagged data from identified infections compiled into travel histories [127].

On the topic of transportation networks, we emphasize analyses of the movement or trajectory of the following: vessels during lockdown in Venice, Italy [128]; public transport and taxis in Qingdao, China [76] ; transportation in Milano, Italy [129].

Other studies are centered on highway networks in Jaipur, India [130] or in Tarragona province, Spain [131] , where there was a significant decline in capacity during the lockdown. Furthermore, Silalahi, Hidayat, Dewi, Purwono, and Oktaviani [132] for Jakarta, Indonesia, and Alasadi, Aziz, and Dhiya [133] for Basora, Iraq, provide important analyses of hospital accessibility.

11.2.8 Web mapping

Thanks to the availability of consistently-updated open data and to the availability of web cartographic templates, plugins, and shared code, there are many geospatial on-line platforms for monitoring COVID-19 around the world [134], which have emerged from organizations, academic institutions, or media organizations.

Here, we present a specific example in Mexico and briefly explain the operation of the university-based platform maintained by the National Autonomous University of Mexico (UNAM) (https://covid19.ciga.unam.mx/). The platform is supported by a package written in R, which runs a daily process to download the current COVID-19 data published by the federal government. The system first cross-checks the data publication date and downloads the database for the current day from the page titled Open Data maintained by the Health Agency.

The data are combined with spatial data for the geographic regions (states and municipalities) and with 2020 population counts for the given regions. Next, the following statistics are calculated at the national, state, and municipal level: total recovered cases, active cases, deaths, and accumulated cases. Combining these totals with the population counts, the process computes rates of incidence, mortality, and fatality. The absolute change between consecutive weeks is calculated for weekly values for positive cases, hospitalizations, and deaths, as well as the weekly change in incidence and mortality rates, and the percent positivity rate [134]. These geographic data are mapped and published through an ArcGIS Online web dashboard using the template created by Dong, Du, and Gardner [135].

In recent months, other notable graphics have been produced using cartograms to represent COVID-19 data world-wide [136], China [137,138], USA [139], and Europe [140].

11.3 Discussion and conclusions

During the first half of 2020, Tobler's first law of geography was not apparent in analyzing COVID-19 at a global scale. The disease evolution indicated by the constellation of the epicenters on the global map had more resemblance to Lévy flight [141], in other words randomly distributed, where human movement seemed to be the sole driving factor in the spatial distribution and intensity of COVID-19. In fact, the dominance of this factor has changed little, but, by reviewing the works from the second half of the year in combination as we have here, we see more evidence of Tobler's law. We find that cartographic scale and enhanced resolution of the units of analysis uncovers the role of geography in disease patterns. Physio-geographic dynamics have a greater role in disease distribution at more local levels. As such, to facilitate the applicability of spatial analysis to decision-making, information must be collected and made available with high spatial resolution. Detailed datasets allow the design of targeted management strategies with greater chances of limiting chains of infection. The likelihood of significantly reducing spread increases further if high resolution data is paired with fieldwork [134]. Whereas aggregated spatial information, even at the municipal level, often serves only anecdotally. Although it may be useful to inform the general public, it offers minimal utility in containing the pandemic. To address this, many studies working at the city level use alternative data sources, such as VGI, mobility data, and remote sensing to break through this limiting factor.

From anthropogenic, economic, and social perspectives, there are conclusions to be drawn related to Tobler's concentual umbrella [48]. In fact, spatial statistics models are among the most widely used and thus could be considered among the most commonly used tools for studying the COVID-19 pandemic during the second half of 2020. Spatial correlation and autocorrelation, and multicriteria analysis are used in the greatest number of GIS-based studies among all the disciplines studying the pandemic, although we find that socio-economic variables have become more common in relation to the body work when compared to the first half of the year [10]. Specifically there has been a significant increase in the production of vulnerability maps of COVID-19 and urban environments.

PR has also exhibited outstanding production. Air pollution thematic has concentrated the largest number of studies. On the other hand, in the previous semester, the main objective of analysis was the climate and its relationship with COVID-19. In this regard, some studies affirm that the climate played a greater role in the first phase of the pandemic, and that its impact has dissipated in the later phases [142].

Web mapping continues to be a principal medium for disseminating public information about COVID-19. Consistent with Chatterjee et al. [57], public health organizations and governments advise many preventive measures such as social distancing and personal hygiene, but one of the dominant strategies remains communicating risks and generating consciousness to break chains of infection. Web maps provide an eloquent means to effect these strategies.

Spatiotemporal analysis of before and during lockdowns was the topic most developed in the second half of 2020. Although not all countries implemented lockdowns, but to a greater or lesser degree, all have reduced the mobility of their residents as well as their economic and production activities. In the geographic dimension of the pandemic, lockdown dynamics were among the most analyzed phenomena, studied by all the disciplines with spatial analyses of COVID-19.

As can be seen, most spatial analyzes use administrative boundaries as their units of studies. From a global perspective, nation-states dictate the management of the pandemic and borders become the land, air and water ports of entry, which fulfill a series of functions (legal, control, security, health) that are decisive in time of crisis: terrorist threats or armed conflict and, in 2020, a pandemic [143]. Mobility is determined by state power and, as the spread of the virus is dictated by the main axes of air transport, the stoppage of more than 90% of traffic is the first effect of this policy of interrupting chains of contagion [143].

On the other hand, the global pandemic is inherently metropolitan in character [67] as proven by the large number of reviewed works in urban environments around the world. The publication by the UN-Habitat of a Response Plan for the mitigation of the externalities based on SARS-CoV-2 in global cities is further evidence. The report expresses concern for the urban nature of the pandemic and thus highlights the role of cities in disease transmission [67].

In this second half of the year, the most spatial analyses are conducted within the territories of China and the U.S. In comparison with the previous six months [10], India accounted for the greatest increase in spatial analysis studies whereas global-scale studies have most declined. There continue to be many studies based in Brazil, Iran, and Italy. Although there are notable studies based in Egypt, Ghana, Morocco, Nigeria, and Oman, we observe a powerful gap in contributions for the whole of Africa, where there are still many unknowns to solve.

GIS has served to monitor, evaluate situations, predict events, and inform policy decisions, all while the global populace awaits a vaccine. We expect that the economic dynamics and study topics will once again change in 2021, especially as vaccines begin to be administered and induce further changes around the globe.

References

[1] T.C. Bailey, A.C. Gatrell, Interactive spatial data analysis, Essex: Longman Scientific & Technical 413 (8) (1995).

[2] R. Kearns, G. Moon, From medical to health geography: Novelty, place and theory after a decade of change, Progress in Human Geography 26 (5) (2002) 605–625. http://doi.org/10.1191/0309132502ph389oa.
[3] M.S. Meade, M. Emch, Medical geography. New York: Guilford Press (2010).
[4] N.S. Walford, Demographic and social context of deaths during the 1854 cholera outbreak in Soho, London: A reappraisal of Dr John Snow's investigation, Health Place 65 (2020). http://doi.org/10.1016/j.healthplace.2020.102402.
[5] E. Mbunge, Integrating emerging technologies into COVID-19 contact tracing: Opportunities, challenges and pitfalls, Diabetes and Metabolic Syndrome: Clinical Research and Reviews 14 (6) (2020) 1631–1636. http://doi.org/10.1016/j.dsx.2020.08.029.
[6] M.N.K. Boulos, M.E. Geraghty, Geographical tracking and mapping of coronavirus disease COVID-19/severe acute respiratory syndrome coronavirus 2 (SARS-CoV-2) epidemic and associated events around the world: How 21st century GIS technologies are supporting the global fight against outbreaks and epidemics, International Journal of Health Geographics (2020). http://doi.org/10.1186/s12942-020-00202-8.
[7] J. Dangermond, Pesaresi C. Vito, Using GIS in the time of the COVID-19 crisis, casting a glance at the future. A joint discussion, J-Reading-Journal of Research and Didactics in Geography 1 (2020).
[8] Buzai G.D.De Wuhan a Luján, Evolución espacial del COVID-19, Posición, Núm 3 (2020).
[9] J.T. Devasia, S. Lakshminarayanan, S.S. Kar, How modern geographical information systems based mapping and tracking can help to combat severe acute respiratory syndrome coronavirus 2 (SARS-CoV-2) pandemic around the World and India, International Journal of Health Systems and Implementation Research 4 (1) (2020) 30–54.
[10] I. Franch-Pardo, B.M. Napoletano, F. Rosete-Verges, L. Billa, Spatial analysis and GIS in the study of COVID-19. A review, Sci. Total Environ. 739 (2020). http://doi.org/10.1016/j.scitotenv.2020.140033.
[11] B. Radojević, L. Lazić, M. Cimbaljević, Rescaling Smart Destinations – The Growing Importance of Smart Geospatial Services during and after COVID-19 Pandemic, Geographica Pannonica 24 (3) (2020) 221–228. http://doi.org/10.5937/gp24-2800.
[12] L. Rosenkrantz, N. Schuurman, N. Bell, O. Amram, The need for GIScience in mapping COVID-19, Health Place (2020). http://doi.org/10.1016/j.healthplace.2020.102389.
[13] C.D. Smith, J. Mennis, Peer Reviewed: Incorporating Geographic Information Science and Technology in Response to the COVID-19 Pandemic, Preventing Chronic Disease (2020).
[14] C. Zhou, F. Su, T. Pei, et al., COVID-19: Challenges to GIS with Big Data, Geography and Sustainability (2020) 77–87. http://doi.org/10.1016/j.geosus.2020.03.005.
[15] J.E. Hanke, A.G. Reitsch, D.W. Wichern, Business Forecasting. 9 (2001).
[16] E. Delmelle, M. Hagenlocher, S. Kienberger, I. Casas, A spatial model of socioeconomic and environmental determinants of dengue fever in Cali, Colombia, Acta Trop. 164 (2016) 169–176. http://doi.org/10.1016/j.actatropica.2016.08.028.
[17] X. Wu, J. Yin, C. Li, H. Xiang, M. Lv, Z. Guo, Natural and human environment interactively drive spread pattern of COVID-19: A city-level modeling study in China, Sci. Total Environ. (2020).
[18] A.E. Iyanda, R. Adeleke, Y. Lu, et al., A retrospective cross-national examination of COVID-19 outbreak in 175 countries: A multiscale geographically weighted regression analysis (January 11-June 28, 2020), Journal of Infection and Public Health 13 (10) (2020) 1438–1445. http://doi.org/10.1016/j.jiph.2020.07.006.
[19] S. Sannigrahi, F. Pilla, B. Basu, A.S. Basu, A. Molter, Examining the association between socio-demographic composition and COVID-19 fatalities in the European region using spatial regression approach, Sustainable Cities and Society 62 (2020). http://doi.org/10.1016/j.scs.2020.102418.

[20] F. Sun, S.A. Matthews, T.C. Yang, M.H. Hu, A spatial analysis of the COVID-19 period prevalence in U.S. counties through June 28, 2020: Where geography matters? Ann. Epidemiol. (2020) 54–59 e1. http://doi.org/10.1016/j.annepidem.2020.07.014 .

[21] A. Stier, M. Berman, L. Bettencourt, COVID-19 attack rate increases with city size, SSRN (2020). https://ssrn.com/abstract=3564464.

[22] H. Tian, Y. Liu, Y. Li, et al., An investigation of transmission control measures during the first 50 days of the COVID-19 epidemic in China, Science 368 (6491) (2020) 638–642. http://doi.org/10.1126/science.abb6105.

[23] T.L. Yip, Y. Huang, C. Liang, Built environment and the metropolitan pandemic: Analysis of the COVID-19 spread in Hong Kong, Build. Environ. (2020). http://doi.org/10.1016/j.buildenv.2020.107471.

[24] I.M. Karaye, J.A. Horney, The impact of social vulnerability on COVID-19 in the U.S.: An analysis of spatially varying relationships, Am. J. Prev. Med. 59 (3) (2020) 317–325. http://doi.org/10.1016/j.amepre.2020.06.006.

[25] B.F. Snyder, V. Parks, Spatial variation in socio-ecological vulnerability to Covid-19 in the contiguous United States, Health Place (2020) 102471. http://doi.org/10.1016/j.healthplace.2020.102471.

[26] C. Brunsdon, A.S. Fotheringham, M.E. Charlton, Geographically weighted regression: A method for exploring spatial nonstationarity, Geographical Analysis 28 (4) (1996) 281–298. http://doi.org/10.1111/j.1538-4632.1996.tb00936.x.

[27] A. Maiti, Q. Zhang, S. Sannigrahi, S. Pramanik, S. Chakraborti, F. Pilla, Spatiotemporal Effects of the Causal Factors on COVID-19 Incidences in the Contiguous United States (2020).

[28] B. Murgante, G. Balletto, G. Borruso, Casas Las, P. Castiglia, M. Dettori, Geographical analyses of Covid-19's spreading contagion in the challenge of global health risks, TeMA-Journal of Land Use, Mobility and Environment. (2020) 283–304.

[29] S. Mansour, A. Al Kindi, A. Al-Said, A. Al-Said, P. Atkinson, Sociodemographic determinants of COVID-19 incidence rates in Oman: Geospatial modelling using multiscale geographically weighted regression (MGWR), Sustainable Cities and Society (2020) 102627. http://doi.org/10.1016/j.scs.2020.102627.

[30] A.S. Fotheringham, W. Yang, W. Kang, Multiscale Geographically Weighted Regression (MGWR), Annals of the American Association of Geographers 107 (6) (2017) 1247–1265. http://doi.org/10.1080/24694452.2017.1352480.

[31] A. Mollalo, B. Vahedi, K.M. Rivera, GIS-based spatial modeling of COVID-19 incidence rate in the continental United States, Sci. Total Environ. 728 (2020). http://doi.org/10.1016/j.scitotenv.2020.138884.

[32] A. Lovett, R. Flowerdew, Analysis of count data using Poisson regression, Professional Geographer 41 (2) (1989) 190–198. http://doi.org/10.1111/j.0033-0124.1989.00190.x.

[33] F.A. Diaz Quijano, Regresiones aplicadas al estudio de eventos discretos en epidemiología. Revista de la Universidad Industrial de Santander Salud. (2016) 9–15. doi: http://doi.org/10.18273/revsal.v48n1-2016001

[34] R. Harris, Exploring the neighbourhood-level correlates of Covid-19 deaths in London using a difference across spatial boundaries method, Health Place 66 (2020). http://doi.org/10.1016/j.healthplace.2020.102446.

[35] L.M. Andersen, S.R. Harden, M.M. Sugg, J.D. Runkle, T.E. Lundquist, Analyzing the spatial determinants of local Covid-19 transmission in the United States, Sci. Total Environ. (2020) 754.

[36] MR Desjardins, A Hohl, EM. Delmelle, Rapid surveillance of COVID-19 in the United States using a prospective space-time scan statistic: Detecting and evaluating emerging clusters, Appl. Geogr. (2020) 102202. http://doi.org/10.1016/j.apgeog.2020.102202.

[37] A. Hohl, E.M. Delmelle, M.R. Desjardins, Y. Lan, Daily surveillance of COVID-19 using the prospective space-time scan statistic in the United States, Spatial and Spatio-temporal Epidemiology 34 (2020). http://doi.org/10.1016/j.sste.2020.100354.

[38] A. Das, S. Ghosh, K. Das, T. Basu, I. Dutta, M. Das, Living environment matters: Unravelling the spatial clustering of COVID-19 hotspots in Kolkata megacity, India, Sustainable Cities and Society. (2020). http://doi.org/10.1016/j.scs.2020.102577.

[39] M.M. Sugg, T.J. Spaulding, S.J. Lane, et al., Mapping community-level determinants of COVID-19 transmission in nursing homes: A multi-scale approach, Sci. Total Environ. (2020) 752.

[40] J.K. Ord, A. Getis, Local Spatial Autocorrelation Statistics: Distributional Issues and an Application, Geographical Analysis 27 (4) (1995) 286–306. http://doi.org/10.1111/j.1538-4632.1995.tb00912.x.

[41] G.E. Barboza, L.B. Schiamberg, L. Pachl, A spatiotemporal analysis of the impact of COVID-19 on child abuse and neglect in the city of, Child Abuse Negl. (2020).

[42] T. Basu, A. Das, S. Pal, Application of geographically weighted principal component analysis and fuzzy approach for unsupervised landslide susceptibility mapping on Gish River Basin, India. Geocarto International. (2020). http://doi.org/10.1080/10106049.2020.1778105.

[43] C.F. Baum, M. Henry, Socioeconomic Factors influencing the Spatial Spread of COVID-19 in the United States, SSRN Electronic Journal (2020). http://doi.org/10.2139/ssrn.3614877.

[44] E. Bopp, J. Douvinet, Spatial performance of location-based alerts in France, International Journal of Disaster Risk Reduction 50 (2020). http://doi.org/10.1016/j.ijdrr.2020.101909.

[45] L. Fang, J. Huang, Z. Zhang, V. Nitivattananon, Data-driven framework for delineating urban population dynamic patterns: Case study on Xiamen Island, China. Sustainable Cities and Society. 62 (2020). http://doi.org/10.1016/j.scs.2020.102365.

[46] A. Mollalo, B. Vahedi, S. Bhattarai, L.C. Hopkins, S. Banik, B. Vahedi, Predicting the hotspots of age-adjusted mortality rates of lower respiratory infection across the continental United States: Integration of GIS, spatial statistics and machine learning algorithms, Int. J. Med. Informatics 142 (2020). http://doi.org/10.1016/j.ijmedinf.2020.104248.

[47] A.M. Oster, G.J. Kang, A.E. Cha, et al., Trends in number and distribution of COVID-19 hotspot counties - United States, March 8-July 15, 2020, Morbidity and Mortality Weekly Report 69 (33) (2020) 1127–1132. http://doi.org/10.15585/MMWR.MM6933E2.

[48] W.R. Tobler, A Computer Movie Simulating Urban Growth in the Detroit Region, Econ. Geogr. 46 (1970) 234–240. http://doi.org/10.2307/143141.

[49] M.F. Goodchild, A spatial analytical perspective on geographical information systems, International Journal of Geographical Information Systems (1987) 327–334. http://doi.org/10.1080/02693798708927820.

[50] G. Huang, P.E. Brown, Population-weighted exposure to air pollution and COVID-19 incidence in Germany, Spatial Statistics (2020).

[51] D. Kang, H. Choi, J. Kim, J. Choi, Spatial epidemic dynamics of the COVID-19 outbreak in China, Int. J. Infect. Dis. (2020) 96–102. http://doi.org/10.1016/j.ijid.2020.03.076.

[52] M.V. Santana Juárez, G.S. Castañeda, C. Carillo, R. Carrillo, R. Alcántara, COVID-19 en México: Asociación espacial de cara a la fase tres, Hygeia - Revista Brasileira de Geografia Médica e da Saúde (2020) 36–48. http://doi.org/10.14393/hygeia0054317.

[53] Z. Xie, Y. Qin, Y. Li, W. Shen, Z. Zheng, S. Liu, Spatial and temporal differentiation of COVID-19 epidemic spread in mainland China and its influencing factors, Sci. Total Environ. (2020) 140929. http://doi.org/10.1016/j.scitotenv.2020.140929.

[54] Y. Yao, J. Pan, W. Wang, et al., Association of particulate matter pollution and case fatality rate of COVID-19 in 49 Chinese cities, Sci. Total Environ. 741 (2020). http://doi.org/10.1016/j.scitotenv.2020.140396.
[55] L. Anselin, Local indicators of spatial association—LISA, Geographical Analysis 27 (2) (1995) 93–115. http://doi.org/10.1111/j.1538-4632.1995.tb00338.x.
[56] Z. Sun, L. Di, W. Sprigg, D. Tong, M. Casal, Community venue exposure risk estimator for the COVID-19 pandemic, Health Place 66 (2020). http://doi.org/10.1016/j.healthplace.2020.102450.
[57] R. Chatterjee, S. Bajwa, D. Dwivedi, R. Kanji, M. Ahammed, R. Shaw, COVID-19 Risk Assessment Tool: Dual application of risk communication and risk governance, Progress in Disaster Science 7 (2020). http://doi.org/10.1016/j.pdisas.2020.100109.
[58] H. Bherwani, S. Anjum, S. Kumar, et al., Understanding COVID-19 transmission through Bayesian probabilistic modeling and GIS-based Voronoi approach: A policy perspective, Environment, Development and Sustainability (2020). http://doi.org/10.1007/s10668-020-00849-0.
[59] R. Tao, J. Downs, T.M. Beckie, Y. Chen, W. McNelley, Examining spatial accessibility to COVID-19 testing sites in Florida, Ann. Gis (2020). http://doi.org/10.1080/19475683.2020.1833365.
[60] M. Coccia, The effects of atmospheric stability with low wind speed and of air pollution on the accelerated transmission dynamics of COVID-19, Int. J. Environ. Stud. (2020) 1–27. http://doi.org/10.1080/00207233.2020.1802937.
[61] M. Marquès, J. Rovira, M. Nadal, J.L. Domingo, Effects of air pollution on the potential transmission and mortality of COVID-19: A preliminary case-study in Tarragona Province, Environ. Res. (2020) 192.
[62] P. Prunet, O. Lezeaux, C. Camy-Peyret, H. Thevenon, Analysis of the NO2 tropospheric product from S5P TROPOMI for monitoring pollution at city scale, City and Environment Interactions (2020) 100051. http://doi.org/10.1016/j.cacint.2020.100051.
[63] S.K. Pani, N.H. Lin, S. RavindraBabu, Association of COVID-19 pandemic with meteorological parameters over Singapore, Sci. Total Environ. 740 (2020). http://doi.org/10.1016/j.scitotenv.2020.140112.
[64] J. Ran, S. Zhao, L. Han, et al., The ambient ozone and COVID-19 transmissibility in China: A data-driven ecological study of 154 cities, J. Infect. (2020) 81.
[65] T.L. Saaty, What Is the Analytic Hierarchy Process?, Springer Science and Business Media LLC, 1988, pp. 109–121. http://doi.org/10.1007/978-3-642-83555-1_5.
[66] W.J. Requia, E.K. Kondo, M.D. Adams, D.R. Gold, C.J. Struchiner, Risk of the Brazilian health care system over 5572 municipalities to exceed health care capacity due to the 2019 novel coronavirus (COVID-19), Sci. Total Environ. 730 (2020). http://doi.org/10.1016/j.scitotenv.2020.139144.
[67] SV Mishra, A Gayen, SM. Haque, COVID-19 and urban vulnerability in India, Habitat International 103 (2020). http://doi.org/10.1016/j.habitatint.2020.102230.
[68] H. Yao, X. Zuo, D. Zuo, H. Lin, X. Huang, C. Zang, Study on soybean potential productivity and food security in China under the influence of COVID-19 outbreak, Geography and Sustainability (2020) 163–171. http://doi.org/10.1016/j.geosus.2020.06.002.
[69] J.T. Wei, Y.X. Liu, Y.C. Zhu, et al., Impacts of transportation and meteorological factors on the transmission of COVID-19, Int. J. Hyg. Environ. Health 230 (2020). http://doi.org/10.1016/j.ijheh.2020.113610.
[70] D.F. Cuadros, Y. Xiao, Z. Mukandavire, et al., Spatiotemporal transmission dynamics of the COVID-19 pandemic and its impact on critical healthcare capacity, Health Place 64 (2020). http://doi.org/10.1016/j.healthplace.2020.102404.
[71] F. Moral García, Aplicación de la geoestadística en las ciencias ambientales, Revista Ecosistemas (1) (2004).

[72] J. Saha, B. Barman, P. Chouhan, Lockdown for COVID-19 and its impact on community mobility in India: An analysis of the COVID-19 Community Mobility Reports, 2020, Child. Youth. Serv. Rev. 116 (2020). http://doi.org/10.1016/j.childyouth.2020.105160.
[73] J. Park, W. Jo, M. Cho, et al., Spatial and Temporal Exposure Assessment to PM2.5 in a Community Using Sensor-Based Air Monitoring Instruments and Dynamic Population Distributions, Atmosphere (2020) 1284. http://doi.org/10.3390/atmos11121284.
[74] A.K. Sarfo, S. Karuppannan, Application of geospatial technologies in the COVID-19 Fight of Ghana, Transactions of the Indian National Academy of Engineering (2020) 193–204. http://doi.org/10.1007/s41403-020-00145-3.
[75] A. Kerimray, N. Baimatova, O.P. Ibragimova, et al., Assessing air quality changes in large cities during COVID-19 lockdowns: The impacts of traffic-free urban conditions in Almaty, Kazakhstan. Science of the Total Environment. 730 (2020). http://doi.org/10.1016/j.scitotenv.2020.139179.
[76] Y. Sui, H. Zhang, W. Shang, et al., Mining urban sustainable performance: Spatio-temporal emission potential changes of urban transit buses in post-COVID-19 future, Appl. Energy 280 (2020). http://doi.org/10.1016/j.apenergy.2020.115966.
[77] S. Fernández, T. Cotos-Yáñez, J. Roca-Pardiñas, C. Ordóñez, Geographically Weighted Principal Components Analysis to assess diffuse pollution sources of soil heavy metal: Application to rough mountain areas in Northwest Spain, Geoderma 311 (2018) 120–129. http://doi.org/10.1016/j.geoderma.2016.10.012.
[78] P. Clifford, S. Richardson, D. Hemon, Assessing the significance of the correlation between two spatial processes, Biometrics 45 (1) (1989) 123–134. http://doi.org/10.2307/2532039.
[79] S. Nomura, D. Yoneoka, S. Shi, et al., An assessment of self-reported COVID-19 related symptoms of 227,898 users of a social networking service in Japan: Has the regional risk changed after the declaration of the state of emergency? The Lancet Regional Health - Western Pacific (2020) 100011. http://doi.org/10.1016/j.lanwpc.2020.100011.
[80] R. Zulkarnain, K.D. Ramadani, Kualitas udara dan potensi transmisi COVID-19 di pulau Jawa, Seminar Nasional Official Statistics 2020-Statistics in the New Normal: A Challenge of Big Data and Official Statistics (2020).
[81] P. Melin, J.C. Monica, D. Sanchez, O. Castillo, Analysis of Spatial Spread Relationships of Coronavirus (COVID-19) Pandemic in the World using Self Organizing Maps, Chaos, Solitons and Fractals. 138 (2020). http://doi.org/10.1016/j.chaos.2020.109917.
[82] Z. Dehghan Shabani, R. Shahnazi, Spatial distribution dynamics and prediction of COVID-19 in Asian countries: Spatial Markov chain approach, Regional Science Policy and Practice (2020). http://doi.org/10.1111/rsp3.12372.
[83] P.M. Buscema, F. Della Torre, M. Breda, G. Massini, E. Grossi, COVID-19 in Italy and extreme data mining, Physica A 557 (2020). http://doi.org/10.1016/j.physa.2020.124991.
[84] M. Kulldorff, SaTScan User Guide (2018).
[85] M.P. Michalak, J. Cordes, A. Kulawik et al., A Systematic Framework for Spatiotemporal Modelling of COVID-19 Disease (2020).
[86] H.R. Pourghasemi, S. Pouyan, B. Heidari, et al., Spatial modeling, risk mapping, change detection, and outbreak trend analysis of coronavirus (COVID-19) in Iran (days between February 19 and June 14, 2020), Int. J. Infect. Dis. 98 (2020) 90–108. http://doi.org/10.1016/j.ijid.2020.06.058.
[87] I. Abou El-Magd, N. Zanaty, Impacts of short-term lockdown during COVID-19 on air quality in Egypt, Egyptian Journal of Remote Sensing and Space Science (2020). http://doi.org/10.1016/j.ejrs.2020.10.003.
[88] M. Ghahremanloo, Y. Lops, Y. Choi, S. Mousavinezhad, Impact of the COVID-19 outbreak on air pollution levels in East Asia, Sci. Total Environ. 754 (2003) 142226, doi:https://doi.org/10.1016/j.scitotenv.2020.142226.

[89] Z. Pei, G. Han, X. Ma, H. Su, W. Gong, Response of major air pollutants to COVID-19 lockdowns in China, Sci. Total Environ. 743 (2020). http://doi.org/10.1016/j.scitotenv.2020.140879.
[90] M.S. Rahman, M.A.K. Azad, M. Hasanuzzaman, et al., How air quality and COVID-19 transmission change under different lockdown scenarios? A case from Dhaka city, Sci. Total Environ. 762 (2021) 143161, doi:https://doi.org/10.1016/j.scitotenv.2020.143161.
[91] E. Tello-Leal, Macías-Hernández BA, Association of environmental and meteorological factors on the spread of COVID-19 in Victoria, Mexico, and air quality during the lockdown, Environ. Res. (2020). http://doi.org/10.1016/j.envres.2020.110442.
[92] K.P. Wyche, M. Nichols, H. Parfitt, et al., Changes in ambient air quality and atmospheric composition and reactivity in the South East of the UK as a result of the COVID-19 lockdown, Sci. Total Environ. (2020) 755.
[93] C. He, L. Yang, B. Cai, Q. Ruan, S. Hong, Z. Wang, Impacts of the COVID-19 event on the NOx emissions of key polluting enterprises in China, Appl. Energy 281 (2021) 116042, doi:https://doi.org/10.1016/j.apenergy.2020.116042.
[94] D.M. Stieb, G.J. Evans, T.M. To, J.R. Brook, R.T. Burnett, An ecological analysis of long-term exposure to PM2.5 and incidence of COVID-19 in Canadian health regions, Environ. Res. 191 (2020). http://doi.org/10.1016/j.envres.2020.110052.
[95] PHS Brancalion, DRA De Almeida, E Vidal, et al., Fake legal logging in the brazilian amazon, Sci. Adv. 4 (8) (2018). http://doi.org/10.1126/sciadv.aat1192.
[96] Y. Wang, D. Peng, L. Yu, et al., Monitoring Crop Growth during the Period of the Rapid Spread of COVID-19 in China by Remote Sensing, IEEE Journal of Selected Topics in Applied Earth Observations and Remote Sensing 13 (2020) 6195–6205. http://doi.org/10.1109/JSTARS.2020.3029434.
[97] M. Renigier-Biłozor, S. Źróbek, M. Walacik, A. Janowski, Hybridization of valuation procedures as a medicine supporting the real estate market and sustainable land use development during the covid-19 pandemic and afterwards, Land Use Policy (2020) 99. http://doi.org/10.1016/j.landusepol.2020.105070.
[98] C.D. Elvidge, T. Ghosh, F.C. Hsu, M. Zhizhin, M. Bazilian, The dimming of lights in china during the COVID-19 pandemic, Remote Sensing 12 (17) (2020).
[99] R. Minetto, M.P. Segundo, G. Rotich, S. Sarkar, Measuring human and economic activity from satellite imagery to support city-scale decision-making during COVID-19 pandemic, IEEE Transactions on Big Data (2020) 1. http://doi.org/10.1109/tbdata.2020.3032839.
[100] I. Okyere, E.O. Chuku, B. Ekumah, et al., Physical distancing and risk of COVID-19 in small-scale fisheries: A remote sensing assessment in coastal Ghana, ResearchSquare (2020).
[101] S. Mobaied, A new method for identifying and mapping areas vulnerable to Covid-19 in an armed conflict zone: Case study north-west Syria, MethodsX 7 (2020). http://doi.org/10.1016/j.mex.2020.101091.
[102] V. Garg, S.P. Aggarwal, P. Chauhan, Changes in turbidity along Ganga River using Sentinel-2 satellite data during lockdown associated with COVID-19, Geomatics, Natural Hazards and Risk. 11 (1) (2020) 1175–1195. http://doi.org/10.1080/19475705.2020.1782482.
[103] M.A. Aman, M.S. Salman, A.P. Yunus, COVID-19 and its impact on environment: Improved pollution levels during the lockdown period – A case from Ahmedabad, India. Remote Sensing Applications, Society and Environment 20 (2020). http://doi.org/10.1016/j.rsase.2020.100382.
[104] A.P. Yunus, Y. Masago, Y. Hijioka, COVID-19 and surface water quality: Improved lake water quality during the lockdown, Sci. Total Environ. (2020) 139012. http://doi.org/10.1016/j.scitotenv.2020.139012.

[105] C.V. Rodríguez-Benito, G. Navarro, I. Caballero, Using Copernicus Sentinel-2 and Sentinel-3 data to monitor harmful algal blooms in Southern Chile during the COVID-19 lockdown, Mar. Pollut. Bull. (2020) 161. http://doi.org/10.1016/j.marpolbul.2020.111722.
[106] J. Honey-Roses, I. Anguelovski, J. Bohigas, et al., The impact of COVID-19 on public space: A review of the emerging questions, 5, Cities & Health, 2021, pp. S263–S279. https://doi.org/10.1080/23748834.2020.1780074.
[107] K. Samuelsson, S. Barthel, J. Colding, G. Macassa, M. Giusti, Urban nature as a source of resilience during social distancing amidst the coronavirus pandemic, OSF Preprints (2020). http://doi.org/10.31219/osf.io/3wx5a.
[108] A. Gorayeb, J. de Oliveira, H.N. da Cunha, et al., Volunteered geographic information generates new spatial understandings of Covid-19 in Fortaleza, Journal of Latin American Geography (2020) 260–271. http://doi.org/10.1353/lag.2020.0048.
[109] H. Rossman, A. Keshet, S. Shilo, et al., A framework for identifying regional outbreak and spread of COVID-19 from one-minute population-wide surveys, Nat. Med. 26 (5) (2020) 634–638. http://doi.org/10.1038/s41591-020-0857-9.
[110] V. Antoniou, E. Vassilakis, M. Hatzaki, Is crowdsourcing a reliable method for mass data acquisition? The case of COVID-19 spread in greece during spring 2020, ISPRS International Journal of Geo-Information 9 (10) (2020). http://doi.org/10.3390/ijgi9100605.
[111] R. Debnath, R. Bardhan, India nudges to contain COVID-19 pandemic: A reactive public policy analysis using machine-learning based topic modelling, PLoS One 15 (9) (2020). http://doi.org/10.1371/journal.pone.0238972.
[112] P.L. Brito, M. Kuffer, M. Koeva, et al., The spatial dimension of COVID-19: The potential of earth observation data in support of slum communities with evidence from Brazil, ISPRS International Journal of Geo-Information 9 (2020).
[113] J. Coven, A. Grupta, Disparities in mobility responses to covid-19. NYU Stern Working Paper (2020).
[114] S. Gao, J. Rao, Y. Kang, et al., Association of mobile phone location data indications of travel and stay-at-home mandates With COVID-19 infection rates in the US, JAMA Network Open 3 (9) (2020) e2020485. http://doi.org/10.1001/jamanetworkopen.2020.20485.
[115] S. Heo, C.C. Lim, M.L. Bell, Relationships between local green space and human mobility patterns during COVID-19 for Maryland and California, USA, Sustainability (Switzerland) 12 (22) (2020) 1–16. http://doi.org/10.3390/su12229401.
[116] Y. Kang, S. Gao, Y. Liang, M. Li, J. Rao, J. Kruse, Multiscale dynamic human mobility flow dataset in the U.S. During the COVID-19 epidemic (2020b). arXiv. https://arxiv.org.
[117] E. Pepe, P. Bajardi, L. Gauvin, et al., COVID-19 outbreak response, a dataset to assess mobility changes in Italy following national lockdown, Scientific Data 7 (1) (2020). http://doi.org/10.1038/s41597-020-00575-2.
[118] T. Yabe, K. Tsubouchi, N. Fujiwara, T. Wada, Y. Sekimoto, V.S. Ukkusuri, Non-compulsory measures sufficiently reduced human mobility in Tokyo during the COVID-19 epidemic, Sci. Rep. (2020). http://doi.org/10.1038/s41598-020-75033-5.
[119] X. Huang, Z. Li, Y. Jiang, X. Ye, C. Deng, J. Zhang, et al., The characteristics of multi-source mobility datasets and how they reveal the luxury nature of social distancing, Int. J. Digit. Earth 14 (4) (2021) 424–442.
[120] A. Kapoor, X. Ben, L. Liu et al., Examining covid-19 forecasting using spatio-temporal graph neural networks. arXiv preprint (2020). doi:http://doi.org/arXiv:2007.03113.

[121] T. Kuchler, D. Russel, J. Stroebel, The geographic spread of COVID-19 correlates with structure of social networks as measured by Facebook (No. w26990), National Bureau of Economic Research (2020).

[122] X. Mu, A.G.O. Yeh, X. Zhang, The Interplay of Spatial Spread of COVID-19 and Human Mobility in the Urban System of China during the Chinese New Year, Environment and Planning B: Urban Analytics and City Science (2020).

[123] V. Pászto, A. Vondráková, Mobility Community Reports data: Geovisual analytics and cartographic synthesis of behaviour changes due to COVID-19 pandemic in Europe, Abstracts of the ICA (2020) 1–2. http://doi.org/10.5194/ica-abs-2-13-2020.

[124] Z. Peng, R. Wang, L. Liu, H. Wu, Exploring Urban Spatial Features of COVID-19 Transmission in Wuhan Based on Social Media Data, ISPRS International Journal of Geo-Information (2020) 402. http://doi.org/10.3390/ijgi9060402.

[125] C. Yang, D. Sha, Q. Liu, et al., Taking the pulse of COVID-19: A spatiotemporal perspective, Int. J. Digital Earth 13 (10) (2020) 1186–1211. http://doi.org/10.1080/17538947.2020.1809723.

[126] F. Pase, F. Chiariotti, A. Zanella, M. Zorzi, Bike Sharing and Urban Mobility in a Post-Pandemic World, IEEE Access (2020) 187291–187306. http://doi.org/10.1109/access.2020.3030841.

[127] U. Saeed, K. Sherdil, U. Ashraf, I. Younas, H.J. Butt, S. Rashid, Identification of potential lock-down area during COVID-19 transmission in punjab, Public Health 190 (2020) 42–51. https://doi.org/10.1016/j.puhe.2020.10.026.

[128] D. Depellegrin, M. Bastianini, A. Fadini, S. Menegon, The effects of COVID-19 induced lockdown measures on maritime settings of a coastal region, Sci. Total Environ. (2020), doi:10.1016/j.scitotenv.2020.140123.

[129] C. Guida, G. Carpentieri, Quality of life in the urban environment and primary health services for the elderly during the Covid-19 pandemic: An application to the city of Milan (Italy), Cities (2020) 103038. http://doi.org/10.1016/j.cities.2020.103038.

[130] S. Kanga, M.G. Sudhanshu, M. Farooq, M.S. Nathawat, S.K. Singh, Reporting the management of COVID-19 threat in India using remote sensing and GIS based approach, Geocarto International (2020) 1–8. http://doi.org/10.1080/10106049.2020.1778106.

[131] Ò Saladié, E. Bustamante, A. Gutiérrez, COVID-19 lockdown and reduction of traffic accidents in Tarragona province, Spain, Transportation Research Interdisciplinary Perspectives 8 (2020). http://doi.org/10.1016/j.trip.2020.100218.

[132] F.E.S. Silalahi, F. Hidayat, R.S. Dewi, N. Purwono, N. Oktaviani, GIS-based approaches on the accessibility of referral hospital using network analysis and the spatial distribution model of the spreading case of COVID-19 in Jakarta, Indonesia. BMC Health Services Research. 20 (1) (2020). http://doi.org/10.1186/s12913-020-05896-x.

[133] H.A.A. Alasadi, M.T. Aziz, M. Dhiya, A Abdulmajed, A network analysis for finding the shortest path in hospital information system with GIS and GPS, Journal of Network Computing and Applications 5 (1) (2020) 10–23.

[134] A. Ghilardi, I. Ruiz, A. Navarrete, et al., Plataforma de información geográfica de la UNAM sobre COVID-19 en México, BioTecnología 4 (3) (2020).

[135] E. Dong, H. Du, L. Gardner, An interactive web-based dashboard to track COVID-19 in real time, Lancet Infect. Dis. 20 (5) (2020) 533–534. http://doi.org/10.1016/S1473-3099(20)30120-1.

[136] M. Yalcin, Mapping the global spatio-temporal dynamics of COVID-19 outbreak using cartograms during the first 150 days of the pandemic, Geocarto International (2020). http://doi.org/10.1080/10106049.2020.1844310.

[137] P. Gao, H. Zhang, Z. Wu, J. Wang, Visualising the expansion and spread of coronavirus disease 2019 by cartograms, Environ. Plan. A. 52 (4) (2020) 698–701. http://doi.org/10.1177/0308518X20910162.
[138] Q. Shi, T. Liu, Should internal migrants be held accountable for spreading COVID-19? Environ. Plan. A. 52 (4) (2020) 695–697. http://doi.org/10.1177/0308518X20916764.
[139] T. Zhang, Integrating geographic information system technique with Google Trends data to analyse COVID-19 severity and public interest, Public Health 189 (2020) 3–4. http://doi.org/10.1016/j.puhe.2020.09.005.
[140] M. Galluzzi, D. Rocchini, R. Canullo, R.E. McRoberts, G. Chirici, Mapping uncertainty of ICP-Forest biodiversity data: From standard treatment of diffusion to density-equalizing cartograms, Ecological Informatics 48 (2018) 281–289. http://doi.org/10.1016/j.ecoinf.2018.06.005.
[141] B. Gross, Z. Zheng, S. Liu et al., Spatio-temporal propagation of COVID-19 pandemics. (2020). medRxiv. doi:http://doi.org/10.1101/2020.03.23.20041517
[142] F. Méndez-Arriaga, The temperature and regional climate effects on communitarian COVID-19 contagion in Mexico throughout phase 1, Sci. Total Environ. (2020) 139560. http://doi.org/10.1016/j.scitotenv.2020.139560.
[143] M. Foucher, La Réaffirmation Des Frontières En Europe et Dans Le Monde. La Pandémie de Covid-19, Regards Croisés de Géographes (2020).

CHAPTER

12

Current clinical testing approach of COVID

Manisha Bhardwaj[a], *and Priyanka Kakkar*[b]

[a]**Department of Biochemistry, Ch. Charan Singh University, Meerut, India,** [b]**Department of Chemistry, Ch. Charan Singh University, Meerut, India**

12.1 Introduction

Human fight with microorganism is not new. Past centuries have been the witness to emergence of number of new micro-organisms. Every time this tiny nonvisible enemy came with new approach by changing their genetic features but the astute human has always conquered and uprooted it from the ground. Innumerable pandemics came in past century; humans confronted against all and surpassed every opponent whichever tried to cease human community. But the year 2019 was different and gave the most awful memories to human community by claiming more than 0.8 million lives across the world so far. The recent emergence of novel coronavirus in the human population has caused dramatic and appalling influence on world economy and prompted mobilization of public health authorities around the world. Having stuck almost the whole world and affecting nearly 213 countries, this vicious creature had a deadly impact on human community and has caused colossal damage to society, swallowed umpteen lives. The aftermath consequences are also equally harrowing that is, the psychological impact on survivors and the social and economic downturn are relatively more disgusting. A recent mathematical model suggested that undocumented infections might be major drivers of SARS-CoV-2 spread in the world [1]. The efficacy of the virus to get adapted in new environment by modifying itself through mutation make a constant threat to population which may turn illnesses into pandemic and testing became more crucial.

Sensing Tools and Techniques for COVID-19: Developments and Challenges in Analysis and Detection of Coronavirus
DOI: https://doi.org/10.1016/B978-0-323-90280-9.00003-1

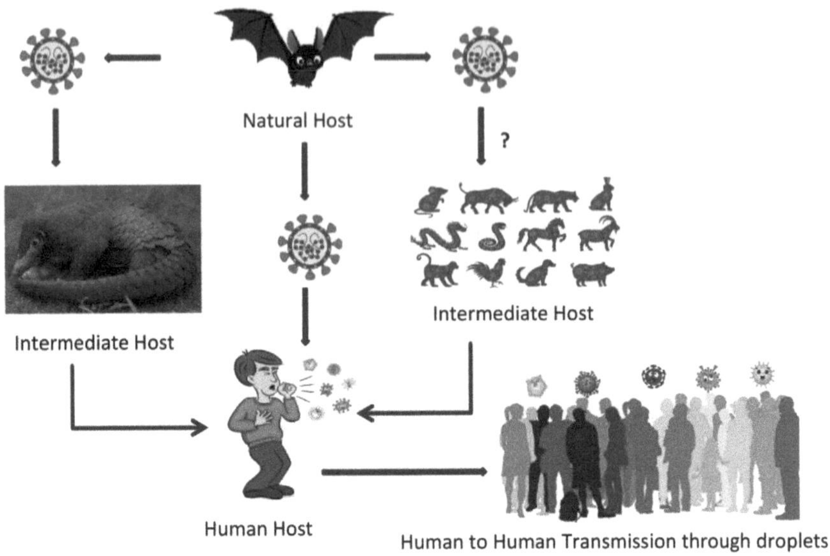

FIGURE 12.1 Mode of transmission of Coronavirus via different mediators (only alpha and beta virus have ability to transmit to humans).

Corona viruses belong to the *Coronaviridae* family in the *Nidovirales* order. They are large, enveloped, single-stranded RNA viruses which have the largest genome among all RNA viruses, typically ranging from 27 to 32 kb (Fig. 12.1) and approx. 50–200 nm in diameter. The subgroups of coronavirus family are alpha (a), beta (b), gamma (c), and delta (d). Among them, alpha- and beta coronaviruses infect mammals, gamma coronaviruses infect avian species, and delta coronaviruses infect both mammalian and avian species.

The genome is packed inside a helical capsid formed by the nucleocapsid protein (N) and further surrounded by an envelope. This N protein holds the RNA genome embraced with three structural proteins that is, M (The membrane protein), E (The envelop protein) and S (The Spike protein) which fabricates the viral envelope together [2]. M and E are responsible for viral assembly whereas the S spike protein is located on the surface of virus and plays a key role in virus entry into host cells. Viruses enter the host cells through Receptor-mediated endocytosis. The spike proteins responsible for the viral entry have N-terminal and C-terminal domains, and two major subunits S1 and S2 which are present in almost all coronaviruses [3]. One of these S1 or S2 subunits binds with the host receptors and acts as a receptor-binding domain (RBD). Among these structural proteins, these spikes form large protrusions from the virus surface, giving coronaviruses the appearance of having crowns (hence their name; corona in Latin means crown).

These viruses were thought to infect only animals until the world witnessed a severe acute respiratory syndrome (SARS) outbreak caused by SARS-CoV, at Foshan municipality, Guangdong Province, China, in November 2002 [4]. Only a decade later, in 2012 another pathogenic coronavirus, known as Middle East respiratory syndrome coronavirus (MERS-CoV) occurred in April 2012 in Jordan and has been diffused within and sporadically outside the Middle East regions [5] causing steady endemics in the countries. The SARS-CoV and MERS-CoV are prone to infect the lower respiratory tract, resulting in acute lung injury (ALI)/acute respiratory distress syndrome (ARDS), septic shock and multiorgan failure, with high case fatality ratio (CFR). In 2019 emergence of new CoV outburst into pandemic has 80% sequence identity to SARS-CoV (with whom it is classified into the species severe acute respiratory syndrome-related coronavirus) [6] and 50% to MERS CoV. [VIII] SARS-CoV-2 belongs to the *Sarbecovirus* sub genus of the *Corona viridae family* and is the seventh coronavirus known to infect humans. The virus has been found 96% nucleotide identity with CoV isolated from bat origin, but not yet clear, that it directly infects human or through an intermediate host [7]. The SARS-CoV-2 full genome has been determined and published in GenBank [8]. Distinct features of SARS-CoV, MERS-CoV, and SARS Cov-2 are described in Table 12.1.

The phylogenetic analysis of SARS-CoV-2 shows similar Beta lineage as that of bat corona virus named as Bat CoV-RaTG13 [6]. Recent studies have suggested that bats may be the potential natural host of SARS-CoV-2 [7,9] and Malayan pangolin the potential intermediate host [10]. However, SARS-CoV-2 has been shown to have much higher human-to-human transmissibility [11]. As for the new virus, it is highly infectious and has already killed over 200,000 people with an estimated sCFR (symptomatic case fatality risk) of 1.4% (0.9–2.1%) [12] on comparison, fatality rate of SARS was roughly 10%, whereas, for MERS, it was approximately 36%, making it one of the deadliest human pathogens [13].

On the basis of recent studies most of the patients develop symptoms against infection of COVID-19 within 3–7 days [14] while clinical feature arises in 1–14 days. In general, the most common clinical manifestations are fever, dry cough, fatigue, sputum production, dyspnea, sore throat, headache, myalgia or arthralgia, and chills. Less common symptoms include nausea or vomiting, nasal congestion, diarrhea, hemoptysis, and conjunctival congestion [15]. In children, SARS-CoV-2 infection is generally mild and, in some case, asymptomatic; however, when presented, the main symptom includes fever (43% on admission, and 88.7% during hospitalization), cough (67.8%), diarrhea (3.8%), and fatigue [16,17]. More recently, several studies have been suggested that COVID-19 infection was associated with cutaneous manifestations in patients [18–20]. Major manifestations observed in COVID-19 patients including different types

TABLE 12.1 Comparative analysis of biological and historical features of SARS-CoV, MERS-CoV, and SARS-CoV 2.

Features	SARS-CoV [32,218–222]	MERS [223,224]	SARS-CoV-2 [7,143,225–229]
Genus	Clade I, lineage B	Clade II, lineage C	Clade I, lineage B
Emergence date	16 November 2002	4 April 2012	7 December 2019
Location of emergence	Guangdong, China	Zarqa, Jordan	Wuhan, China
Current status	Controlled	Sporadic	Still ongoing
Potential animal hosts	Bats, palm civets and Raccon dogs	Bats, Dromedry camels	Bats, Pangolins
Countries infected	26	27	213[a]
Total infected cases	8098	2519	102 million[a] (as of 31 January 2021)
Total attributed deaths	776	876	Above
Mortality rate	9.6%	35%	4.4%
Entry receptor in humans	ACE2 receptor	DPP4 receptor	ACE2 receptor
Sign and symptoms	Fever, cough and shortness of breath, ARDS	Fever, cough and shortness of breath, ARDS	Fever, myalgia, cough, shortness of breath, sneezing, headache, diarrhea, shivering, ARDS

of lesions such as purpuric, papulovesicular, livedoid, urticarial, maculopapular, and thrombotic ischemic [21]. Symptoms of severe COVID-19 disease include shortness of breath, loss of appetite, confusion, persistent pain, or pressure in the chest.

According to WHO criteria, a person is suspected of being infected with COVID-19 in three scenarios: (i) a patient with acute respiratory illness (fever and at least one sign/symptom of respiratory disease, e.g., cough, shortness of breath) and a history of travel to or residence in a location reporting community transmission of COVID-19 disease during the 14 days prior to symptom onset; (ii) a patient with any acute respiratory illness who has been in contact with a confirmed or probable COVID-19 case in the last 14 days prior to symptom onset; and (iii) a patient with severe acute respiratory illness (fever and at least one sign/symptom of respiratory disease, for example, cough, shortness of breath, and requiring hospitalization) and in the absence of an alternative diagnosis that fully explains the clinical presentation. If a patient then tests positive with a laboratory diagnostic, then the infection becomes a confirmed case of COVID-19, irrespective of clinical signs and symptoms [22].

12.2 Mechanism of infection: Key features and entry mechanism of human coronaviruses

The human-to-human spreading of the virus occurs due to close contact with an infected person through coughing, sneezing, respiratory droplets, or aerosols. These aerosols can penetrate the human body (lungs) via inhalation through the nose or mouth [23–26]. Animal models play a vital role to uncover the mechanisms of viral pathogenicity from the entrance to the transmission and designing therapeutic strategies. As the entire genome of the 2019-novel coronavirus is more than 80% similar to the SARS-like bat CoV, previously used animal models for SARS-CoV can be utilized to study the infectious pathogenicity of SARS-CoV-2. The sequence analysis revealed a mutation at spike glycoprotein in SARS CoV [27]. Thus, it could be another suitable option to develop spike glycoprotein targeting therapeutics against novel coronaviruses. Recently, mice models and clinical isolates were used to develop any therapeutic strategy against SARS-CoV-2 induced COVID-19 [28,29]. In a similar study, artificial intelligence prediction was also used to investigate the inhibitory role of the drug against SARS-CoV-2 [28]. SARS-CoV-2 infected patients were also used to conduct randomized clinical trials [28,30,31]. The rate of infection or the average number of people getting infected by an individual (R0) was 2.75 in the case of SARS pandemic in 2003. The R0 value of MERS, was around 1. The SARS-CoV-2 R0 value was estimated to be in the range of 1.5–3.5. Comparative study of SARS, MERS and SARS-CoV-2 has been displayed in Table 12.1. However, the difficulties arise for investigating therapeutic reagent against SARS-CoV-2 infection due to (1) basic mechanism of viral infection and incubation periods were uncertain, (2) most of the asymptomatic individuals are equally efficient in expansion of infection, (3) alteration in responsiveness of the society in proliferation of infection remains unanswered. In addition, there were no control measures for this spread [30]. With the ability to infect people through asymptomatic carriers, it can remain unnoticed and quickly disseminate itself, making the disease containment a confounding public health challenge [32]. It is now important that the scientists worldwide collaborate to design a suitable model and investigate the in vivo mechanisms associated with pathogenesis of SARS-CoV-2.

The respiratory droplets are the main routes of transmissions; SARS-CoV-2 can be transmitted to a healthy person if they encounter with infected person or any of his belongings, including clothes, doorknobs, etc.

Studies have been reported that aerosol transmission (Airborne transmission) is also possible for SARSCoV-2, but there is no clear study on neonatal infections (mother–to-child) [17,20,21,33,34]. During the initial phase of the COVID-19 outbreak, a dataset was obtained from 1099 patients with laboratory confirmed COVID-19 from 552 hospitals in 30 provinces of China on January 29, 2020. Only 2% of the patients had a history of contact with animals; more than three quarters have either visited the Wuhan city or are residents. Hence, the outbreak patterns or the source of infection could not be predicted from their study. The SARS-CoV-2 was detected in saliva, blood, sputum, and urine before the development of viral pneumonia, and some patients do not develop pneumonia at all. Asymptomatic persons are potential sources of SARS-CoV-2 infection, which control the transmission dynamics of the current outbreak [2,7,35,36]. Spikes comprises of three segments: (1) large ectodomain, (2) transmembrane domain, and (3) intracellular tail. The receptor-binding subunits S1 and S2 are placed in the ectodomain region. During the infection, the S1 binds with the host receptor, and S2 fuses the host and viral membranes, thereby releasing the viral genome into the cell. The spike protein is a clove-shaped trimer with three S1 heads and a trimeric S2 stalk [9]. During viral infection, spike protein (~1300 amino acid residues) is cleaved by host proteases into receptor binding subunit S1 and membrane fusion subunit S2. During cell entry, the S1 subunit binds directly to the sugar receptors [37] and ACE2 of the host cell surface, and the S2 subunit undergoes conformational changes and obtains postfusion state [38]. During this state, the three pairs of heptads repeat region HR-N and HR-C in trimeric S2 form a six-helix bundle structure [39]. The buried hydrophobic fusion peptides become exposed and insert into the target host membrane. These fusion peptides and the transmembrane anchors are positioned at the end of a six-helix bundle structure, bringing the viral, and host membranes to fuse [39,40]. During this process, a large amount of energy is released, which accelerates the membrane fusion forward. Along with this, receptor binding and low pH can also trigger this membrane fusion. Since the spike protein has a good binding affinity for sugar receptors of human cells, it uses them as a mechanism of cell entry [3]. Notably, the SARS-CoV-2 has a higher affinity to human ACE2 than the SARS-CoV virus strain. The ectodomain of the SARS-CoV-2 spike protein binds to the peptidase domain (PD) of ACE2 with a Kd (equilibrium dissociation constant) of ~ 15 nM [41]. Spike protein priming is done by transmembrane protease serine 2 (TMPRSS2), which is also essential for the entry of SARS-CoV-2 [42]. The glycoprotein spikes on the outer surface of coronaviruses are responsible for the attachment and entry of the virus to host cells. The receptor-binding domain (RBD) is loosely attached among virus; therefore, the virus may infect multiple hosts [43,44]. The spike protein of SARS-CoV-2 contains a 3-D structure in the RBD region to maintain the van der Waals forces [45]. The 394-glutamine residue in the RBD region of SARS-CoV-2 is recognized by the critical lysine 31 residue on the human ACE2 receptor [46].

Throughout this viral entry process, the Lys31 and Lys 353 of human ACE2 receptors are termed as "hot spot" residues, which consist of a salt bridge buried in a hydrophobic environment and contribute critically for the virus and host cell receptor binding. Thus, this key residue comparison of SARS-CoV-2 with the civet and human SARs-CoV explains how actively SARS-CoV-2 is choosing and binding with the human ACE2 receptors, which is likely to cause the human-to-human transmission [46]. Hussain et al. reported that the mutations S19P and E329G in ACE2 disrupt the intermolecular interactions and have low binding affinity with viral spike protein. In addition to the variations in the viral spike protein, ACE2 allelic variants can also drive the potential resistance against SARS-CoV-2 infection [47].

A diagnostic test method should have sufficient sensitivity and accuracy to make appropriate clinical decisions rapidly during a pandemic [48]. The most effective way to curb outbreak of pandemic is early detection and isolate patients [49,50]. The laboratory-based diagnosis assumes a role for the clinical management of patients and the implementation of disease control measures. Here, we review the clinical features, laboratory methods, and imaging findings that are used for COVID-19 diagnosis.

12.3 Types of specimen

The right type of sample, appropriate collection procedure, and reliable transportation must be placed to minimize the risk of inaccurate results. The difference in the sensitivity of different types of swabs may depend on disease progression. Therefore, it is important to identify the appropriate type of sample considering the medical condition of the patient and diagnostic facility available for the test.

Two types of sample specimens are being primarily used for the diagnosis of COVID-19. Respiratory specimens are used for direct detection of virus and serum samples are used for identification of antiviral antibodies [51]. For specimen collection, the sensitivity of nasal, nasopharyngeal (NP), and throat swabs was found to be 80%, 90%, and 87%, respectively [18]. The respiratory specimens [52] are collected most frequently from the upper respiratory tract, that is, nasopharyngeal and oropharyngeal swab which is considered with regard to upper respiratory specimens, collected in the acute phase of infection—ideally within 7 days, and for lower respiratory specimens, sputum and endotracheal aspirate or bronchoalveolar lavage fluid (BLF) are considered [33] which should be obtained from patients remain symptomatic after more than a week [53,54]. A case study with a pneumonia patient in Thailand showed a negative test with nasal or oropharyngeal swab samples but a positive test with bronchoalveolar lavage fluid [20]. A study reported that the highest rates of positive were found from bronchoalveolar lavage fluid (93%), followed by sputum (72%), nasal swab (63%), fibrobronchoscope brush biopsy (46%), pharyngeal swabs (32%), feces (29%), and blood (1%) [55]. Additionally, blood and stool specimen could be collected for clinical inspection which requires less challenging sampling procedures than respiratory specimens [56].

Serum samples are collected for immunoassay methods. Volume of blood sample for immunoassays ranges from 5 to 10 mL for lab assays to capillary draws of 50–200 μL blood for lateral flow immunoassays (LFIA) [57]. The virus samples should be processed and tested as soon as possible. If immediate testing is not possible, the sample can be stored up to 72 h at 2–8°C. However, for more than 72 h storage, the specimens should be frozen at -70°C as soon as possible after collection [53]. It is recommended to avoid repeated freezing and thawing of the specimen [33].

12.4 Clinical testing methods

It is of critical importance to rationally choose specific diagnostic methods to fight against viral outbreak, any negligence or compromise in the diagnosis may lead to devastating consequences. As part of the quality assessment of each assay, the FDA requires demonstration of specificity and exclusivity. Exclusivity means that no other viruses or bacteria from a specified list are detected by the test. Depending on the performance of the internal controls of the assays, the test results are reported as positive, negative, inconclusive, or invalid. Tests that detect 2 or more viral genes are interpreted differently in various assays. Some assays require that all viral genes be detected for the test result to be interpreted as positive, whereas others rely on the detection of 1 or 2 viral genes for a positive interpretation. These test results are positive during the incubation period, which is several days before the onset of symptoms of the disease and remain positive for the duration of symptoms. The tests detect parts of the viral RNA that can be present after the virus fragments. Therefore, the results can continue to be positive after the resolution of symptoms, even though a complete infectious virus may no longer be present [14,58–60].

Regulatory agencies such as the World Health Organization (WHO) and the US Food and Drug Administration (FDA) have approved the use of a number of diagnostic methods, while some new methods are receiving conditional approval under emergency use authorization (EUA) [61]. These diagnostic methods have varying throughput, batching capacity, requirement of infrastructure setting, analytical performance, and turnaround times [62]. In addition to depending on the equipment and method itself, the result from a method also relies on sample collection protocol, reagents used, potential for cross-contamination, and sample/reagent storage requirements. These factors must be considered while selecting a reliable and rapid diagnostic method to help make an appropriate decision and prompt public health actions. Nucleic acid amplification using the reverse transcription polymerase chain reaction (RT-PCR) is the most widely used method for direct SARS-CoV-2 diagnosis [48]. Immunoassays are used to measure the antibodies against the SARS-CoV-2. CT imaging includes use of plain x-rays, computed tomography (CT) scanning, magnetic resonance imaging (MRI), nuclear scanning, and ultrasonography. Number of scientists also modified these traditional approaches and developed new accurate, error-free and faster techniques in order to avert any obstruction in spotting of this meticulous enemy. Despite of having these conventional procedures new emerging techniques like CRISPR (Clustered Regularly Interspaced Short Palindromic Repeats), RT-LAMP, DPCR, RCA, LFA, PLA have also been reported. Apart from laboratory-based CT imaging, RT-PCR, immunoassay and various emerging techniques, several point-of-care (POC) and rapid test methods have become available in the last few months. This Chapter aims to summarize the successive development of the current clinical approaches for SARS-CoV-2 diagnosis methods and provide recently presented information on various tests till date and compare their analytical efficiency in terms of limit of detection (LOD), sensitivity, and specificity. In addition, their limitations, ease of use, affordability, and availability of accessories have also been evaluated.

12.4.1 Molecular testing (nucleic acid test)

After identification of SARS-CoV-2 as the causative virus for this pandemic, the SARS-CoV-2 genome was quickly sequenced [63] from which unique sequences have been identified for COVID-19 diagnosis. PCR-based tests are widely used for the detection of viruses in human disease and are currently the most commonly used nucleic acid tests (NATs) performed in clinical laboratories. eRT-PCR-based diagnostic is highly sensitive, sequence specific, and useful in the early detection of COVID-19. Since the test accuracy varies depending on the disease stage and viral multiplication, the sensitivity can range from 71 to 98%, whereas the specificity is recorded to be 95% [64]. The same technique has been used in the diagnosis and surveillance of various other viral diseases including SARS-CoV and MERS-CoV [16,35,65]. The RT-PCR tests take less than an hour to a couple of days to give results, depending on the version of the PCR. The RT-PCR assay can be carried out in one- or two-step approaches. PCR instruments and techniques are in widespread use in both clinical and research laboratories and the basis of the tests is well known. The

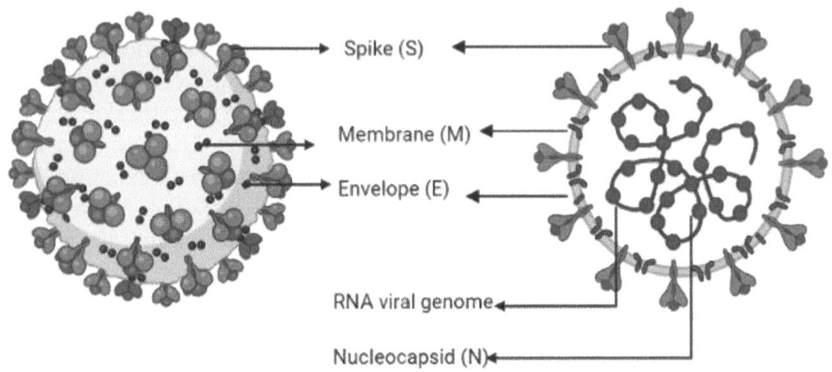

FIGURE 12.2 (A) Schematic presentation of Coronavirus showing viral structural proteins, spike (S), envelope (E), and membrane (M). (B) The single-stranded viral RNA is associated with the nucleocapsid protein having structural and accessory genes.

tests consist of nucleic acid extraction and purification from the human specimen using authorized extraction methods/instruments followed by real-time RT-PCR, where the RNA is reverse transcribed into cDNA and then amplified using the primer sets and detected using specific probes. Laboratory RT-PCR tests the RT-PCR assays in centralized laboratories are generally performed in 96-well plates for signal reading in batches. The high-throughput 384-well assay system using lower volume was reported recently with detection sensitivity down to 5 copies of viral genome per microliter [66] which yields 100% sensitivity and specificity. Real time reverse transcriptase polymerase chain reaction (rRT-PCR), or quantitative reverse transcriptase PCR (qRT-PCR), uses nucleic acid amplification test (NAAT) that has long been used routinely to amplify the DNA for the detection of viral RNA in clinical settings (Accelerated Emergency Use Authorization [[67] Summary; [68]]. Heat treatment prior to RNA extraction is not recommended as studies suggest that thermal inactivation in samples with low viral loads could result in potential false-negative nucleic acid test [69]. The SARS-CoV-2 genes most frequently used for detection so far include theORF1ab/RdRP gene, Nucleocapsid (N) gene, Envelope (E) gene, and Spike protein (S) gene (Fig. 12.2). One study reported that PCR amplification of the E and SARS coronavirus RdRp genes is 95% sensitive111. Virus-specific PCR-based tests required generating primers and probes unique to SARS-CoV-2 but not to the other closely related corona viruses [70–73]. In silico analysis using the many sequences available on publicly available databases (e.g., GenBank, the European Molecular Biology Laboratory (EMBL), Global Initiative on Sharing All Influenza Data (GISAID) to discriminate the SARS-COV-2 from other respiratory viruses is a hallmark widely employed to generate a specific primer for COVID-19 detection.

An amplification of each of these genes can be accomplished by the supply of proper forward and reverse primers. Studies also showed that the clinical sensitivity of a RT-PCR assay is under the influence of specimen type, amount of virus in a swab and the specimen collection time in relation to the onset of symptoms and the test result relies on sample collection, primers and probes used, analysis of fluorescence curves, use of suitable controls, and reliability of the temperature control. Respiratory specimens may contain different genera of coronaviruses along with other major viral pathogens. In the last six decades, before SARS-CoV-2, the human population was already infected with six other members (229E, OC43, SARS-CoV, NL63, HKU1, and MERS-CoV) of the CoV family [70]. False positive results occurring due to the cross reactivity with these viruses, human genome, and microflora can be obliterated with the sequence fidelity Fig. 12.3. Tsang et al. revealed that the saliva samples were more promising for use in RT-PCR [74], while Yam et al. concluded that testing more than one specimen could significantly maximize the sensitivity of the RT-PCR testing [73]. These findings suggest that it is rather important to apply nucleic acid-based kits with optimized conditions to maximize their diagnostic potency. Repeat testing, using various biospecimens (respiratory secretions, sputum, stool, rectal swabs, and serum), should be considered in patients with a high clinical suspicion of COVID-19 [51,59,72,75,76].

In a study of 205 patients, RT-PCR sensitivity was 93% for BLF, 72% for sputum, 63% for nasal swabs, and only 32% for throat swabs [51]. The presence of viral load that is below the assay's LOD (limit of detection) will also elicit false negative results. Therefore, a judicious way to increase viral load is to collect combined nose and throat swabs. Viral kinetics of SARS-CoV-2 showed that the viral load in respiratory specimens often peaks in the first week of illness and decreases as the disease progresses [75]. This posited appropriate sample collection times to enhance sensitivity. These findings are similar to a report that showed a 100% positive RT-PCR result by week 1 after onset of symptoms, followed by 89.3%, 66%, and 32% at week 2, week 3, and week4, respectively. By week 5, the positive detection rate plummeted down to only 5.4% [77]. In contrast to the widely used NP swabs, a study more recently showed an increase in the sensitivity by 13% when saliva samples were used [78]. Virus titers from saliva samples were found significantly higher than NP swabs and more importantly, unlike NP (nasopharyngeal) swabs, less temporal variation in viral titer was observed with longitudinally collected saliva samples. There are variants of RT-PCR methods that share the same mechanism while differing in the detection strategy. Real-time RT-PCR reads fluorescent signals during PCR amplification [68] to quantify the target, whereas nested RT-PCR uses two sets of primers to avoid nonspecific PCR amplifications [79].

The FDA has already issued several other molecular in vitro diagnostics under EUA [80]. In many protocols, RT-PCR assay of more than one gene

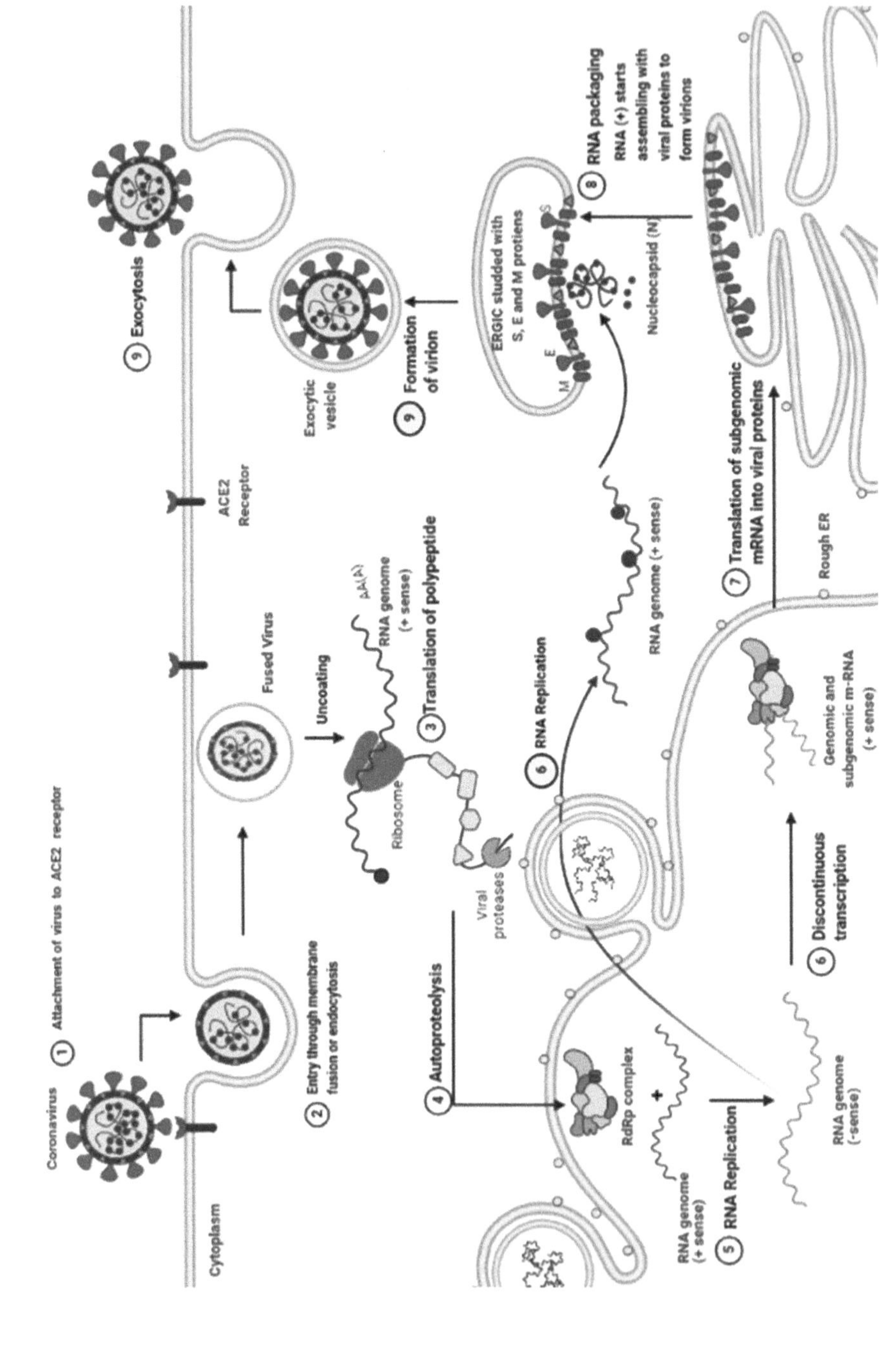

FIGURE 12.3 The schematic diagram of the mechanism of COVID-19 entry led to viral replication and viral RNA packing in the human cell.

target is performed for the positive authenticity of COVID-19. The CDC Real-Time RT-PCR Diagnostic Panel was the first to receive approval and used the N1 and N2 genome segment. The CDC generated the primers and probes and made the materials available for other laboratories to use with the same test. The US CDC real-time RT-PCR diagnostic panel under EUA targets two different loci of the N gene126 and considers positive results only when both gene targets (N1 and N2) are positive with LOD of 1 copy of RNA per microliter for both N1 and N2 primer sets [81]. However, sensitivity was affected by the use of different RNA extraction and purification protocols. If any of the two assays are negative, the result is inconclusive, and the assay has to be repeated following strict guidelines. Positive confirmation with a single gene target is possible if the amplicons are subjected to deep sequence analysis. In addition to these controls, the US CDC recommends the use of a human specimen control (HSC) [81] to ensure successful lysis and integrity of extraction reagents and to minimize false negative results by ensuring collection of enough human cellular material [2].

A PCR assay with absolute exclusiveness to SARS-CoV-2 was made possible in the Charité protocol with the inclusion of additional probe (RdRp_P2) that anneals to only SARS-CoV-2 mRNA transcript. Nonetheless, this protocol does not discriminate between clades of the Sarbeco viruses like SARS-related CoVs from bats. This overlap is corroborated with the sim plot of SARS-CoV-2 showing more than 96% identity to a bat coronavirus [7]. The Charité protocol uses the E gene as the screening assay followed by confirmatory assay with the RdRp gene [82]. Further improvement in the sensitivity was found [65] while conducting a proficiency test for the sensitivity with E Sarbeco and RdRp genes using in vitro transcribed

RNA derived from SARS-CoV strain Frankfurt-1, where they found LOD of 5.2 and 3.8 copies per reaction, respectively, which was in good agreement with other participating laboratories. They also found that N gene assay was slightly less sensitive than the RdRp and the E Sarbecogene. However, a comparative study showed that the N2 primer/probe set developed by US CDC was highly sensitive comparable with the E Sarbeco set described in Charité assay at low viral copy number [83].

The protocol from Pasteur Institute [82] utilizes IP2 and IP4 gene targets as the first-line screening tool, while confirmatory testing utilizes the E gene target. Primer sets (IP2 and IP4), when used individually in an assay, can detect about 100 copies of RNA genome equivalent per reaction at 95% detection probability. A lower LOD of 10 copies was obtained with a multiplex assay using these primer sets [82]. The PCR assays developed by Pasteur Institute have been claimed that there is no cross-react with respiratory viruses like influenza A (H1N1, H3N2), enterovirus, adenovirus, human coronaviruses (HKU1, OC43, 22E, NL63), and MERS-CoV, indicating 100% specificity.

Assays from HKU (Hong Kong University) are specific to only subgenus Sarbecoviruses [84] where N gene was used as the first-line screening while the ORF1b as the confirmatory testing [85]. This assay did not distinguish between SARS-CoV and SARS-CoV-2 transcripts because SARS-CoV-2 shares a 79.6% sequence identity with SARS-CoV BJ01 [7], and many of the regions homologous with the primers are conserved nucleotide regions [13]. Primers based on the receptor binding domain of the S gene developed by Zhou et al. could discriminate SARS-CoV-2 from bat SARSr-CoV WIV1 [7].

Some studies also shown that primer/probe sets E Sarbeco from the Charité protocol and ORF1b-nsp14 from the Hong Kong University (HKU) protocol were the most sensitive with an LOD of 10 virus genome equivalents per microliter at 75% detection frequency [62].

The Chu et al. protocol recommends the N gene for screening, while ORF1b provides a confirmatory test [86]. They used the cloned DNA plasmid containing SARS-CoV ORF1b and N gene for one step real time RT-PCR approach to calculate the LOD of their assay. With a known titer of viral RNA, their preliminary study showed the LOD of <10 copies per reaction [84]. A dynamic range of0.0002–20 TCID50 (50% tissue culture infective dose) per reaction and detection limit below 10 RNA copies per reaction was obtained using RNA extracted from cells infected by SARS coronavirus as a positive control using one-step RT-PCR assays to detect two different regions (ORF1b and N) of the viral genome [84].

Scientists from Germany used the E gene for the first-line screening and the RdRP gene for confirmatory testing [85]. This method further increased sensitivity to detect as few as 5.6 RNA copies per reaction for the E gene and 3.8 RNA copies per reaction for the RdRP gene. In France, two RdRP genes were used for initial screening followed by the confirmatory E gene testing [85]. While in Japan, nested RT-PCR was used [85], which significantly reduced nonspecific target amplification, leading to decreased false positive results (i.e., increased specificity). In general, the sensitivity of these assays ranges from 3.8 to 10 RNA copies per reaction, with high specificities. In a public health emergency, highly sensitive methods are desirable. Indeed, the Chinese authority adopted Wang et al [87] approach for combining RT-PCR with other methods to diagnose SARS-CoV-2, in Wuhan by combining RT-PCR with CT scans [88].

Later, WHO developed a technical guidance including the protocols from different countries to aid COVID-19 diagnosis [85]. According to this compilation, in the US, CDC developed a real-time RT-PCR diagnostic kit with detection limits as low as 4–10 RNA copies per μL.

In addition, rapid test design for the detection of viral protein is also in progress. The viral nucleocapsid (N) and spike (S) are the main immunogenic proteins. While nucleocapsid protein is the most abundant and 90% similar to SARS-CoV, viral spike (S) protein is divergent and elicits strong immune response [89]. However, the S1 subunit of spike protein was found to be specific for SARS-CoV-2, but the S2 subunit of spike protein was conserved across coronaviruses [90]. Since many people have antibodies to the four endemic human coronaviruses, targeting specific part of the spike protein could avoid cross-reactivity.

12.4.2 Limitations

While having the advantage of assisting large mass of samples at a time, its turnaround time is measured in days. The standard RT-PCR test requires, high-cost facilities and machines and sample preparation and assay procedures need well-trained manpower. These shortcomings limit the wider use of the technology during viral pandemics. Since RT-PCR assays amplify specific target loci, but due to behavioral frequent change of SARS CoV, changes in its genome due to insertion or deletion, recombination, and interchange are swift [91,92] so even a single-nucleotide polymorphism due to mutation at the primer or probe binding site could vitiate the true RT-PCR result. The sensitivity may not be enough to detect early infections due to low concentrations of the virus, especially in asymptomatic cases and may result in false negative results. For areas afflicted with COVID-19, a negative PCR result does not imply the absence of the virus since a multitude of factors including viral mutation, PCR inhibition, improper handling of the sample, specimen collection time, low viral RNA, inappropriate shipment, or poor specimen quality can also lead to a negative result in an infected individual [93]. Although studies have shown that RT-PCR may be less sensitive than CT imaging at certain stages of COVID-19, its specificity makes it superior to other methods to detect SARSCoV-2.

12.4.3 Chest imaging

Imaging techniques such as chest X-rays, pulmonary computed tomography (CT) scans, and lung ultrasounds are important tools in the early diagnosis of pneumonia in patients with COVID-19. Apart from detecting the presence of the virus, chest CT scan can demonstrate the disease status and severity. During the early stage of pneumonia, there are multiple small patchy shadows seen with interstitial changes, which is unusual in the lung periphery [94,95]. Some cases could develop bilateral multiple ground-glass opacity, infiltrating shadows, and pulmonary consolidation with infrequent pleural effusion [96]. Chest CT imaging studies showed that chest CT images contained characteristic features for COVID-19 patients. The hallmarks of these CT images include ground glass opacities, crazy-paving pattern, consolidative opacities, septal thickening, and the reverse-halo sign [74,97–99]. These features demonstrate a highly organized pattern of pneumonia. Reports indicate that COVID-19 patients submitted to chest CT scans on admission presented abnormal results (about 90%), showing bilateral multiple ground-glass and patchy opacity [100]. A study with 101 patients in China showed that COVID-19 pneumonia displayed typical imaging features, such as ground glass opacities (GGO) (86.1%) or mixed GGO and consolidation (64.4%), vascular enlargement in the lesion (71.3%), and traction bronchiectasis (52.5%). A recent study suggested that CT scan is more sensitive than the PCR procedure [101]. Importantly, it has a sensitivity that is greater than that of RT-qPCR (97.2 vs. 83.3%) for the diagnosis of COVID-19 patients [101].

However, studies on the sensitivity of CT imaging over RT-PCR, showed that CT imaging is more sensitive and reliable in detecting SARS-CoV-2 infections during certain stages of the COVID-19. Fang et al. studied 51 patients with COVID-19 symptoms based on their clinical manifestations and epidemiological histories [101]. They found that the chest CT scan was more sensitive (98%) than the RT-PCR method (71%). This study was limited by the number of subjects involved. However, another study involving more than 1000 patients reached similar conclusions [88]. Among 1014 patients, 59% were RT-PCR positive, from which 97% showed positive CT features. In some cases, chest CT imaging showed positive SARS-CoV-2 infection, while RT-PCR testing was negative [102]. In addition, 75% of RT-PCR negative patients showed positive CT features. To further validate this, Ai et al. studied multiple RT-PCR testing and serial CT imaging in a selected group. They found that 60–93% people who were RT-PCR negative showed initial positive CT images consistent with SARS-CoV-2 infections. From the patients in the recovery stage, 42% showed improvement in CT features before their RT-PCR results turned negative. According to these diagnostic studies, RT-PCR assays were not as sensitive and reliable as CT images in certain stages of the COVID-19.

Although various types of pneumonias have certain imaging features, COVID-19 and other viral pneumonias, bacterial pneumonia, and some lesions share some common imaging features [103]. In some cases, it is difficult to differentiate COVID-19 from them by imaging alone, and first clinical manifestations, contact history, and laboratory tests should also be considered to make the final diagnosis. In the initial screening, computed tomography (CT) examination is needed for the auxiliary diagnosis [104]. The diagnosis is then confirmed by the positive results of the nucleic acid amplification test (NAAT) of the respiratory tract or blood specimens using reverse transcription real-time fluorescence polymerase chain reaction (RT-PCR) [65]. Chinese researchers strongly recommend CT imaging as the main basis for the diagnosis of COVID-19 in the current situation [46]. An academician of the American Society for Radiation Oncology called for the immediate establishment of a CT-based diagnostic method for COVID-19 and improvement of the detection rate of the severe acute respiratory syndrome coronavirus 2 (SARS-CoV-2) [105]. If a patient with clinically suspected COVID-19 has negative NAAT results but positive imaging results, the patient should be isolated and treated as soon as possible. The advantage of CT examination in the diagnosis of COVID-19 is obvious. Below, we analyze some typical patients who were examined and diagnosed by our institution.

12.4.4 Limitations

Although chest X-rays are less expensive and more convenient for follow up in pneumonia cases, the technique has low-resolution and projection overlapping, which could lead to many false-negative COVID-19 cases [106]. However, CT scans has drawbacks as the mechanism cannot differentiate pneumonia and other pulmonary anomalies from COVID-19 [107]. Despite its high sensitivity in diagnosing COVID-19, chest CT findings in COVID 19 share similarities with other viral pneumonia findings, resulting in false positives [108]. This is because features of the chest imaging from COVID-19 patients may overlap with other infections caused by influenza, H1N1, or SARS-CoV [109,110]. The caveat for the CT scans is that, at an early stage of infection, the lungs of a patient may not develop damaging features that can be picked up by CT scans, increasing its false negative rate.

Despite giving reliable results, the Centres for Disease Control (CDC) in the US does not currently recommend CT to diagnose COVID-19. Laboratory testing of the virus remains the reference standard, even if the CT findings are suggestive of SARS-CoV-2 infections [111].

12.4.5 Immunological test

Immunoassay is another widely used bioanalytical methods to diagnose and/or quantify an analyte using an antigen-antibody interaction. Sensitivity of the immunoassays for the presence of antibodies in human samples may depend on the viral titer and time of sample collected after viral infection; both factors impact circulating antibody concentration [72]. These methods detect viral protein antigens or serum antibodies in patients who have been exposed to the SARS-CoV-2. Several immunoassay kits are already on the market for emergency detection of COVID-19 specific antibodies. These antibody tests are important in detecting prior infections. In the SARS-CoV-2 infection, studies have shown that the seroconversion in the patient generally starts after a week of the first symptom [112]. In a study of postsymptomatic patients, Amanat et al. detected high IgA and IgM immune responses [113]. Most immune assays used for testing individuals for COVID-19 exposure detect specific antibodies (IgG and IgM) in the serum that react with SARS-CoV-2 proteins. Using recombinant viral proteins, this immunoassay could detect antibodies as early as 3 days after the development of the first symptom. IgM is expressed earlier during an infection (~ 3–6 days) and IgG is the late antibody detectable only after ~ 8 days. IgG is generally more specific for a protein antigen than IgM and affinity generally increase with continued exposure to the (viral) antigen [114]. Liu et al. reported that the accuracy of the ELISA for IgG and IgM antibodies was more than 80% [115]. The efficacy of the immunoassay also depends on the specificity of the antigens used to capture the antibodies from the patients. Between the spike (S) proteins and nucleocapsid (N) proteins, the sensitivity of the S proteins is higher for the antibody capture. Among various spike proteins, the S1 protein has shown more capability to bind to SARS-CoV-2 antibodies [116].

Enzyme-linked immunosorbent assays (ELISAs) and lateral flow immunoassays (LFIAs) are the most widely practiced techniques to detect the antibodies against SARS-CoV-2 [117]. In a comparative study, both ELISA and colloidal gold immunochromatographic kits showed equal sensitivity with 100% specificity for the SARS-CoV-2 detection [86]. A two-stage ELISA protocol was described for measuring human antibody responses to the recombinant receptor-binding domain (RBD) of the spike protein or full-length spike protein of SARS-CoV-2 by Krammer's group from the Icahn School of Medicine in New York [118]. The first stage included a high through put screening of samples in a single-serum dilution against the RBD, followed by a second stage in which positive samples from the first stage underwent a confirmatory ELISA against the full-length spike protein.

Recently, a peptide-based magnetic chemiluminescence enzyme immunoassay was also developed for the diagnosis of COVID-19 [119]. The test has high throughput and can perform simultaneous clinical tests for other biomarkers, such as C-reactive protein (CRP), which should also be tracked in COVID-19 suspects [120].

ELISA and chemiluminescence assays based on the antibodies have shown a sensitivity of 70–95% and 82–97% respectively [121–123]. A peptide luminescence method was developed with recombinant S protein to detect IgM and IgG antibody. The total positive rate of detection was 81% with < 6% coefficient of variation. The immunoassay-based rapid diagnostic test (RDT) kits, depending on the vendor type, are reported to have sensitivity in the range of 60–80% and selectivity of 85–100% at a confidence interval of 95% [124]. The RDTs that are LFIAs have shown 86–88% sensitivity and 90–99% specificity to detect total antibodies [121,122,125,126]. A rapid (10 min) and sensitive LFIA that uses lanthanide-doped polystyrene nanoparticles to detect anti-SARS-CoV-2 IgG has also been reported [127]. A study by Bendavid et al. demonstrated 82% sensitivity and 99.5% specificity for detecting antibodies to SARS-CoV-2 tested in 3330 adults and children in Santa Clara County, CA, using LFIAs [57].

A test that detects both IgM and IgG produced a sensitivity, specificity, and accuracy of 57%, 100%, and 69% for IgM and 81%, 100%, and 86% for IgG, respectively. Combining the results from both IgM and IgG yielded a clinical sensitivity of 82% [86].

The Single Molecule Array (SIMOA) multiplexed method developed by Quanterix targets three immunoglobulin responses (IgG, IgM, and IgA) to four viral proteins (spike protein, S1 subunit, receptor binding domain, and nucleocapsid) in a single sample, thereby enabling the quantification of 12 antibody isotype-viral protein interactions. It gives a high-resolution profile of immune response of SARS-CoV-19 when compared with a traditional ELISA where only a single interaction can be interrogated. The SIMOA method demonstrated a sensitivity of 86% and a specificity of 100% during the first week of infection, and 100% sensitivity and specificity thereafter [128]. RDTs employing serological immunoassays are less complex, cost less than the molecular tests, and can give results in a short time period. Immunoassays are good tools to track and study past infections, especially in asymptomatic cases. Serological assays can be used to determine the infection rate and to estimate the population extent and prevalence of infection. Results from a serological survey can also be used to project mortality rates in a community. Furthermore, they are useful to characterize the immune response to the virus.

12.4.6 Limitations

Nevertheless, immunoassays are faster [129] and cheaper than the RT-PCR methods. They can be used for rapid screening of previous SARS-CoV-2 infections. This is particularly useful in the reopening stages of society at which people recovered from previous COVID-19 infections, and therefore immune to the virus, can safely re-engage with society. This method also has a unique advantage of identifying individuals who have strong immune responses against the virus and, therefore, can serve as potential donors for therapeutic and research purposes. However, immunoassays are not as specific, as the antibody tests may cross-react with other pathogens similar to human coronavirus and give false positive results [75,130].

In the past, molecules like interferon, rheumatoid factor, and nonspecific IgM have been shown to cause problems in immunoassays [123], and levels of such potential interferents can be highly variable in COVID-19 patients. There are a few reports where affinity of antibodies used in a biosensor [131], is not sufficient for direct detection of small numbers of virus particles. Results may also be biased due to prior COVID-like illnesses that confound the specificity of the antibody response. One of the difficulties in validating an assay for antiviral antibodies is the availability of appropriate negative and positive controls. Other drawbacks of immunoassay include changes in viral load over the course of infection [132], potential cross reactivity (less specific) [133], and low sensitivity with respect to nucleic acid-based methods. To improve the sensitivity, it is paramount to include other biomarkers of the early stage ofSARS-CoV-2 infection [120]. In addition, the complexity of interpreting results from diagnostics for antiviral antibodies, the antibody response in a patient depends on age, nutritional status, and existing medical conditions and medications [75,116]. Most of the patients develop antibodies only after the second week of infection, that is, in the recovery phase of COVID [134,135]. It loses the main course of period for probability to intervene the disease. WHO has not recommended the use of antibody-based POC systems in clinical decision-making [136].

12.4.7 Serological testing

Serological tests are fundamental to determine the acquired immunity of patients who had the disease and to establish the level of immunity in the general population. This assay overcomes two important limitations of SARS-CoV-2 PCR-based techniques. One that it could be done under biosafety level 2 and second is that no weight given to sampling issues. Serologic testing helps to determine the abundance of an outbreak or extent of infection in society under study. It provides more accurate data of infected population with SARS-CoV-2 and can also assess unknown cases which have not been identified through routine or active surveillance. To analyze the explosion of COVID-19 society-based serosurveys has been done to determine the specificity and sensitivity of SARS-CoV-2 ubiquity and transmission dynamics [137–139]. Moreover, serology can also be useful in situations where RT-qPCR is negative and there is a strong epidemiological link to COVID-19 infection [140]. In these cases, paired serum samples collected in the acute and convalescent phase can be of diagnostic value. Several studies have reported hematologic and blood chemistry alterations in patients infected by SARS-CoV-2 [141,142] which allows us to understand the pattern of infections and help us to opt better defensive measures. It also helps in recognizing asymptomatic individuals who are at higher risk of infections.

Major laboratory findings in COVID-19 patients identified by meta-analysis include leukopenia, leukocytosis, decreased albumin levels, increased levels of C-reactive protein, lactate dehydrogenase (LDH), creatinine kinase, and bilirubin, and a high erythrocyte sedimentation rate (ESR) [143]. Development of an impeccable antibody testing is a major challenge, and as of now, hundreds of trials are going on. There are important considerations like immune status, previous infections of infected person, and cross reaction and time of testing which could differ the test result for antibody testing.

Most serological assays are based on the SARS-CoV-2 nucleocapsid protein (N), transmembrane spike protein (S), or S receptor-binding domain (RBD) because of their high antigenicity [144–146]. As mentioned before, SARS-CoV-2 enters host cells using its spike protein and represents the main target of neutralizing antibodies produced by the host [147]. Several studies have demonstrated that serological tests using the S antigen are more sensitive than those using N-antigen-based antibody assays [148]. Instead of a single antigen, some companies have combined N- and S-based antigens to develop serological assays for SARS-CoV-2 detection [148]. Cross-reactivity between SARS-CoV-2 and SARS-CoV-1 has been reported against the N protein but not against the S1 subunit of the S protein [140]. Cross-reactive antibodies between SARS-CoV-2 and SARSCoV-1 have been detected in both the RBD region located in the S1 subunit and also in non-RBD regions [149].

In the past few months, a variety of serological tests have been designed to detect SARS-CoV-2, mainly neutralization tests, enzyme-linked immunosorbent assays (ELISA), immunofluorescence assays (IFA), chemiluminescence enzyme immunoassays (CLIA), and lateral flow assays (LFA). [148] and are used for the first-line screening of patient samples with COVID-19.

1. Neutralization tests specifies about vital and effective antibodies in the patient by measuring the proportion of viral growth inhibition by antibodies in the lab. This can be used with SARS-CoV-2 virus in a BSL-3 setting, or pseudo viruses that express certain SARS-CoV-2 proteins in a lower BSL setting.
2. Chemiluminescent immunoassay (CLIA) interprets the presence of antibody by exhibiting fluorescent signals during the interaction of viral protein with antibodies in infected patient.
3. Enzyme-linked immunosorbent assays (ELISAs): It is a most common test performed in labs to determine the antigen antibody interaction. Sandwich Elisa is carried out in which antibodies elicited from patients are "sandwiched" between the viral protein and detector antibody to identify active antibody against virus.
4. Lateral flow assays (LFAs) Mostly used in point of care settings providing the qualitative presence of antibodies via color reflection. In this sample from patient passed through a membrane coated with target antigen. Colored band appears if sample is having complementary antibodies interpreting the formation of complex between target antigens with specific antibody. These are similar to pregnancy tests.

However, a few antibody-based tests got emergency approval by the FDA, for instance, COVID-19 ELISA IgG Antibody Test from Cellex, which has a sensitivity of 93%, and qSARSCoV-2 IgG/IgM Rapid Test from Mount Sinai Laboratory with 92.5% sensitivity [13,150]. Another method includes lateral flow chromatographic immunoassay that qualitatively assesses the presence of an analyte (e.g., antibody) from a patient's whole blood, serum, or plasma specimen. Recently, Abbott launched an IgG antibody test received CE mark with its 99.6% specificity [94]. TheElecsyAntiSARS-CoV-2 antibody test from Roche also received FDA approval that employs in-solution double-antigen sandwich format. This test can detect antibodies in human serum or plasma samples with specificity greater than 99.8% and sensitivity of 100% after 14 days of PCR confirmation [151].

The adaptive immune response to SARS-CoV-2 infection has also been studied [60,140,152,153] by Long et al. in 285 patients with COVID-19 in China. Seroconversion for IgM and IgG occurred simultaneously or sequentially, and the median day of seroconversion for both immunoglobulins was 13 days after COVID-19 onset. The seroconversion rate for IgG reached 100% 19 days after symptom onset [140].

In a related work, Zhao et al. showed that the median seroconversion times for total antibodies, IgM, and IgG were 11, 12, and 14 days after SARS-CoV-2 infection [154]. In addition, several studies have been used to evaluate the clinical sensitivity of serological tests with COVID-19 patient samples collected on different days after the onset of symptoms [155–158]. Pan et al. evaluated a commercial lateral flow immunochromatographic test targeting viral IgM or the IgG antibody and compared it with RT-qPCR. The sensitivity of the assay for SARS-CoV-2 detection was 11.1% in early-stage patients (1–7 days after onset), 92.9% in intermediate-stage patients (8–14 days after onset), and 96.8% in late-stage patients (more than 15 days after onset) [157]. In similar study, Tang et al. compared the diagnostic performance of two SARS-CoV-2 commercial serologic tests using 103 specimens from PCR-confirmed SARS-CoV-2 patients and 153 control specimens from different days after disease onset. The IgG antibodies bind to the recombinant SARS-CoV-2 antigen-coated gold nanoparticles (AuNP); the rabbit IgG gold-conjugates are used as control that binds to the anti-rabbit antibodies. The assay resembles a lateral flow pregnancy test but detects antibodies instead of a human glycoprotein.

The serological assays are relatively quick and easy to use which requires no highly trained staff and sophisticated equipment. Therefore, they are being increasingly used for diagnosis, contact tracing, herd immunity assessment, and vaccine efficacy evaluation. Serology testing emerges into a useful tool which may have several more applications in the future. It could be useful for complementing RNA tests, for the rapid identification of cases and for recommending quarantine to remarked raceme. It could be a reliable way to identifying the route of transmission to monitor the spread of the pandemic and facilitate epidemiological studies. Serology assay may also be useful to check immune status after vaccination. Seroepidemiologic studies can assist in the investigation of the ongoing pandemic and the retrospective assessment to determine the attack rate or the progress of the pandemic through antibody detection. These studies can assist health authorities and governments in making sound decisions with respect to the implementation of public health measures during the course of the current pandemic

12.4.8 Emerging test

As discussed in Ideal Characteristics of Diagnostic Methods, diagnostic tests developed so far rarely meet all the ASSURED criteria. The most important features for the SARS-CoV-2 detection are sensitivity, specificity, and efficiency (throughput and cost-effective ness). In addition to the modification of conventional methods approaches like RT-PCR, isothermal amplification for nucleic acid targets, lateral flow based detection of nucleic acid targets sections, and immunoassay other emerging methods named CRISPR technology, differential diagnostics, AI assays have further been developed to fasten and specified diagnostic features. Besides all these techniques single molecule detection capability is used to improve the sensitivity [159–161].

12.4.8.1 Isothermal amplification for nucleic acid targets

Although RT-PCR is a widely used method in the confirmatory screening of COVID-19 infections (see Nucleic Acid Based Methods section), it is time-consuming and requires a sophisticated laboratory facility and trained personnel to operate [68]. Apart from conventional RT-PCR, additional molecular diagnostic tools are emerging for SARS-CoV-2.

Whole-genome sequencing of SARS-CoV-2 has the potential to overcome the limitations of RT-PCR. Genomic sequencing can detect fragments even when a complete genome is not present in the sample. Notably, the SARS-CoV-2 genome is free of repeats, making it susceptible to complete characterization using short sequence reads. The genome sequencing involves the construction of RNA library, next-generation sequencing (NGS) of the RNA construct, de novo assembly of the quality trimmed reads to generate a contig map and phylogenetic analysis [162]. The genome sequencing may overcome the false results arising from specious priming. However, due to cost, technical complexities associated with the instrument, data analysis, and higher turnaround time of 24–48 h, this method may not be a routine use for clinical diagnosis during pandemic [163].

Droplet-based digital PCR (dPCR) methods have also been sought as a more sensitive method to test for the RNA of SARS-CoV-2. This method, when tested in 77 patient samples, showed an improved sensitivity from 44 to 94% and the same specificity when compared with the RT-PCR method [164]. As the LOD reported for RT-PCR has been 2.1 copies/reaction for ORF1ab and 1.8 copies/reaction for N primer/probe set and in d PCR method, it was found as 2 copies/reaction [165]. When tested in 103 fever-suspected patients, the d PCR method improved the sensitivity from 28.2% with RT-PCR to 87.4%.

Dual-functional plasmonic biosensor combined the plasmonic photothermal effect and localized surface plasmon resonance sensing transduction was reported as an alternative for the clinical diagnosis of COVID-19 [136]. Two-dimensional gold nanoislands functionalized with complementary DNA receptors were used to detect SARS-CoV-2 through nucleic acid hybridization. The biosensor was able to detect 0.22 pM and allowed precise detection of the specific target in a multigene mixture. Gold nanoparticles (AuNPs) functionalized with antisense oligonucleotides were developed for the simultaneous detection of two regions of the SARS-CoV-2 N gene [166]. In the presence of RNA target, AuNPs agglomerate, and the agglomeration is further enhanced by the addition of RNAse H enzyme to generate a distinguishable visible precipitate.

To simplify the testing procedures, isothermal nucleic acid amplifications have been developed. These methods do not require any thermal cycler to perform the amplification and, therefore, can be carried out in a simple water bath at a constant temperature of 40–65°C [167].

Reverse transcription loop mediated isothermal amplification (RT-LAMP) is an amplification approach in which RNA genome of SARS-CoV-2 is first reverse transcribed to cDNA and is then amplified using four to six target-specific primers. Prior to the LAMP amplification, a dumbbell-shaped single-stranded DNA (ssDNA) is formed through the annealing and the strand-displacing cycle on both ends of the target sequence with the help of the primers and a strand-displacing polymerase. The looped ssDNA on each end then serves as a seed for the LAMP amplification cycle [168–171]. As a result, the target sequence is amplified exponentially, which is detected by turbidometry [172] or fluorescence/colorimetry [171]. As an example, RNA extraction and LAMP amplification have been performed in the same tube [168,173]. This method has LOD ranging 80–500 SARS-CoV-2 RNA copies per milliliter, which is comparable to the RT-PCR assay.

Rolling circle amplification (RCA) [174] is another isothermal amplification method that gives sensitive detection of nucleic acids. In this method, a segment of the target genome is circularized and amplified by a highly processive strand displacing DNA polymerase. Wang et al. used this method to develop a highly sensitive and efficient assay for SARS-CoV [175]. Compared to the LAMP assay, the RCA method is simpler since it requires fewer steps and can be performed at room temperature. The method offers high sensitivity comparable to RT-PCR [176] since it amplifies the target sequence by ~10,000. In addition, it presents high specificity, as the RCA is initiated only after the formation of a circular template upon which a specific primer is hybridized [176]. Therefore, RCA reduces false-positive results often encountered in PCR-based assays.

A major difficulty in this method is that it requires a circular template whose preparation is dependent on the length of a linear template and the ligation efficiency of the DNA circularization. Inappropriate design of complementary sequences therefore results in failure of amplifications.

Penn-RAMP was developed by, El Tholoth et al.to improve the LOD. It is a two-stage closed tube test combines LAMP with recombinase polymerase amplification [177]. In the Penn-RAMP, each amplification was performed at a separate compartment in a single tube followed by mixing. The method demonstrated 10 times higher sensitivity than LAMP or RT-PCR alone. A new platform with engineered complementary recombinant RNA (crRNA) has been reported with LOD as low as ~700 fM cDNA from HIV, 290 fM RNA from HCV, and 370 fM cDNA from SARS-CoV-2 in 30 min without a need for target amplification. In this case, the isothermal amplification of SARS-CoV-2 RNA was performed using RT-LAMP, and the modified crRNAs were incorporated in a paper-based lateral flow assay that could detect the target within 40–60 min [178].

Some commercial COVID-19 diagnostic kits based on isothermal RT-LAMP assays are already on the market (see Table 12.1). Abbott ID Now is such an example. This method only requires 5 min to give positive results. Recently, however, issues on false negativity have been raised for the Abbott ID Now because of its relatively high LOD [179]. This may be attributed to the compromised performance of the RdRP target [180,181] used in this assay, which is found to be mutating and evolving [182].

12.4.8.2 *Lateral flow-based detection of nucleic acids and proteins (LFA)*

The nucleic acid based isothermal amplifications discussed above partially overcome the limitations of conventional RT-PCR assays, as they do not require sophisticated laboratory facilities while their turnaround time is short. However, these methods still require trained staff to operate various sample collection and processing steps. In LFAs, both nucleic acid detection methods and immunoassays can be utilized. The device is often made of papers with immobilized capture probes. Upon binding with nucleic acid targets, the probes give a visible signal [183–185].

Paper-based lateral flow assays have gained interest because of their low cost, easy manufacture, and full compatibility with POCT, which allow them to be conveniently performed by anyone at home. Byers et al. developed a 2D paper network to perform immunoassay for the detection of nucleic acids of SARS-CoV-2 with the POCT format [186]. Such methods still require initial nucleic acid extraction and amplification steps, the latter of which can be accomplished by the PCR or isothermal amplifications as discussed above. On the POCT platform, all those steps are integrated in a single device.

TABLE 12.2 Screening and confirmatory genes with their limit of detection of various research organizations during COVID-19 pandemic.

S. No.	Agency	Screening gene	Confirmatory gene	LOD
1.	CDC	N1-N2		1 micro lt
2.	Charite protocol	E gene	RdRp-P2	5.2 and 3.8 /rexn
3.	Pasteur Institute	IP2-IP4	E gene	10/rexn
4.	Chu et al.	N gene	ORF1b	<10
5.	HKU (Hon Kong University)			10/microlitre
6.	Germany	E gene	RdRp	5.6 & 3.8
7.	France	RdRp gene	E gene	

A novel method based on the Sanger sequencing was developed by the Chandler-Brown group that targets the SARS-CoV-2 N protein [187]. This method uses COVID-19 spike-in DNA as an internal control that provides quantification of the viral mRNA. The major advantages of this method are the omission of RNA extraction procedure so there will be no impedance in the testing capacity due to shortage of RNA extraction kits and potential for a very high performance of one million tests per day with proper customization of the sequencer.

12.4.8.3 Immunoassay

Similarly, Byers et al. developed a 2D paper network to perform immunoassay for the detection of nucleic acids of SARS-CoV-2 with the POCT format [186]. Although nucleic acid based lateral flow assays are sensitive, lateral flow immunoassays have gained interest in the massive surveillance of the COVID-19 pandemic because of their simplicity and cheap cost. An electrochemical ultrasensitive POC device named eCovSens detects the spike protein of SARS-CoV-2 within 10–30 s [188]. This device was fabricated by immobilizing antibody against S1 protein on screen-printed carbon electrodes. With spiked saliva samples, the LOD of the device was found to be 90 fM.

Currently, IgM/IgG rapid test kits are available for qualitative antibody testing of COVID-19. Many such commercial devices have already been developed (see Table 12.2). One problem associated with the immunoassay based lateral flow assay is the weak signal, which results in reduced sensitivity [183]. Various signal enhancement strategies therefore have been proposed. A promising signal amplification strategy in lateral flow assays is the use of colloidal gold nanoparticles conjugated with the probes. Upon binding with the target, the gold nanoparticles linked to the capture probe aggregate to change the color, enhancing the signal [189]. Other signal amplification strategies include solvent evaporation for analyte preconcentrations nanoparticle, catalyzed nanoparticle labeled assays, and ion concentration polarization methods [190]. Due to the low-cost requirement of the POCT, detection in the LFA is usually achieved by visual inspection. To improve detection sensitivity, cameras in smartphones have been used [191]. These cameras are sensitive to subtle color changes and hence provide more effective color detection than traditional RGB sensors or the naked eye [192]. For improved readout of the results and data processing, a machine learning algorithm could also be used [193]. Smartphones can also be coupled with external adapters to integrate external biosensor platforms for more versatile POC testing [194].

The single molecule enzyme-linked immunosorbent assay (ELISA) has been developed to offer a detection limit of subfemto molar protein concentrations [195]. In this method, each microscopic bead decorated with specific antibodies is loaded into individual femtoliter wells. Sensing was accomplished by the ELISA on each bead, whereas the excellent concentration detection limit was achieved by a large array of such beads. To be applicable in clinical setting, however, this method requires special equipment, increasing its cost.

12.4.8.4 Proximity ligation assay (PLA)

The method utilizes two or more DNA tagged aptamers or antibodies for binding of multiple targets. [196] The DNA tags on the probes are amplified only when the two different targets are in close proximity. The multiple targets ensure the specificity of target detection. However, this method requires intact SARS-CoV-2 virus particles from which two different targets are present for positive detection. This demands stringent sample processing steps to increase the throughput.

12.4.8.5 Next generation sequencing [197] and DNA microarray [198]

In the case of COVID-19, evidence has suggested that the SARS-CoV-2 is rapidly evolving while infecting people. Therefore, it is critical to rapidly identify the genome of the causative agent [199]. The DNA microarray has been used in high-work rate identification of mutations in SARS-CoV-2 [200]. However, for these methods, the time limiting step becomes the sample collection, which must be performed one-at-a-time. In addition, these methods involve rather advanced equipment with high cost; therefore, they may not be appropriate for the economic and rapid screening in the COVID-19 pandemic.

12.4.8.6 CRISPR/Cas-based diagnostic methods

The clustered regularly interspaced short palindromic repeats (CRISPR)/Cas machinery has recently been adapted as a POC tool for the rapid detection of nucleic acids (DNA or RNA) [154,201]. Overall, this CRISPR machinery is programmed to cleave specific sequences in the DNA/RNA target where the results can be easily observed by combination with a lateral-flow strip. CRISPR gene editing tool has been employed to construct an accurate, faster, and simple-to-use SARS-CoV-2 detection test. A rapid (<40 min), easy-to-implement, and accurate CRISPR–Cas12-based lateral flow assay for the detection of SARS-CoV-2 from respiratory swab RNA extracts has been recently reported.

DNA endonuclease-targeted CRISPR trans reporter (DETECTR) developed by Mammoth Bioscience Company based on the CRISPR/Casto detect any RNA or DNA target, which has now been used to detect the SARS-CoV-2 RNA genome from respiratory swab RNA extracts [202]. It is based on CRISPR–Cas12 and it can distinguish SARS-CoV-2 with no cross-reactivity for related coronavirus strains using N gene gRNA within 40 min [202]. The result is visualized by using a FAM-biotin reporter molecule and lateral flow strips to capture labeled nucleic acids. The suitability of DETECTR technology for the detection of SARS-CoV-2 was evaluated using 78 patient specimens, including 36 patients with COVID-19 infection and 42 patients with other viral respiratory infections, and then compared with the CDC RT-qPCR as a reference method to confirm COVID-19 infection. Clinical sensitivity and specificity were 95% and 100%, respectively, for the detection of the coronavirus in 83 total respiratory swab samples [202]. The SARS-CoV-2 DETECTR test had 95% positive predictive agreement and 100% negative predictive agreement when compared with RT-qPCR results. The LOD of this method was 10 copies per microliter reaction.

Specific high-sensitivity enzymatic reporter unlocking (SHERLOCK), was developed by Zhang's team on the basis of CRISPR/Cas-based platform that combined isothermal preamplification to detect strains of single-strand RNA viruses, identifies mutations and human genotype DNA, and distinguishes pathogenic bacteria [154,201]. More recently, using the same knowledge, they adapted a protocol using the SHERLOCK system for SARS-CoV-2 detection [203], which incorporates a thermostable Cas12b enzyme from alicyclobacillus acidiphilus and targets S and ORF1ab gene fragments of SARS-CoV-2 was described by Sherlock Biosciences. The test recently got FDA approval under emergency use authorizations [204]. The test can be conducted by extracting RNA from patient samples and can be read out in less than an hour using a dipstick, without requiring extensive instrumentation. This method was able to detect synthetic COVID-19 RNA sequence in a range between 10 and 100 copies per microliter and could be completed within 1 h [203].

FnCas9 Editor Linked Uniform Detection Assay (FELUDA) is a different approach taken to build a CRISPR-based tool in India. It is a highly accurate enzymatic readout for detecting nucleotide sequences [205]. The assay is quick to provide output and can be used in rapid diagnosis.

In other developments, Zhang's group [203] and Chiu's group [202] integrated the LAMP with the CRISPR-based SHERLOCK (see Table 12.1) [206] and CRISPR-Cas12 based methods, respectively, to detect the SARS-CoV-2 RNA with a detection limit as low as 10 copies/μL on a point-of-care testing (POCT) format. Taken together, these results highlight the great potential of CRISPR-based diagnostic methods as a rapid, specific, portable, and accurate detection platform for the detection of the SARS-CoV-2 genome in patient samples. Despite these promising results, CRISPR/Cas-based diagnostic methods are not widely used by diagnostic laboratories and need further implementation.

12.4.8.7 Artificial intelligence

Though relatively slow and expensive, computed tomography (CT) of the chest has also been explored in the diagnosis of COVID-19 as a complementary tool to molecular techniques when combined with medical history and clinical observations [46,187]. In a recent study, artificial intelligence (AI) algorithms were used to integrate chest CT findings with clinical symptoms, exposure history, and laboratory testing to rapidly diagnose patients who are positive for COVID-19 [207]. Moreover, the machine learning approach shows potential to predict criticality in patients [208]. The AI system had equal sensitivity as compared with a senior thoracic radiologist. In addition, it improved the detection of patients who were positive for COVID-19 via RT-PCR who presented with normal CT scans, correctly identifying 17 of 25 (68%) patients, whereas radiologists classified all of these patients as negative for COVID-19. When CT scans and associated clinical history are available, the proposed AI system can help to diagnose COVID-19 patients. The supplementary CT scan imaging may help to rule out negative RT-PCR results [88]. Some AI inspired mobile application-based tools are in development to preliminary detect suspected COVID-19 patients [13].

12.4.8.8 Differential diagnosis

Differential diagnosis is the process by which a single disease or condition is differentiated from those having similar clinical features. Patients with COVID-19 can have coinfection or superimposed infection by other viruses or bacteria simultaneously. Differential diagnosis is therefore important to differentiate SARS-CoV-2 induced infection from other viral or bacterial and mycoplasma pneumonia. Since the clinical manifestations of COVID-19 are similar to those of other respiratory diseases, differential diagnosis is of paramount importance in assisting physicians in the therapeutic management of patients and health officials in establishing disease control policies.

In an effort to do so, respiratory pathogens in patients suspected of COVID-19 in Italy have been investigated [109]. Here, researchers tested the nasopharyngeal swabs of 126 suspected causes using SARS-CoV-2 RT-qPCR and a commercial multiplex respiratory cartridge that detects and differentiates viral and bacterial pathogens. Only 3 patients were confirmed to be infected with SARS-CoV-2, and 67 (53.2%) were positive for other respiratory pathogens, including 26 (20.6%) positive for influenza A and 10 (7.9%) positive for influenza B. Other pathogens detected in the patient samples were common cold CoV (H-CoV 229 E, H-CoV NL63, and H-CoV HKU1), rhinovirus, enterovirus, metapneumovirus, adenovirus, Mycoplasma pneumoniae, Streptococcus pneumoniae, and Legionella pneumophila [109]. These results highlight the diagnostic differential and demonstrate the importance of using a spectrum of molecular kits for the rapid detection of respiratory pathogens in order to improve the clinical management and treatment of patients infected with COVID-19. In another study, Yan et al. reported the cases of two patients in Singapore who initially had moderate symptoms including myalgia, a mild cough, and diarrhea and presented with thrombocytopenia and a normal chest radiograph [209]. Dengue was suspected, and a rapid serological test for dengue produced false-positive results. As patient symptoms worsened, they were later diagnosed with COVID-19 by RT-qPCR. Taken together, these findings suggest that special attention is needed in the differential diagnosis between arboviruses and SARS-CoV-2 infection, especially in countries where is there a circulation of DENV, ZIKV, and CHIKV. Moreover, coinfection with SARS-CoV-2 and other respiratory viruses, such as common cold CoV, influenza, and metapneumovirus, have been reported, highlighting the need for repeat testing based on clinical indications [210–213].

References

[1] R. Li, S. Pei, B. Chen, Y. Song, T. Zhang, W. Yang, et al., Substantial undocumented infection facilitates the rapid dissemination of novel coronavirus (SARS-CoV2), Science 368 (2020) 489.

[2] F. Wu, S. Zhao, B. Yu, Y.M. Chen, W. Wang, Z.G. Song, et al., A new coronavirus associated with human respiratory disease in China, Nature 265 (9) (2020) 579.

[3] F. Li, Structure, function, and evolution of coronavirus spike proteins, Annu Rev Virol (3) (2016) 237–261.

[4] N. Zhong, B. Zheng, Y. Li, L. Poon, Z. Xie, K. Chan, et al., Epidemiology and cause of severe acute respiratory syndrome (SARS) in Guangdong, People's Republic of China, in February,, The Lancet 362 (9393) (2003) 1353–1358.

[5] N. Wang, X. Shi, L. Jiang, S. Zhang, D. Wang, P. Tong, et al., Structure of MERS-CoV spike receptor-binding domain complexed with human receptor DPP4, Cell Research 23 (8) (2013) 986.
[6] A.E. Gorbalenya, S.C. Baker, R.S. Baric, et al., The species Severe acute respiratory syndrome-related coronavirus: Classifying 2019-nCoV and naming it SARS-CoV-2, Nat. Microbiol 5 (4) (2020) 536–544.
[7] P. Zhou, X.L. Yang, X.G. Wang, B. Hu, L. Zhang, W. Zhang, et al., A pneumonia outbreak associated with a new corona virus of probable bat origin, Nature 270 (3) (2020) 579.
[8] Gen (Bank). (https://www.ncbi.nlm.nih.gov/genbank/)
[9] R. Lu, X. Zhao, J. Li, P. Niu, B. Yang, H. Wu, et al., Genomic characterization and epidemiology of 2019 novel coronavirus: Implications for virus origins and receptor binding, Lancet 395 (2020) 565–574.
[10] T.T.Y. Lam, M.H.H. Shum, H.C. Zhu, Y.G. Tong, X.B. Ni, S. LiaoY, et al., Identifying SARS-CoV-2 related coronaviruses in Malayan pangolins, Nature 1 (6) (2020) 579.
[11] S. Sanche, Y.T. Lin, C. Xu, E. Romero-Severson, N. Hengartner, R. Ke, High contagiousness and rapid spread of severe Acute respiratory syndrome coronavirus 2, Emerging Infectious Diseases 26 (7) (2020).
[12] J.T. Wu, K. Leung, M. Bushman, et al., Estimating clinical severity of COVID-19 from the transmission dynamics in Wuhan, China, Nature Medicine 26 (4) (2020) 506–510.
[13] E. De Wit, N. Van Doremalen, D. Falzarano, V.J. Munster, SARS and MERS: Recent insights into emerging coronaviruses, Nature Reviews Microbiology 14 (8) (2016) 523–534.
[14] W. Guan, Z. Ni, Y. Hu, W. Liang, C. Ou, J. He, et al., Clinical characteristics of coronavirus disease 2019 in China, New England Journal of Medicine 382 (2020) 1708.
[15] WHO Report of the WHO-China Joint Mission on Coronavirus Disease (2019). (COVID-19) 2020. https://www.who.int/docs/default-source/coronaviruse/who-china-joint-mission-oncovid-19-final-report.pdf (accessed April 10, 2020).
[16] I.M. Mackay, K.E. Arden, A. Nitsche, Real-time PCR in virology, Nucleic Acids Res. 30 (2002) 1292–1305.
[17] X. Wang, L. Tan, X. Wang, W. Liu, Y. Lu, L. Cheng, et al., Comparison of nasopharyngeal and oropharyngeal swabs for SARS-CoV-2 detection in 353 patients received tests with both specimens simultaneously, International Journal of Infectious Diseases 107 (9) (2020) 94.
[18] Berenger, B.M., Fonseca, K., Schneider, A.R., Hu, J., & Zelyas, N. (2020).Sensitivity of nasopharyngeal, nasal and throat swab for the detection of SARS-CoV-2, preprint, Infectious Diseases [except HIV/AIDS].
[19] S.A. Irving, M.F. Vandermause, D.K. Shay, E.A. Belongia, Comparison of nasal and nasopharyngeal swabs for influenza detection in adults, Clin Med Res 10 (2012) 215–218.
[20] P. Winichakoon, R. Chaiwarith, C. Liwsrisakun, P. Salee, A. Goonna, A. Limsukon, et al., Negative nasopharyngeal and oropharyngeal swabs do not rule out COVID-19, Journal of Clinical Microbiology 58 (5) (2020) E00297–20.
[21] OASH. Nasal [anterior nasal] specimen collection for SARS-CoV2 diagnostic testing. https://www.cdc.gov/coronavirus/2019-ncov/downloads/OASH-nasal-specimen-collection-fact-sheet.pdf [accessed May 20, 2020].
[22] WHO Coronavirus Disease 2019 (COVID-19), Situation Report*83*; (2020 e).
[23] Q. Li, X. Guan, P. Wu, X. Wang, L. Zhou, Y. Tong, et al., Early transmission dynamics in Wuhan, China, of novel coronavirus–infected pneumonia, New England Journal of Medicine (2020).

[24] J. Parry, China coronavirus: Cases surge as official admits human to human transmission, British Medical Journal Publishing Group, 2020.
[25] L.T. Phan, T.V. Nguyen, Q.C. Luong, T.V. Nguyen, H.T. Nguyen, H.Q. Le, et al., Importation and human-to-human transmission of a novel coronavirus in Vietnam, New England Journal of Medicine (2020).
[26] J. Riou, CL. Althaus, Pattern of early human-to-human transmission of Wuhan 2019 novel coronavirus (2019-nCoV), December 2019 to January 2020, Euro Surveillance 25 (4) (2020).
[27] N. Nagata, N. Iwata, H. Hasegawa, S. Fukushi, M. Yokoyama, A. Harashima, et al., Participation of both host and virus factors in induction of severe acute respiratory syndrome (SARS) in F344 rats infected with SARS coronavirus, Journal of Virology 81 (4) (2007) 1848–1857.
[28] T.P. Sheahan, A.C. Sims, S.R. Leist, A. Schäfer, J. Won, A.J. Brown, et al., Comparative therapeutic efficacy of remdesivir and combination lopinavir, ritonavir, and interferon beta against MERS-CoV, Nature Communications 11 (1) (2020) 1–14.
[29] Z. Wang, X. Chen, Y. Lu, F. Chen, W. Zhang, Clinical characteristics and therapeutic procedure for four cases with 2019 novel coronavirus pneumonia receiving combined Chinese and Western medicine treatment, BioScience Trends (2020).
[30] M.L. Holshue, C. DeBolt, S. Lindquist, K.H. Lofy, J. Wiesman, H. Bruce, et al., First case of 2019 novel coronavirus in the United States, New England Journal of Medicine (2020).
[31] C.S. Ng, D.M. Kasumba, T. Fujita, H. Luo, Spatio-temporal characterization of the antiviral activity of the XRN1-DCP1/2 aggregation against cytoplasmic RNA viruses to prevent cell death, Cell Death and Differentiation (2020) 1–20.
[32] D.S.C. Hui, A. Zumla, Severe acute respiratory syndrome: Historical, epidemiologic, and clinical features, Infect Dis Clin N Am 33 (2019) 869–889.
[33] WHO Laboratory testing for coronavirus disease 2019 [COVID19] in suspected human cases: Interim guidance, 2 March 2020, World Health Organization. (2020 d). https://apps.who.int/iris/handle/10665/331329?show=full. Accessed 2 July 2020.
[34] L. Zou, F. Ruan, M. Huang, L. Liang, H. Huang, Z. Hong, et al., SARS CoV-2 viral load in upper respiratory specimens of infected patients, New England Journal of Medicine 382 (2020) 1177–1179. https://doi.org/10.1056/NEJMc2001737.
[35] C. Drosten, M. Seilmaier, V.M. Corman, W. Hartmann, G. Scheible, S. Sack, et al., Clinical features and virological analysis of a case of Middle East respiratory syndrome coronavirus infection, The Lancet Infectious Diseases 13 (2013) 745–751.
[36] SARS-(CoV-2). possesses the typical coronavirus structure with spike protein and also expressed other polyproteins, nucleoproteins, and membrane proteins, such as RNA polymerase, 3- chymotrypsin-like protease, papain-like protease, helicase, glycoprotein, and accessory proteins.
[37] X. Huang, W. Dong, A. Milewska, A. Golda, Y. Qi, Q.K. Zhu, et al., Human coronavirus HKU1 spike protein uses O-acetylated sialic acid as an attachment receptor determinant and employs hemagglutinin-esterase protein as a receptor-destroying enzyme, Journal of Virology 89 (2015) 7202–7213.
[38] J.K. Millet, GR. Whittaker, Host cell proteases: critical determinants of coronavirus tropism and pathogenesis, Virus Research 202 (2015) 120–134.
[39] H. Hofmann, S. Pöhlmann, Cellular entry of the SARS coronavirus, Trends in Microbiology 12 (2004) 466–472.
[40] M. Gui, W. Song, H. Zhou, J. Xu, S. Chen, Y. Xiang, et al., Cryoelectron microscopy structures of the SARS-CoV spike glycoprotein reveal a prerequisite conformational state for receptor binding, Cell Research 27 (2017) 119–129.

[41] D. Wrapp, N. Wang, K.S. Corbett, J.A. Goldsmith, C.L. Hsieh, O. Abiona, et al., Cryo-EM structure of the 2019-nCoV spike in the prefusion conformation, Science 367 (2020) 1260–1263.
[42] L.E. Gralinski, VD. Menachery, Return of the Coronavirus: 2019nCoV, Viruses 12 (2020) 135.
[43] S. Perlman, J. Netland, Coronaviruses post-SARS: Update on replication and pathogenesis, Nature Reviews Microbiology 7 (6) (2009) 439–450.
[44] V.S. Raj, H. Mou, S.L. Smits, D.H. Dekkers, M.A. Müller, R. Dijkman, et al., Dipeptidyl peptidase 4 is a functional receptor for the emerging human coronavirus-EMC, Nature 495 (7440) (2013) 251–254.
[45] X. Xu, P. Chen, J. Wang, J. Feng, H. Zhou, X. Li, et al., Evolution of the novel coronavirus from the ongoing Wuhan outbreak and modelling of its spike protein for risk of human transmission, Science China Life Sciences 63 (3) (2020) 457–460.
[46] Y. Wan, J. Shang, R. Graham, R.S. Baric, F. Li, Receptor recognition by novel coronavirus from Wuhan: An analysis based on decade-long structural studies of SARS, Journal of Virology (2020), doi:10.1128/JVI.00127-20.
[47] M. Hussain, N. Jabeen, F. Raza, S. Shabbir, A.A. Baig, A. Amanullah, et al., Structural variations in human ACE2 may influence its binding with SARS-CoV-2 spike protein, Journal of Medical Virology (2020). https://doi.org/10.1002/jmv.25832.
[48] M.P. Cheng, J. Papenburg, M. Desjardins, S. Kanjilal, C. Quach, M. Libman, et al., Diagnostic testing for severe acute respiratory syndrome–related coronavirus-2: A narrative review, Annals of Internal Medicine (2020). https://doi.org/10.7326/M20-1301.
[49] S. Knobler, A. Mahmoud, S. Lemon, A. Mack, L. Sivitz, K. Oberholtzer, Learning from SARS: Preparing for the next disease outbreak. Workshop summary, National Academies Press, United States, 2004, doi:10.17226/10915.
[50] WHO Managing epidemics: Key facts about major deadly diseases; (2018). https://www.who.int/publications/i/item/managingepidemicskeyfactsaboutmajordeadly diseases.
[51] W. Wang, Y. Xu, R. Gao, R. Lu, K. Han, G. Wu, et al., Detection of SARS-CoV-2 in different types of clinical specimens, JAMA 323 (2020) 1843–1844.
[52] A. See, S.T. Toh, Respiratory sampling for severe acute respiratory syndrome coronavirus 2: An overview, Head & Neck 42 (2020) 1652–1656.
[53] CDC. Coronavirus disease 2019 [COVID-19] https://www.cdc.gov/coronavirus/2019-ncov/lab/guidelines-clinical-specimens.html. Accessed 28 Apr (2020).
[54] P.H.L.N. PHLN., Guidance on laboratory testing for SARS-CoV-2 [the virus that causes COVID-19], PHLN, 2020.
[55] W. Wang, Y. Xu, R. Gaoetal, Detection of SARS-CoV-2 in different types of clinical specimens, Journal of the American Medical Association 323 (18) (2020) 1843–1844.
[56] W. Feng, A. Newbigging, C. Le, B. Pang, H. Peng, Y. Cao, et al., Molecular diagnosis of COVID-19: Challenges and research needs, Analytical Chemistry 92 (2020) 10196–10209.
[57] E. Bendavid, B. Mulaney, N. Sood, S. Shah, E. Ling, R. BromleyDulfano, et al., COVID-19 antibody seroprevalence in Santa Clara County, California, medRxiv (2020). https://doi.org/10.1101/2020.04.14.20062463.
[58] J.A. Backer, D. Klinkenberg, J. Wallinga, Incubation period of 2019 novel coronavirus (2019-nCoV) infections among travellers from Wuhan, China, 20-28 January 2020, Euro Surveillance 25 (2020).
[59] Z. Yongchen, H. Shen, X. Wang, X. Shi, Y. Li, J. Yan, et al., Different longitudinal patterns of nucleic acid and serology testing results based on disease severity of COVID-19 patients, Emerg Microbes Infect 9 (2020) 833–836.
[60] J. Zhao, Q. Yuan, H. Wang, W. Liu, X. Liao, Y. Su, et al., Antibody responses to SARS-CoV-2 in patients of novel coronavirus disease, Clinical Infectious Diseases (2019). https://doi.org/10.1093/cid/ciaa344.

[61] U.S. FDA. FAQs on emergency use authorizations [EUAs] for medical devices during the COVID-19 pandemic. FDA. (2020). https://www.fda.gov/medical-devices/emergency-situationsmedical-devices/faqs-emergency-use-authorizations-euas-medicaldevices-during-covid-19-pandemic. Accessed 2 July 2020.
[62] C.B.F. Vogels, A.F. Brito, A.L. Wyllie, J.R. Fauver, I.M. Ott, C.C. Kalinich, et al., Analytical sensitivity and efficiency comparisons of SARS-CoV-2 RT-qPCR primer-probe sets, Nat Microbiol (2020) Epub 2020/07/12. pmid:32651556.
[63] D.L. Heymann, N. Shindo, COVID-19: What is next for public health? Lancet 395 (2020) 542–545.
[64] J. Watson, F. Whiting, J.E. Brush, Interpreting a COVID-19 test result, BMJ (Clinical Research Ed.) 369 (2020) 1–7.
[65] V.M. Corman, O. Landt, M. Kaiser, R. Molenkamp, A. Meijer, D.K. Chu, et al., Detection of 2019 novel coronavirus [2019-NCoV] by real-time RT-PCR, Euro Surveillance 25 (2020) 2000045.
[66] A.C. Nelson, B. Auch, M. Schomaker, D.M. Gohl, P. Grady, D. Johnson, et al., Analytical validation of a COVID-19 QRT-PCR detection assay using a 384-well format and three extraction methods, BioRxiv (2020). https://doi.org/10.1101/2020.04.02.022186.
[67] Accelerated Emergency Use Authorization (EUA) (Summary). COVID-19 RT-PCR Test (Laboratory Corporation of America), https://www.fda.gov/media/136151/download.109.
[68] S.A.& Bustin, R. Mueller, Real-time reverse transcription PCR (qRT-PCR) and its potential use in clinical diagnosis, Clinical Science 109 (2005) 365–379.
[69] Y. Pan, L. Long, D. Zhang, et al., Potential false-negative nucleic acid testing results for severe acute respiratory syndrome coronavirus 2 from thermal inactivation of samples with low viral loads, Clinical Chemistry 66 (6) (2020) 794–801.
[70] S. Su, G. Wong, W. Shi, J. Liu, A.C. Lai, J. Zhou, et al., Epidemiology, genetic recombination, and pathogenesis of coronaviruses, Trends in Microbiology 24 (2016) 490–502.
[71] K.K. To, O.T. Tsang, C.C. Yip, K. Chan, T. Wu, J.M.C. Chan, et al., Consistent detection of 2019 novel coronavirus in saliva, Clinical Infectious Diseases (2020), doi:10.1093/cid/ciaa149.
[72] K.K. To, O.T. Tsang, W.S. Leung, A.R. Tam, T.C. Wu, D.C. Lung, et al., Temporal profiles of viral load in posterior oropharyngeal saliva samples and serum antibody responses during infection by SARS-CoV-2: An observational cohort study, The Lancet Infectious Diseases 20 (2020) 565–574.
[73] W.C. Yam, K.H. Chan, L.L.M. Poon, K.Y. Yuen, W.H. Seto, J.S.M. Peiris, Evaluation of reverse transcription-PCR assays for rapid diagnosis of severe acute respiratory syndrome associated with a novel coronavirus, Journal of Clinical Microbiology 41 (2003) 4521–4524.
[74] A. Bernheim, X. Mei, M. Huang, Y. Yang, Z.A. Fayad, N. Zhang, et al., Chest CT findings in coronavirus disease-19 (COVID-19): Relationship to duration of infection, Radiology 295 (2020) 200463.
[75] R. Wolfel, V.M. Corman, W. Guggemos, M. Seilmaier, S. Zange, M.A. M¬uller, et al., Virological assessment of hospitalized patients with COVID-2019, Nature (2020). https://doi.org/10.1038/s41586-020-2196-x.
[76] S. Zheng, J. Fan, F. Yu, B. Feng, B. Lou, Q. Zou, et al., Viral load dynamics and disease severity in patients infected with SARS-CoV-2 in Zhejiang province, China, BMJ (Clinical Research Ed.) 369 (2020) m1443.
[77] A.T. Xiao, Y.X. Tong, S. Zhang, Profile of RT-PCR for SARS-CoV2: A preliminary study from 56 COVID-19 patients, Clinical Infectious Diseases (2020) 460. https://doi.org/10.1093/cid/ciaa.
[78] A.L. Wyllie, J. Fournier, A. Casanovas-Massana, M. Campbell, M. Tokuyama, P. Vijayakumar, et al., Saliva is more sensitive for SARS-CoV-2 detection in

COVID-19 patients than nasopharyngeal swabs, Med Rxiv (2020) 20067835. https://doi.org/10.1101/2020.04.16.

[79] H. Chen, J. Guo, C. Wang, F. Luo, X. Yu, W. Zhang, et al., Clinical characteristics and intrauterine vertical transmission potential of COVID-19 infection in nine pregnant women a retrospective review of medical records, Lancet 395 (2020) 809–815.

[80] CDC, U. Emergency use authorizations. FDA. 2020. https://www.fda.gov/emergency-preparedness-and-response/mcm-legalregulatory-and-policy-framework/emergency-use-authorization. Accessed 2 July 2020.

[81] CDC. CDC 2019-Novel coronavirus [2019-nCoV] real-time RTPCR diagnostic panel https://www.fda.gov/media/134922/download. Accessed 2 July (2020).

[82] WHOMolecularassaystodiagnoseCOVID-19: Summary table of available protocols, World Health Organization, 2020 chttps://.

[83] A.K. Nalla, A.M. Casto, M-L.W. Huang, G.A. Perchetti, R. Sampoleo, L. Shrestha, Comparative performance of SARS-CoV-2detection assays using seven different primer/probe sets and one assay kit, Journal of Clinical Microbiology 58 (6) (2020) E00557 -00520.

[84] D.K. Chu, Y. Pan, S.M. Cheng, K.P. Hui, P. Krishnan, Y. Liu, et al., Molecular diagnosis of a novel coronavirus [2019-NCoV] causing an outbreak of pneumonia, Clinical Chemistry 66 (2020) 549–555.

[85] WHO Summary of available protocols to detect SARS-CoV-2 (2020 b). https://www.who.int/docs/default-source/coronaviruse/whoinhouseassays.Pdfsfvrsn=de3a76aa_2.

[86] J. Xiang, M. Yan, H. Li, T. Liu, C. Lin, S. Huang, et al., Evaluation of enzyme-linked immunoassay and colloidal gold immunochromatographic assay kit for detection of novel coronavirus (SARS-Cov-2) causing an outbreak of pneumonia (COVID-19), medRxiv (2020), doi:10.1101/2020.02.27.20028787.

[87] C. Wang, P.W. Horby, F.G. Hayden, G.F. Gao, A novel corona virus outbreak of global health concern, Lancet 395 (2020) 470–473 a.

[88] Ai, T., Yang, Z., Hou, H., Zhan, C., Chen, C., Lv, W., et al. (2020).Correlation of chest CT and RT-PCR testing in coronavirusdisease2019(COVID-19) in China: A report of 1014 cases. Radiology, 200642.

[89] S.F. Ahmed, A.A. Quadeer, M.R. McKay, Preliminary identification of potential vaccine targets for the COVID-19 coronavirus (SARS-CoV-2) based on SARS-CoV immunological studies, Viruses 12 (3) (2020) 254.

[90] N.M.A. Okba, M.A. Mu¨ller, W. Li, et al., Severe acute respiratory syndrome coronavirus 2-specific antibody responses in coronavirus disease 2019 patients, Emerging Infectious Diseases 26 (7) (2020).

[91] D. Benvenuto, M. Giovanetti, A. Ciccozzi, S. Spoto, S. Angeletti, M. Ciccozzi, The 2019-new coronavirus epidemic: Evidence for virus evolution, Journal of Medical Virology 92 (2020) 455–459.

[92] J. Cui, F. Li, Z-L. Shi, Origin and evolution of pathogenic coronaviruses, Nature Reviews Microbiology 17 (2019) 181–192.

[93] False Negatives and ReinfectionsThe Challenges of SARSCoV-2 RT-PCR Testing, American Society for Microbiology, Washington,DC, USA, 2020 https://asm.org/Articles/2020/April/False-Negatives-and-Reinfections-the-Challenges-of.

[94] J.F.-W. Chan, S. Yuan, K.-H. Kok, et al., A familial cluster of pneumonia associated with the 2019 novel coronavirus indicating person-to-person transmission: A study of a family cluster, The Lancet 395 (no.10223) (2020) 514–523.

[95] C. Huang, Y. Wang, X. Li, et al., Clinical features of patients infected with 2019 novel corona virus in Wuhan, China, The Lancet 395 (10223) (2020) 497–506.

[96] D. Wu, T. Wu, Q. Liu, Z. Yang, The SARS-CoV-2 outbreak:wHatweknow, International Journal of Infectious Diseases 94 (2020) 44–48.

[97] W. Kong, P.P. Agarwal, et al., Chest imaging appearance of COVID-19 infection, Radiol. Cardiothorac. Imaging 2 (2020) e200028.
[98] F. Pan, T. Ye, P. Sun, S. Gui, B. Liang, L. Li, et al., Timecourseoflungchangesonchest CT during recovery from 2019 novel coronavirus (COVID-19) pneumonia, Radiology 295 (2020) 715.
[99] H. Shi, X. Han, N. Jiang, Y. Cao, O. Alwalid, J. Gu, et al., Radiological findings from 81 patients with COVID-19 pneumonia in Wuhan, China: A descriptive study, The Lancet Infectious Diseases 20 (2020) 425.
[100] J.J. Zhang, X. Dong, Y.Y. Cao, Y.D. Yuan, Y.B. Yang, Y.Q. Yan, et al., Clinical characteristics of 140 patients infected with SARS-CoV-2 in Wuhan, China. Allergy 75 (2020) 1730.
[101] C. Long, H. Xu, Q. Shen, X. Zhang, B. Fan, C. Wang, et al., Diagnosis of the Coronavirus disease (COVID-19): RRT-PCR or CT? European Journal of Radiology 126 (2020) 108961.
[102] Y. Fang, H. Zhang, J. Xie, M. Lin, L. Ying, P.& Pang, et al., Sensitivity of chest CT for COVID-19: Comparison to RT-PCR, Radiology (2020) 200432.
[103] W. Wang, J. Tang, F. Wei, Updated understanding of the outbreak of 2019 novel coronavirus (2019-nCoV) in Wuhan, China, J Med Virol. 92 (4) (2020) 441–447.
[104] H.J. Koo, S. Lim, J. Choe, S.H. Choi, H. Sung, K.H. Do, Radiographic and CT features of viral pneumonia, Radiographics 38 (3) (2018) 719–739.
[105] M. Garg, N. Prabhakar, A. Gulati, R. Agarwal, S. Dhooria, Spectrum of imaging findings in pulmonary infections. Part 1: Bacterial and viral, Pol J Radiol 84 (2019) e205–ee13.
[106] W. Zhao, Z. Zhong, X. Xie, Q. Yu, J. Liu, Relation between chest CT findings and clinical conditions of coronavirus disease (COVID-19) pneumonia: A multicenter study, AJR, Am. J. Roentgenol. 214 (2020) 1072.
[107] X. Li, M. Geng, Y. Peng, L. Meng, S. Lu, Molecular immunepathogenesisanddiagnosis of COVID-19, Journal of Pharmaceutical Analysis 10 (2) (2020) 102–108.
[108] H.X. Bai, B. Hsieh, Z. Xiong, K. Halsey, J.W. Choi, T.M.L. Tran, et al., Performance of radiologists in differentiating COVID-19 from viral pneumonia on chest CT, Radiology 296 (2020) E46.
[109] L. Bordi, E. Nicastri, L. Scorzolini, A. Di Caro, M.R. Capobianchi, C. Castilletti, et al., On Behalf of Inmi Covid-Study Group and Collaborating Centers, Differential diagnosis of illness in patients under investigation for the novel coronavirus (SARS-CoV-2), Italy, February 2020, Eurosurveilance 25 (8) (2020), doi:10.2807/1560-7917.ES.2020.25.8.2000170.
[110] M. Hosseiny, S. Kooraki, A. Gholamrezanezhad, S. Reddy, R. Myers, adiologyperspectiveofcoronavirusdisease2019(COVID19): Lessons from severe acute respiratory syndrome and Middle East respiratory syndrome, AJR, Am. J. Roentgenol. 214 (2020) 1078.
[111] American Roentgen Ray Society. (Novel). coronavirus (COVID19) imaging features overlap with SARS and MERS: COVID-19's imaging features are variable and nonspecific, but the imaging findings reported thus far do show; Science Daily; https://www.sciencedaily.com/releases/2020/02/200228142018.htm
[112] Y. Gao, Y. Yuan, T.T. Li, W.X. Wang, Y.X. Li, A. Li, et al., Evaluation the auxiliary diagnosis value of antibodies assays for detection of novel coronavirus (SARS-Cov-2) causing an outbreak of pneumonia (COVID-19), medRxiv (2020), doi:10.1101/2020.03.26.20042044.
[113] F. Amanat, D. Stadbauer, S. Strohmeier, T.H.O. Nguyen, V. Chromikova, M. McMahon, et al., A serological assay to detect SARSCoV-2 seroconversion in humans, Nature Medicine 26 (2020) 1.

[114] C.A. Janeway Jr, P. Travers, M. Walport, M.J. Shlomchik, The distribution and functions of immunoglobulin isotypes, Immunobiology: The immune system in health and disease In, 5th ed, Garland Science, 2001.
[115] W. Liu, L. Liu, G. Kou, Y. Zheng, Y. Ding, W. Ni, et al., Evaluation of nucleocapsid and spike protein-based ELISAs for detecting antibodies against SARS-CoV-2, Journal of Clinical Microbiology (2020), doi:10.1128/JCM.00461-20.
[116] N.M. Okba, M.A. Mulle, W. Li, C. Wang, C.H. Geurtsvan Kessel, V.M. Corman, et al., SARS-CoV-2 specific antibody responses in COVID-19 patients, Emerging Infect. Dis. 26 (2020) 1478.
[117] X. Li, M. Geng, Y. Peng, L. Meng, S. Lu, Molecular immune pathogenesis and diagnosis of COVID-19, J Pharm Anal 10 (2020) 102–108.
[118] D. Stadlbauer, F. Amanat, V. Chromikova, K. Jiang, S. Strohmeier, G.A. Arun kumar, et al., SARS-CoV-2 seroconversion in humans: A detailed protocol for a serological assay, antigen production, and test setup, Curr Protoc Microbiol 57 (2020) e100.
[119] X.F. Cai, J. Chen, Q.X. Long, H.J. Deng, P. Liu, K. Fan, et al., A peptide-based magnetic chemiluminescence enzyme immunoassay for serological diagnosis of coronavirus disease 2019, The Journal of Infectious Diseases (2020). https://doi.org/10.1093/infdis/jiaa243.
[120] SK. Vashist, In vitro diagnostic assays for COVID-19: recent advances and emerging trends, Diagnostics 10 (2020) 202. https://doi.org/10.3390/diagnostics10040202.
[121] D. Lin, L. Liu, M. Zhang, Y. Hu, Q. Yang, J. Guo, et al., Evaluations of the serological test in the diagnosis of 2019 novel coronavirus (SARS-CoV-2) infections during the COVID-19 outbreak, European Journal of Clinical Microbiology & Infectious Diseases (2020) 1–7. https://doi.org/10.1007/s10096-020-03978-6.
[122] P. Zhang, Q. Gao, T. Wang, Y. Ke, F. Mo, R. Jia, et al., Evaluation of recombinant nucleocapsid and spike proteins for serological diagnosis of novel coronavirus disease 2019 [COVID-19], MedRxiv. (2020). https://doi.org/10.1101/2020.03.17.20036954.
[123] L. Zhong, J. Chuan, B. Gong, P. Shuai, Y. Zhou, Y. Zhang, et al., Detection of serum IgM and IgG for COVID-19 diagnosis, Sci China Life Sci 63 (2020) 777–780.
[124] G.J. Gorse, M.M. Donovan, G.B. Patel, Antibodies to coronaviruses are higher in older compared with younger adults and binding antibodies are more sensitive than neutralizing antibodies in identifying coronavirus-associated illnesses, Journal of Medical Virology 92 (2020) 512–517.
[125] Z. Li, Y. Yi, X. Luo, N. Xiong, Y. Liu, S. Li, et al., Development and clinical application of a rapid IgM-IgG combined antibody test for SARS-CoV-2 infection diagnosis, Journal of Medical Virology (2020). https://doi.org/10.1002/jmv.25727.
[126] Scott, A. (2020).Is a reagents shortage delaying European COVID-19 testing? https://cen.acs.org/analytical-chemistry/diagnostics/reagentsshortage-delaying-European-COVID/98/i13. Accessed 30 Apr
[127] Z. Chen, Z. Zhang, X. Zhai, Y. Li, L. Lin, H. Zhao, et al., Rapid and sensitive detection of anti-SARS-CoV-2 IgG using lanthanide doped nanoparticles-based lateral flow immunoassay, Analytical Chemistry 92 (2020) 7226–7231.
[128] M. Norman, T. Gilboa, A.F. Ogata, A.M. Maley, L. Cohen, Y. Cai, et al., Ultrasensitive high-resolution profiling of anti-SARS-CoV-2 antibodies for detecting early seroconversion in COVID-19 patients, medRxiv (2020). https://doi.org/10.1101/2020.04.28.20083691.
[129] C. Sheridan, Fast, portable tests come online to curb coronavirus pandemic, Nature Biotechnology 38 (2020) 515.
[130] WHO. Advice on the use of point-of-care immunodiagnostic tests for COVID-19. https://www.who.int/news-room/commentaries/detail/advice-on-the-use-of-point-of-care-immunodiagnostic-testsfor-covid-19. Accessed 30 Apr (2020a).

[131] G. Seo, G. Lee, M.J. Kim, S-H. Baek, M. Choi, K.B. Ku, et al., Rapid detection of COVID-19 causative virus [SARS-CoV-2] in human nasopharyngeal swab specimens using field-effect transistor-based biosensor, ACS Nano 14 (2020) 5135–5142.
[132] R. Kassanjee, C.D. Pilcher, M.P. Busch, G. Murphy, S.N. Facente, S.M. Keating, et al., Viral load criteria and threshold optimization to improve HIV incidence assay characteristics-a CEPHIA analysis, Aids 30 (2016) 2361.
[133] J. Tate, G. Ward, Interferences in immunoassay, Clin. Biochem. Rev. 25 (2004) 105.
[134] N. Wang, S.Y. Li, X.L. Yang, H.M. Huang, Y.J. Zhang, H. Guo, et al., Serological evidence of bat SARS-related coronavirus infection in humans, China, Virol Sin 33 (2018) 104–107.
[135] W. Zhang, R.H. Du, B. Li, X.S. Zheng, X.L. Yang, B. Hu, et al., Molecular and serological investigation of 2019-NCoV infected patients: Implication of multiple shedding routes, Emerg Microbes Infect 9 (2020) 386–389 d.
[136] G. Qiu, Z. Gai, Y. Tao, J. Schmitt, G.A. Kullak-Ublick, J. Wang, Dualfunctional plasmonic photothermal biosensors for highly accurate severe acute respiratory syndrome coronavirus 2 detection, ACS Nano 14 (2020) 5268–5277.
[137] E. Percivalle, G. Cambie, I. Cassaniti, E.V. Nepita, R. Maserati, A. Ferrari, et al., Prevalence of SARS-CoV-2 specific neutralising antibodies in blood donors from the Lodi Red Zone in Lombardy, Italy, as at 06 April 2020, Eurosurveillance 25 (24) (2020), doi:10.2807/15607917.ES.2020.25.24.2001031.
[138] S. Stringhini, A. Wisniak, G. Piumatti, A.S. Azman, S.A. Lauer, H. Baysson, et al., Seroprevalence of anti-SARS-CoV-2 IgG antibodies in Geneva, Switzerland (SEROCoV-POP): A population-based study, Lancet 396 (2020) 313.
[139] X. Xu, J. Sun, S. Nie, H. Li, Y. Kong, M. Liang, et al., Seroprevalence of immunoglobulin M and G antibodies against SARS-CoV-2 in China, Nature Medicine (2020), doi:10.1038/s41591-020-0949-6.
[140] Q.X. Long, B.Z. Liu, H.J. Deng, G.C. Wu, K. Deng, Y.K. Chen, et al., Antibody responses to SARS-CoV-2 in patients with COVID19, Nature Medicine 26 (2020) 845.
[141] F. Zhou, T. Yu, R. Du, G. Fan, Y. Liu, Z. Liu, et al., Clinical course and risk factors for mortality of adult inpatients with COVID-19 in Wuhan, China: A retrospective cohort study, Lancet 395 (2020) 1054.
[142] W. Zhu, K. Xie, H. Lu, L. Xu, S. Zhou, S. Fang, Initial clinical features of suspected Coronavirus Disease 2019 in two emergency departments outside of Hubei, China. J. Med. Virol. (2020), doi:10.1002/jmv.25763.
[143] A.J. Rodriguez-Morales, J.A. Cardona-Ospina, E. Gutierrez-Ocampo, R. Villamizar Pena, et al., Clinical, laboratory and imaging features of COVID-19: A systematic review and meta-analysis, Travel Med Infect Dis 34 (2020) 101623.
[144] Y.W. Tang, J.E. Schmitz, D.H. Persing, C.W. Stratton, Laboratory diagnosis of COVID-19: Current issues and challenges, Journal of Clinical Microbiology 58 (6) (2020), doi:10.1128/JCM.00512-20.
[145] A.C. Walls, Y.J. Park, M.A. Tortorici, A. Wall, A.T. McGuire, D. Veesler, Structure, function, and antigenicity of the SARS-CoV-2 spike glycoprotein, Cell 181 (2020) 281.
[146] M. Zheng, L. Song, Novel antibody epitopes dominate the antigenicity of spike glycoprotein in SARS-CoV-2 compared to SARS-CoV, Cell. Mol. Immunol. 17 (5) (2020) 536–538.
[147] K.H.D. Crawford, R. Eguia, A.S. Dingens, A.N. Loes, K.D. Malone, C.R. Wolf, et al., Protocol and Reagents for Pseudotyping Lentiviral Particles with SARS-CoV-2 spike protein for neutralization assays, Viruses 12 (5) (2020) 513.
[148] P.I. Kontou, G.G. Braliou, N.L. Dimou, G. Nikolopoulos, PG. Bagos, Antibody tests in detecting SARS-CoV-2 infection: A meta-analysis, Diagnostics 10 (5) (2020) 319.
[149] H. Lv, N.C. Wu, O.T. Tsang, M. Yuan, R. Perera, W.S. Leung, et al., Cross-reactive antibody response between SARS-CoV-2 and SARS-CoV infections, Cell reports 31 (9) (2020) 107725.

[150] S.K. Lau, K.S. Li, A.K. Tsang, C.S. Lam, S. Ahmed, H. Chen, et al., Genetic characterization of Betacoronavirus lineage C viruses in bats reveals marked sequence divergence in the spike protein of pipistrellus bat coronavirus HKU5 in Japanese pipistrelle: Implications for the origin of the novel Middle East respiratory syndrome coronavirus, Journal of Virology 87 (15) (2013) 8638–8650.

[151] J.F.-W. Chan, K-H. Kok, Z. Zhu, H. Chu, K.K.-W. To, S. Yuan, et al., Genomic characterization of the 2019 novel human-pathogenic coronavirus isolated from a patient with atypical pneumonia after visiting, Wuhan. Emerging Microbes & Infections 9 (1) (2020) 221–236.

[152] S. Jiang, C. Hillyer, L. Du, Neutralizing antibodies against SARS-CoV-2 and other human coronaviruses, Trends in Immunology 41 (5) (2020) 355–359.

[153] L. Ni, F. Ye, M.L. Cheng, Y. Feng, Y.Q. Deng, H. Zhao, et al., Detection of SARS-CoV-2-specific humoral and cellular immunity in COVID-19 convalescent individuals, Immunity 52 (6) (2020) 971–977.

[154] J.S. Gootenberg, O.O. Abudayyeh, J.W. Lee, P. Essletzbichler, A.J. Dy, J. Joung, et al., Nucleic acid detection with CRISPR-Cas13a/C2c2, Science 356 (6336) (2017) 438–442.

[155] N. Kohmer, S. Westhaus, C. Rühl, S. Ciesek, H.F. Rabenau, Clinical performance of different SARS-CoV-2 IgG antibody tests, Journal of Medical Virology (2020), doi:10.1002/jmv.26145.

[156] I. Montesinos, D. Gruson, B. Kabamba, H. Dahma, W.S. Van den, S. Reza, Evaluation of two automated and three rapid lateral flow immunoassays for the detection of antiSARS-CoV-2 antibodies, Journal of Clinical Virology 128 (2020) 104413.

[157] Y. Pan, X. Li, G. Yang, J. Fan, Y. Tang, J. Zhao, et al., Serological immunochromatographic approach in diagnosis with SARS-CoV-2 infected COVID-19 patients, Journal of Infection 81 (2020) 28.

[158] M.S. Tang, K.G. Hock, N.M. Logsdon, J.E. Hayes, A.M. Gronowski, Clinical performance of two SARS-CoV-2 serologic assays, Clinical Chemistry 66 (2020) 1055.

[159] A. Kaur, K. Sapkota, S. Dhakal, Multiplexed nucleic acid sensing with single-molecule FRET, ACS Sens 4 (2019) 623–633.

[160] D. Koirala, P. Shrestha, T. Emura, K. Hidaka, S. Mandal, M. Endo, et al., Single-molecule mechanochemical sensing using DNA origami nanostructures, Angew. Chem., Int. Ed. 53 (2014) 8137–8141.

[161] D. Koirala, Z. Yu, S. Dhakal, H. Mao, Detection of single nucleotide polymorphism using tension-dependent stochastic behaviour of a single-molecule template, Journal of the American Chemical Society 133 (2011) 9988–9991.

[162] L. Chen, W. Liu, Q. Zhang, K. Xu, G. Ye, W. Wu, et al., RNA based MNGS approach identifies a novel human coronavirus from two individual pneumonia cases in 2019 Wuhan outbreak, Emerg Microbes Infect 9 (2020) 313–319.

[163] J.-W. Ai, Y. Zhang, H.-C. Zhang, T. Xu, W.-H. Zhang, Era of molecular diagnosis for pathogen identification of unexplained pneumonia, lessons to be learned, Emerg Microbes Infect 9 (2020) 597–600.

[164] T. Suo, X. Liu, J. Feng, M. Guo, W. Hu, G. Dong, et al., ddPCR: A more accurate tool for SARS-CoV-2 detection in low viral load specimens, Emerg Microbes Infect 9 (2020) 1259–1268.

[165] L. Dong, J. Zhou, C. Niu, Q. Wang, Y. Pan, S. Sheng, et al., Highly accurate and sensitive diagnostic detection of SARS-CoV-2 by digital PCR, medRxiv (2020) 20036129. https://doi.org/10.1101/2020.03.14.

[166] P. Moitra, M. Alafeef, K. Dighe, M. Frieman, D. Pan, Selective naked eye detection of SARS-CoV-2 mediated by N gene targeted antisense oligonucleotide capped plasmonic nanoparticles, ACS Nano 14 (2020) 7617–7627.

[167] S.J.R. Silva, K. Pardee, L. Pena, Loop-mediated isothermal amplification (LAMP) for the diagnosis of Zika virus: A review, Viruses 12 (2020) 19.

[168] W.E. Huang, B. Lim, C. Hsu, D. Xiong, W. Wu, Y. Yu, et al., RT-LAMP for rapid diagnosis of coronavirus SARS-CoV-2, Microbial Biotechnology 13 (2020) 950.
[169] G. Park, K. Ku, S. Baek, S. Kim, S. Kim, B. Kim, et al., Development of reverse transcription loop-mediated isothermal amplification (RT-LAMP) assays targeting SARS-CoV-2, The Journal of Molecular Diagnostics 22 (2020) 729.
[170] L. Yu, S. Wu, X. Hao, X. Dong, L. Mao, V. Pelechano, Rapid detection of COVID-19 coronavirus using a reverse transcriptional loop-mediated isothermal amplification (RT-LAMP) diagnostic platform, Clinical Chemistry 66 (2020) 975.
[171] Y. Zhang, N. Odiwuor, J. Xiong, L. Sun, R.O. Nyaruaba, H. Wei, et al., Rapid molecular detection of SARS-CoV-2 (COVID-19) virus RNA using colorimetric LAMP, medRxiv (2020), doi:10.1101/2020.02.26.20028373.
[172] X. Yuan, Y. Wang, K. Meng, Y. Zhang, H. Xu, W. Ai, LAMP real-time turbidity detection for fowl adenovirus, BMC Vet. Res. 15 (2019) 256.
[173] M. Jiang, W. Pan, A. Aratehfar, W. Fang, et al., Development and validation of a rapid single-step reverse transcriptase loop-mediated isothermal amplification (RT-LAMP) system potentially to be used for reliable and high-throughput screening of COVID-19, medRxiv (2020), doi:10.1101/2020.03.15.20036376.
[174] D. Liu, S.L. Daubendiek, M.A. Zillman, K.& Ryan, E.T. Kool, Rolling circle DNA synthesis: Small circular oligonucleotides as efficient templates for DNA polymerases, Journal of the American Chemical Society 118 (1996) 1587–1594.
[175] B. Wang, S.J. Potter, Y. Lin, A.L. Cunningham, et al., Rapid and sensitive detection of severe acute respiratory syndrome coronavirus by rolling circle amplification, Journal of Clinical Microbiology 43 (2005) 2339–2344.
[176] W. Zhao, M. Ali, M. Brook, M.A.Y. Li, Rolling circle amplification: Applications in nanotechnology and biodetection with functional nucleic acids, Angew. Chem., Int. Ed. 47 (2008) 6330–6337.
[177] M. El-Tholoth, H.H. Bau, J. Song, A single and two-stage, closed-tube, molecular test for the 2019 Novel Coronavirus (COVID19) at home, clinic, and points of entry, chem Rxiv (2020) 10.26434/chem rxiv. 11860137. v1.
[178] L.T. Nguyen, B.M. Smith, P.K. Jain, Enhancement of trans-cleavage activity of Cas12a with engineered CrRNA enables amplified nucleic acid detection, Bio Rxiv (2020). https://doi.org/10.1101/2020.04.13.036079.
[179] A. Basu, T. Zinger, K. Inglima, K. Woo, O. Atie, L. Yurasits, et al., Performance of the rapid nucleic acid amplification by Abbott ID NOW COVID-19 in nasopharyngeal swabs transported in viral media and dry nasal swabs, New York City Academic Institution. bioRxiv. (2020), doi:10.1101/2020.05.11.089896.
[180] M.J. Loeffelholz, D. Alland, S.M. Butler-Wu, U. Pandey, C.F. Perno, A. Nava, et al., Multicenter evaluation of the cepheid Xpert Xpress SARS-CoV-2 test, Journal of Clinical Microbiology (2020), doi:10.1128/JCM.00926-20.
[181] B. Visseaux, Q.L. Hingrat, G. Collin, D. Bouzid, S. Lebourgeois, D.L. Pluart, et al., Evaluation of the QIAstat-Dx Respiratory SARS-CoV-2 Panel, the first rapid multiplex PCR commercial assay for SARS-CoV-2 detection, Journal of Clinical Microbiology (2020), doi:10.1128/JCM.00630-20.
[182] M. Pachetti, B. Marini, F. Benedetti, F. Giudici, E. Mauro, P. Storici, et al., Emerging SARS-CoV-2 mutation hot spots include a novel RNAdependent-RNA polymerase variant, Journal of Translational Medicine 18 (2020) 1–9.
[183] K.M. Koczula, A. Gallotta, Lateral flow assays, Essays in Biochemistry 60 (2016) 111–120.
[184] J. Reboud, G. Xu, A. Garrett, M. Adriko, Z. Yang, E.M. Tukahebwa, et al., Paper-based microfluidics for DNA diagnostics of malaria in low resource underserved rural communities, PNAS 116 (2019) 4834–4842.

[185] D. Zhang, D. Broyles, E.A. Hunt, E. Dikici, S. Daunert, S.K. Deo, A paper-based platform for detection of viral RNA, Analyst 142 (2017) 815–823.

[186] K.M. Byers, A.R. Bird, H.D. Cho, J.C. Linnes, Fully dried two-dimensional paper network for enzymatically enhanced detection of nucleic acid amplicons, ACS Omega 5 (2020) 4673–4681.

[187] D. Chandler-Brown, A.M. Bueno, O. Atay, D.S. Tsao, A highly scalable and rapidly deployable RNA extraction-free COVID-19 assay by quantitative Sanger sequencing, bioRxiv (2020). https://doi.org/10.1101/2020.04.07.029199.

[188] S. Mahari, A. Roberts, D. Shahdeo, GS. eCov, Sensul tra sensitive novel in-house built printed circuit board-based Review of analytical performance of COVID-19 detection methods 47electrochemical device for rapid detection of nCovid-19 antigen, a spike protein domain 1 of SARS-CoV-2. bioRxiv for Enzymatically Enhanced Detection of Nucleic Acid Amplicons, ACS Omega 5 (2020) 4673–4681.

[189] R. Tanaka, T. Yuhi, N. Nagatani, T. Endo, K. Kerman, Y. Takamura, et al., A novel enhancement assay for immune chromatographic test strips using gold nanoparticles, Analytical and Bioanalytical Chemistry 385 (2006) 1414–1420.

[190] B. Giri, A perspective on the sensitivity of paper-analytical devices for bioanalysis, J. Appl. Bionalysis 2 (2016) 6.

[191] A.W. Martinez, S.T. Phillips, E. Carrilho, S.W. Thomas, H. Sindi, G.M. Whitesides, Simple telemedicine for developing regions: Camera phones and paper-based microfluidic devices for realtime, off-site diagnosis, Analytical Chemistry 80 (2008) 3699–3707.

[192] A. Priye, S.W. Bird, Y.K. Light, C.S. Ball, O.A. Negrete, R.J. Meagher, A smartphone-based diagnostic platform for rapid detection of Zika, chikungunya, and dengue viruses, Scientific Reports 7 (2017) 1–11.

[193] B. Berg, B. Cortazar, D. Tseng, H. Ozkan, S. Feng, Q. Wei, Cell phone based hand-held microplate reader for point-of-care testing of enzyme linked immunosorbent assays, ACS Nano 9 (2015) 7857–7866.

[194] T. Laksanasopin, T.W. Guo, S. Nayak, A.A. Sridhara, S. Xie, O.O. Olowookere, smartphone dongle for diagnosis of infectious diseases at the point of care, Science Translational Medicine 7 (2015) 273re1.

[195] D.M. Rissin, C.W. Kan, T.G. Campbell, S.C. Howes, D.R. Fournier, L. Song, et al., Single-molecule enzyme-linked immunosorbent assay detects serum proteins at subfemtomolar concentrations, Nature Biotechnology 28 (2010) 595.

[196] C. Greenwood, D. Ruff, S. Kirvell, G. Johnson, H.S. Dhillon, S.A. Bustin, Proximity assays for sensitive quantification of proteins, Biomol. Detect. Quantif. 4 (2015) 10–16.

[197] M.L. Metzker, Sequencing technologies the next generation, Nature Reviews Genetics 11 (2010) 31–46.

[198] M.J. Khan, A.C. Trabuco, H.L. Alfonso, M.L. Figueiredo, W.C. Batista, S.J. Badra, DNA microarray platform for detection and surveillance of viruses transmitted by small mammals and arthropods, PLoS Neglected Trop. Dis. 10 (2016) E0005017.

[199] Z. Shen, Y. Xiao, L. Kang, W. Ma, L. Shi, L. Zhang, et al., Genomic diversity of severe acute respiratory syndrome–Coronavirus 2 in patients with coronavirus disease, Clinical Infectious Diseases 2020 (2019), doi:10.1093/cid/ciaa203.

[200] W. Long, H. Xiao, X. Gu, Q. Zhang, H. Yang, G. Zhao, et al., A universal microarray for detection of SARS coronavirus, Journal of Virological Methods 121 (2004) 57–63.

[201] J.S. Gootenberg, O.O. Abudayyeh, M.J. Kellner, J. Joung, J.J. Collins, F. Zhang, Multiplexed and portable nucleic acid detection platform with Cas13, Cas12a, and Csm6, Science 360 (6387) (2018) 439–444.

[202] J.P. Broughton, X. Deng, G. Yu, C.L. Fasching, V. Servellita, J. Singh, et al., CRISPR-Cas12-based detection of SARS-CoV-2, Nature Biotechnology 38 (2020) 870.

[203] Zhang F., Abudayyeh O., Gootenberg J., A protocol for detection of COVID-19 using CRISPR diagnostics 2020 a. https://www.broadinstitute.Org/files/publications/special/COVID19%20detection%20(updated).pdf.

[204] J. Joung, A. Ladha, M. Saito, et al., Point-of-care testing for COVID-19 using SHERLOCK diagnostics, MedRxiv (2020) 2020.

[205] M. Azhar, R. Phutela, A.H. Ansari, et al., Rapid, field-deployable nucleobase detection and identification using FnCas9, BioRxiv 2020 (2020).

[206] M.J. Kellner, J.G. Koob, J.S. Gootenberg, O.O. Abudayyeh, F. Zhang, SHERLOCK: Nucleic acid detection with CRISPR nucleases, Nature Protocols 14 (2019) 2986–3012.

[207] X. Mei, H.-.C. Lee, K. Diao, M. Huang, B. Lin, C. Liu, et al., Artificial intelligence–Enabled rapid diagnosis of patients with COVID-19, Nature Medicine (2020) 1–5. https://doi.org/10.1038/s41591-020-0931-3.

[208] A. Wu, Y. Peng, B. Huang, X. Ding, X. Wang, P. Niu, et al., Genome composition and divergence of the novel coronavirus (2019-nCoV) originating in China, Cell Host Microbe (2020).

[209] G. Yan, C.K. Lee, L.T.M. Lam, B. Yan, Y.X. Chua, A.Y.N. Lim, et al., Covert COVID-19 and false-positive dengue serology in Singapore, The Lancet Infectious Diseases 20 (2020) 536.

[210] J. Chaung, D. Chan, S. Pada, P.A. Tambyah, Coinfection with COVID-19 and Coronavirus HKU1 - the critical need for repeat testing if clinically indicated, Journal of Medical Virology (2020), doi:10.1002/jmv.25890.

[211] E. Cuadrado-Payan, E. Montagud-Marrahi, M. Torres-Elorza, M. Bodro, M. Blasco, E. Poch, et al., SARS-CoV-2 and influenza virus co-infection, Lancet 395 (10236) (2020) 84.

[212] F. Touzard-Romo, C. Tape, J.R. Lonks, Coinfection with SARS-CoV-2 and human metapneumovirus, R I Med. J. 103 (2) (2013) 75–76 (2020).

[213] D. Wu, J. Lu, X. Ma, Q. Liu, D. Wang, Y. Gu, et al., Coinfection of influenza virus and severe acute respiratory syndrome coronavirus 2 (SARS-COV-2), Pediatric Infectious Disease Journal 39 (2020) 79.

Further Reading

[1] E.I. Azhar, D.S.C. Hui, Z.A. Memish, C. Drosten, A. Zumla, The Middle East respiratory syndrome (MERS), Infect Dis Clin N Am 33 (2019) 891–905.

[2] M.E. Killerby, H.M. Biggs, C.M. Midgley, S.I. Gerber, J.T. Watson, Middle East respiratory syndrome coronavirus transmission, Emerging Infectious Diseases 26 (2020) 191–198.

[3] S. Shalhoub, AS. Omrani, Middle East respiratory syndrome, BMJ (Clinical Research Ed.) 355 (2016) i5281.

[4] World Health Organization. MERS situation update, January (2020 c). 2020. http://www.emro.who.int/pandemic-epidemic-diseases/mers-cov/merssituation-update-january-2020.html. Accessed 13 July 2020.

[5] G.M. Leung, A.J. Hedley, L.M. Ho, P. Chau, I.O. Wong, T.Q. Thach, et al., The epidemiology of severe acute respiratory syndrome in the 2003 Hong Kong epidemic: An analysis of all 1755 patients, Annals of Internal Medicine 141 (2004) 662–673.

[6] W. Li, M.J. Moore, N. Vasilieva, J. Sui, S.K. Wong, M.A. Berne, et al., Angiotensin-converting enzyme 2 is a functional receptor for the SARS coronavirus, Nature 426 (2003) 450–454.

[7] W. Ren, W. Li, M. Yu, P. Hao, Y. Zhang, P. Zhou, et al., Full-length genome sequences of two SARS-like coronaviruses in horseshoe bats and genetic variation analysis, Journal of General Virology 87 (2006) 3355–3359.

[8] M. Wang, M. Yan, H. Xu, W. Liang, B. Kan, B. Zheng, et al., SARS-CoV infection in a restaurant from palm civet, Emerging Infectious Diseases 11 (2005) 1860–1865.
[9] World Health Organization. Summary of probable SARS cases with onset of illness from 1 November 2002 to 31 July 2003. (2003). https://www.who.int/csr/don/2003_07_05/en/. Accessed 13 July 2020.
[10] G. Lu, Y. Hu, Q. Wang, J. Qi, F. Gao, Y. Li, et al., Molecular basis of binding between novel human coronavirus MERSCoV and its receptor CD26, Nature 500 (2013) 227–231.
[11] Z.A. Memish, S. Perlman, M.D. Van Kerkhove, A. Zumla, Middle East respiratory syndrome, Lancet 395 (2020) 1063–1077.
[12] M. Hoffmann, H. Kleine-Weber, S. Schroeder, N. Krüger, T. Herrler, et al., SARS-CoV-2 cell entry depends on ACE2 and TMPRSS2 and is blocked by a clinically proven protease inhibitor, Cell. 181 (2020) 271–280.
[13] X. Ou, Y. Liu, X. Lei, P. Li, D. Mi, L. Ren, et al., Characterization of spike glycoprotein of SARS-CoV-2 on virus entry and its immune cross-reactivity with SARS-CoV, Nature Communications 11 (2020) 1620.
[14] World Health Organization Statement on the second meeting of the International Health Regulations (2005). Emergency Committee regarding the outbreak of novel coronavirus (2019-nCoV). 2020. https://www.who.int/news-room/detail/30-01-2020 statement on the second meeting of the international health regulations (2005) emergency committee regarding the outbreak of novel coronavirus (2019-ncov). Accessed 13 July 2020.
[15] World Health Organization. Coronavirus disease 2019 (COVID-19) Situation Report –*175*. (2020 a). https://www.who.int/docs/default-source/coronaviruse/situation-reports/20200423-sitrep-94-covid-19.pdf?sfvrsn=b8304bf0_4. Accessed 13 July 2020.
[16] World Health Organization. WHO Director-General's opening remarks at the media briefing on COVID-19 - 11 March 2020. (2020 b). https://www.who.int/dg/speeches/detail/who-director-general-s-opening-remarks-at-the-mediabriefing-on-covid-19%2D%2D-11-march-2020. Accessed 13 July 2020.
[17] https://labblog.uofmhealth.org/rounds/how-scientists-quantifyintensity-of-an-outbreak-like-covid-19. Accessed 30 June (2020).

SECTION 6

Future of Biosensing for COVID-19 and Conclusion

CHAPTER

13

Overcoming the after effect of coronavirus

Anchit Gugnani[a], *and Charu Mehandiratta*[b]

[a]Department of Physiotherapy, Nims University, Jaipur, Rajasthan, India, [b]Human Consciousness and Yogic Science, Department of Yoga and Naturopathy, Apex University, Jaipur, Rajasthan, India

Although we do not have much data as many researches are still underway, one thing common in all of researches is the "long lasting impact on the lungs." Even for the people who were not hospitalized, the respiratory issue still remains a point of concern.

Researchers have theorized that the virus responsible for COVID-19 is very similar to earlier viruses like SARS and MERS which caused severe acute respiratory distress syndrome (ARDS) which lasted for a very long time causing irreversible damage to the lungs and quality of life of individuals being affected by it.

Therefore it is necessary to address this problem by finding suitable rehabilitation techniques for the lungs. The main goal in respiratory physiotherapy is to mobilize secretions and ease the work of breathing while working on increasing the inspiratory and expiratory capacities. Interventions may include techniques such as autogenic drainage, positioning, breath stacking, deep breathing exercises, active cycle of breathing, mobilization and manual techniques (e.g., percussion, vibrations, and assisted cough) to aid sputum clearance and increase the lung efficiency. These interventions can be performed at any stage of the disease as they are appropriate and safe to perform.

Patients recovering from an acute COVID-19 event may present with a disability or functional damage (CIM, respiratory function, PICS, CIP), decreased participation and deterioration in their quality of life (short-term as well as long-term postdischarge). Variable recovery time – depending

Sensing Tools and Techniques for COVID-19: Developments and Challenges in Analysis and Detection of Coronavirus
DOI: https://doi.org/10.1016/B978-0-323-90280-9.00002-X

on the degree of normocapnic respiratory failure, emotional dysfunction, associated physical dysfunction (asthenia, muscle weakness); the presence of other co morbidities. Clinical parameter evaluating protocols are also indicated on a daily basis even after testing negative – temperature, SaO2, respiratory rate, dyspnea, thoracoabdominal dynamics, and cough.

Techniques such as diaphragmatic breathing, pursed-lip breathing, incentive spirometry, lung expansion techniques (positive expiratory pressure) bronchial hygiene, manual mobilization of the ribcage, aerobic exercise, respiratory muscle training, and are useful in rehabilitating the lungs. In the event of comorbidities like secondary pneumonia, bronchiectasis, or increasing secretions, postural drainage, and standing may help with secretion management.

The aim of pulmonary rehabilitation is to not only improve the patient's physical and mental conditions but also to aid the patient return to family and society more promptly. In addition, because COVID-19 has caused a public emergency, patients with COVID-19 may demonstrate different degrees of psychological disorders, such as depression anxiety insomnia anger, fear and loneliness, as well as a lack of cooperation and abandonment of treatment due to fear of the disease. Patients are experiencing post-traumatic stress syndrome also. Therefore, prompt introduction and continuous availability of pulmonary rehabilitation services is critical for patients with COVID-19.So it is important to ensure we address the after effects of COVID-19.

- How it started all
- Impact of COVID-19 and physical inactivity on the immune system
- The lungs- after COVID
- How to increase lung efficiency with various methods
- Diaphragmatic breathing/deep breathing
- Chest/lung physiotherapy
- Fighting COVID-19 with proper nutrition
- Quality of life/mental health
- Long-term effects of lung rehabilitation.

13.1 How it started all/a quick recap

Severe acute respiratory syndrome coronavirus (SARS-CoV)-2, a novel coronavirus RNA in the same family as SARS-CoV and Middle East respiratory syndrome coronavirus (MERS-CoV), was identified at the beginning of January 2020 as the cause of a pneumonia epidemic affecting the city of Wuhan, the capital of Hubei province, from where it spreads rapidly across China [24]. After infecting and killing thousands of people in China, the virus has spread to Italy and other European countries and the USA, with the number of confirmed new cases increasing every day.

The World Health Organization named coronavirus disease in 2019 (COVID-19) and subsequently declared it a pandemic due to widespread infectivity and high contagion rates. Human coronaviruses usually cause respiratory and enteric infections [1].

Although serious lung injury has been reported at all ages, in some high-risk individuals, such as the elderly or those affected by multimorbidity, the virus is more likely to cause severe interstitial pneumonia, acute respiratory distress syndrome (ARDS) and subsequent multiorgan failure, which are responsible for severe acute respiratory failure and high death rates. Typically, affected individuals display a variable degree of dyspnea and radiological signs.

13.1.1 Impact of COVID-19 on the immune system and physical inactivity

COVID-19 characterized by major symptoms of fever, dry cough, myalgia, and fatigue. Currently, neither vaccines nor clinically proven effective treatments are available. Convalescent plasma and antiviral drugs have shown some promise in the treatment of COVID-19 patients, but their widespread use is awaiting statistical rigor. Behavioral strategies for social distancing and hygiene are currently the best and only means of limiting spread and reducing morbidity and mortality [2]. As this strain of the virus is novel to the human immune system, we depend on aspects of our innate immunity to deal with the initial infection.

Like most viral infections, if we survive the infection, we develop virus-specific antibody and cell-mediated immune responses over weeks. In most cases, this exposure-related 'training' of our immune systems provides long-lasting protection against re-infection or, if we are re-infected, the symptoms of disease are much milder. However, we do not currently know whether our response to-CoV-2 is sufficient to be protective and long-lasting. Along with tests for the presence of viral particles and plasma antibodies, there is a clear need for research related to vaccine development and research to determine whether our immune response is adequate to protect us. Public health recommendations (i.e. residence orders, park closures, gymnasiums and fitness centers) to prevent SARS-CoV-2 from spreading have the potential to reduce daily physical activity (PA).

These recommendations are unfortunate because day-to-day exercise can help combat the disease by boosting our immune systems and counteracting some of the co-morbidities such as obesity, diabetes, hypertension, and serious heart conditions that make us more susceptible to serious COVID-19 disease [1].

Studies have clarified some understanding of the mechanisms responsible for these observations. Early epidemiological studies suggested that intense, prolonged exercise was associated with an increase in upper respiratory tract infections. This led to the concept of inverted J theory, where moderate exercise reduces and prolonged, high-intensity exercise increases the susceptibility to infection [3]. Since then, many studies have supported the theory of individual immune parameters, including those specific to viral defense. For example, salivary lactoferrin and its rate of secretion increased for up to 2 hours after moderate exercise. Mucosal lactoferrin is important because it can prevent DNA and RNA virus from forming infecting cells by binding and blocking host receptors.

Conversely, low levels or secretions of salivary immunoglobulin A, which can bind to and inactivate viruses, have been shown to be associated with upper respiratory tract infection in some athletes under intense training. Moreover, because PA and exercise result in profound movement of leukocytes in blood and tissues, many researchers predict that physical activity increases [1].

13.1.2 The lungs after the COVID

Coronavirus 2019 (COVID-19), more recently known as SARS-COV-2, is a coronavirus that belongs to the b-corona cluster and is largely spread via droplets. When an infection occurs, the virus enters the lungs and is received by angiotensin-converting enzyme 2 (ACE2), which is expressed in normal humans in alveolar cells of type I and type II. When the virus binds to ACE2, it damages the alveolar cells Under normal circumstances, alveolar cells act to synthesize and secrete surfactants, perform xenobiotic metabolism, aid in the transepithelial movement of water, and regenerate alveolar epithelium after lung injury. These functions help with normal lung functions. As a result, damage to alveolar cells may result in respiratory problems, other systemic manifestations, and eventually death. As a result, the clinical manifestations of COVID-19 disease include fever, cough, myalgia or fatigue, pneumonia, and complicated dyspnea.

Many people exposed to the SARS-CoVID- virus have no symptoms or are relatively mild [25]. Some patients, on the other hand, experience catastrophic lung damage characterized by rapid-onset fibrosis. Pulmonary fibrosis involves thickening the small airways and surrounding tissues [4]. It can develop as a result of COVID-19 and a variety of other conditions. Specifically, milder symptoms of COVID-19 appear to occur more commonly in younger ages—probably because lower age poses a comparatively lower risk of "cytokine storm" causing the most extensive fibrosis seen in COVID-19 patients. Yet, even among the many older patients, many survive the initial disease [26]. And among younger patients, many survivors may still have persistent pulmonary fibrosis and other respiratory complications from SARS-CoV-2 [5].

Severe respiratory complications appear to occur in about 20 percent of people who have been tested positive for SARS-CoV-2 for polymerase chain reaction tests of throat swabs and blood samples [36]. Severe disease has been defined according to six criteria:

(a) Rapid deterioration of the patient's condition
(b) Lower levels of white blood cells that fight infection
(c) Extreme inflammation throughout the body
(d) Damage to components of the immune system such as spleen and lymph nodes (e) lesions in the lungs with particular pathology
(e) Clotting and blood damage.

Studies suggests that clotting and inflammation components play a key role in causing extensive, permanent lung damage. However, only about 3% of total people tested positive for SARS-CoV-2 are killed by COVID-19. As a result, approximately 17 percent of the total SARS-CoV-2 patients develop severe clinical symptoms but survive the initial infection. Among older survivors of COVID-19, the proportion of patients affected by extensive fibrosis is likely to be higher than the 17% estimate for the total population tested positive for SARS-CoV-2.

13.2 How to increase lung efficiency with different methods

- Diaphragm breathing/deep breathing
- Physiotherapy of the chest/lung
- Fight against COVID-19 with proper nutrition
- Diaphragm breathing/deep breathing

While the clinical course of the COVID-19 pandemic continues, researched, many patients with COVID-19 develop respiratory failure and require mechanical ventilation (MV) to maintain adequate pulmonary function. The exchange of gas in this regard, a recent report shows that ~54% of patients are hospitalized due to COVID-19 respiratory failure and >30% require MV. Although MV is often a life-saving intervention, the unwelcome consequence of prolonged MV is the rapid development of Respiratory muscle weakness due to atrophy of the muscle diaphragm and contractile dysfunction (collectively referred to as ventilator-induced diaphragm Dysfunction) VIDD.

VIDD is clinically significant because of diaphragmatic Weakness is a major contributor to weaning inability Patients with the ventilator. Many patients with COVID-19 often require ventilation. Prolonged time on the ventilator to increase the risk of weaning Issues. Patients with difficult weaning experience suffer higher morbidity and mortality as patients weaned quickly in their first attempts to separate from the ventilator and unfortunately too many COVID-19 patients succumb to ICU-related complications (e.g., sepsis).

Because respiratory muscle weakness is the primary risk factor for failure Developing strategies to protect the diaphragm from the ventilator [6].

In critical cases MV-induced weakness has become a priority Medicinal care. Interestingly, studies on the effects of endurance exercise Training on the respiratory system has led the way. Although many organ systems are adapted in response to endurance exercise Training, the structural and functional properties of the lungs and Airways are not affected by training exercises. However, while airways are not affected by training [7]. The gas-exchange side of the respiratory system is not adapted for exercise training the "pump" side of the respiratory system undergoes adaptive training. Changes in response to the exercise of endurance. Specifically, stamina Exercise training promotes a number of biochemical alterations in diaphragm muscle resulting in a phenotype that is protected against a few.

Challenges including prolonged MV. Indeed, as few as ten consecutive MV. Days of endurance training results in significant protection Against VIDD. As a result, endurance trained individuals are expected to be trained. Developing COVID-19 and requiring support for the ventilator benefit from the exercise-induced conditioning of the diaphragm. In addition, when respiratory symptoms are severe, Can progress to respiratory failure (acute respiratory distress) Syndrome), which could lead to death if it is not managed, prompt use of invasive ventilation [7]. For those, however, With mild to moderate symptoms, non-invasive techniques such as It can be used as a chest physiotherapy.

13.2.1 Physiotherapy of the chest/lung

Physiotherapy plays a vital role in the management of patients who have been infected with COVID or recovered. It plays a pivotal role in the management of acute and chronic respiratory conditions that help to improve the quality of life following acute illness [8].

It will also be important to explore the value and impact of physiotherapy in patients aged with chronic lung fibrosis caused by COVID-19. Physiotherapy with and without devices is often extremely beneficial for people with pulmonary fibrosis [9]. In a variety of different fibrosis-causing conditions, patients can maintain optimal respiratory health by actively clearing mucus from small airways [7].

Percussion therapy is a common component of keeping your fibrotic lungs healthy throughout your life. This can be achieved by manual clapping on the chest and back and/or by means of assistive devices such as oscillating vests and hand-held percussors.

Respiratory muscle training is a different approach. This can be done manually through an active cycle of breathing exercise and/or through the use of assistive devices such as flutter pipes and inspiration/expiratory trainers [10]. In older patients, it will be particularly important to consider the relative value of each type of therapy given the safety concerns associated with percussive pressure in people with lower bone density and/or more fragile skin.

Physiotherapy has been shown to be necessary for the physical rehabilitation of post-COVID-19 patients. These patients have variable symptoms, including either dry or productive cough. Those with dry cough may not benefit from physiotherapy, but patients with productive cough and underlying respiratory conditions or other comorbidities associated with hypersecretion and ineffective cough (e.g. neuromuscular disease, respiratory disease, obesity, etc.) may benefit from the same. Many COVID-19 patients also presented a stroke that increased their morbidity. There is sufficient evidence to suggest the development of ICU-acquired weakness in patients admitted to ICU for a prolonged period of time.

Patient education and training will be critical across the spectrum of clinical interventions for healthy aging with COVID-19 lung fibrosis [27]. These supports should include attention to both biomedical and psychosocial aspects of living with advanced lung damage. Because patients of all ages will be very new to managing the condition if it has been caused by SARS-CoV-2 infection, they may have unique needs on this front. These needs and associated patient preferences may vary slightly depending on age as well as other social locations and experiences [27]. Newer patients with lung fibrosis may need time to determine their needs and preferences in these areas. More experienced patients whose pulmonary fibrosis originated with other conditions can therefore provide valuable support in developing both best clinical practices and responsive community resources for aging and older adults with long-term respiratory impacts from COVID-19.

Chest physiotherapy has been used in many different respiratory states [11]. It has been said to improve the exchange of gas, to reverse Pathological progression, reducing or avoiding the need for Artificial ventilation when provided very early [11]. However, there is still a lack of evidence for COVID-19 patients In addition to its effects, especially during the acute stage, aside from Some position papers or anecdotal recommendations Evidence of this. This is because of the characteristics of the respiratory system the problems in COVID-19 patients differ significantly from in other respiratory conditions [12].

For example, patients with COVID-19 do not usually have exudation during the acute stage. In addition, dyspnea in COVID patients-19 may progress rapidly towards acute respiratory failure. As a result, the timely use of mechanical ventilation in these areas Situations are strongly recommended [13].

COVID-19 is highly infectious and spreads rapidly. Concerns have been raised about the use of chest physiotherapy in Infectious diseases. This is because it was argued that Physiotherapy in the chest may cause aerosolization. This could be Increase the rate at which COVID-19 spreads out. Later, however the findings in similar circumstances refuted this view [14].

According to that- evaluation of the dispersion of droplets Influenza pandemic and other airborne infections have shown Physiotherapy in the chest significantly and predominantly produced Droplets $>10\mu m$ [28]. Droplets of this size are not breathable, only droplets within an inspirable range (about 5 μm) can play an important role in the transmission of infections [29].

Similarly, a review of the transmission of aerosol from influenza A virus disputes whether it is even possible for the droplets to come from Chest physiotherapy for transmitting infections. Moreover, A disease that shared similar pathophysiology with SARS COVID-19, thoracic physiotherapy was later recommended.

In addition, the overall management of COVID- 19 is still symptomatic, as scientists are still trying to do, Understanding its pathophysiology and viral behavior. As a result, the disease can kill in days to days, especially in the elderly and those with disabilities, weak immunity, we can cause our patients to sneeze or cough out the sputum in disposable plastic bags, and after thoracic physiotherapy to prevent or reduce thoracic disease the chance for aerosolization. Chest physiotherapy may improve respiratory function and quality of life in COVID-19 patients, especially after Release [15].

There is still a lack of evidence during the acute stage Its usefulness, apart from some professional recommendations, On the basis of anecdotal evidence. It should be noted, however, Physiotherapy for the chest is an individualized treatment Particular presentations of the patient. When, therefore, Patients with symptoms that may benefit from the chest physiotherapy may be given while the patients are in close proximity observed for any adverse event [16]. In addition, when administered physiotherapy of the chest for patients in the acute stage, measures of this kind Use of surgical masks, if available, should be taken as appropriate prevent cross-infection.

13.2.2 Fighting COVID 19 with proper nutrition

Nutrition is an important factor in human health, including maintains a strong immune system. However, up-to-date research indicates that no single nutrient or dietary supplement can prevent or treat COVID-19 patient. On the contrary, inadequate intake, in particular overdoses Dietary supplements may be more harmful than beneficial. Clinical information demonstrate that patients dying from COVID-19 are mostly elderly people [30]. Complication of other diseases and malnutrition problems ageing development of COVID-19 from mild to severe symptoms, it is closely related to nutritional status [17]. Therefore, the assessment of nutritional status is considered. The status is necessary and important during infection with COVID-19 [17].

Patients with COVID-19 have increased proinflammatory activity Cytokines, high sensitivity C-reactive protein (hsCRP) and increased Risk to sepsis and ARDS; experiences in the treatment of SARS, MERS, and ARDS. Other infectious diseases of the virus and clinical trials in patients with COVID-19 Show the beneficial effects of nutritional support on COVID-19. By reducing oxidative stress and increasing immunity, nutritional support helps people to reduce the risk of infection by the virus or to reduce the symptoms of COVID-19 [4].

In patients with COVID, plasma, hsCRP, a marker of inflammation and oxidative stress is significantly increased. As a result, antioxidant status increases and proinflammatory cytokine release decreases [31].

Fruit and vegetable intakes have been investigated for potential benefits in combination with respiratory and inflammatory conditions due to their nutrient profile consisting of antioxidants, vitamins, minerals and phytochemicals, including phenolic compounds that may have antioxidant, anti-inflammatory, and other beneficial effects.

Vitamin C (ascorbic acid) is a water-soluble vitamin that has been used as part of cultural practice in cold or flu for decades. This is due to research published by Nobel Prize winner Linus Pauling (circa 1970) on how vitamin C helps treat colds.

However, regular vitamin C supplements had a number of benefits, including:

- Reduced cold severity: Reduced symptoms of the cold, making it less severe.
- Reduced cold duration: Supplements reduced recovery time by 8%in adults and 14%in children on average [18].

Regular treatments are likely to be an effective strategy for lowering ARDS And the risk of COVID-19. Vitamin C is a commonly used scavenge antioxidant. ROS and to protect cells against oxidative stress. Intravenous (i.e.) or oral administration of high doses of vitamin C has been reported to be safe Protects against viral infection without major adverse events [19]. In addition, high-dose vitamin- C supplementation with i.v. Administration of the Intensive Care Unit (ICU) stay shortened by 7.8% and significantly decreased mortality rate.

The findings on vitamin D and immune function are also erroneous. Vitamin D is often referred to as sunshine vitamin; it is also found in eggs, mushrooms, fatty fish such as salmon, milk and dairy products, or vitamin D fortified foods [31].

Vitamin D itself may have antiviral effects by inhibiting viral replication and by its immunomodulatory and anti-inflammatory properties. It is assumed that increased vitamin D may benefit from SARS-CoV-2 infection.

Higher doses of vitamin D are certainly recommended for vulnerable individuals. It is also particularly important to increase dietary sources of vitamin D at this time, as many people worldwide are exposed to less sunlight due to 'stay at home' mitigation strategies.

Vitamin E is a group of fat-soluble antioxidants that contains molecules such as tocopherols and tocotrienols [38].

Nuts, seeds and vegetable oils are important contributors to dietary intakes, as are green leafy vegetables and fortified cereals. Numerous studies have shown that vitamin E deficiency affects both humoral and cell-mediated immune functions. Vitamin E is generally accepted to exert its immune-enhancing effects by scavenging oxygen species to reduce oxidative stress and to induce anti-inflammatory effects. [37] Vitamin E may also protect polyunsaturated fatty acids (PUFAs) in cell membranes from oxidation, regulate the production of ROS and reactive nitrogen species (RNS) and modulate signal transduction.

Vitamin E is also present at high concentrations in immune cells, which protects them from oxidative damage due to their high metabolic activity and PUFA content [38]. In particular, aging is associated with immune system deregulation, predisposing people to increased oxidative stress and inflammation. This leads to an increase in the incidence of elderly infections, such as influenza. Although elderly people have comparable vitamin E levels to younger people, increasing vitamin E intakes may benefit their immune function, provide resistance to infection, and reduce morbidity due to infections [31]. As elderly people are predisposed to infection due to immune sensitivity, it is worth investigating the potential benefits of vitamin E against COVID-19. In fact, it has been suggested that a combination of vitamins C and E may be a useful antioxidant therapy for cardiac complications of COVID-19.

Zinc is a trace element of the diet that is critical to the development of immune cells and an important cofactor of many enzymes. Zinc deficiency may contribute to defective cell-mediated immunity and increased susceptibility to various infections, including pneumonia. Despite various zinc-containing foods, including meat, dairy and legumes, etc., the majority of zinc-containing dietary research focused on zinc supplementation. It is therefore not clear from high zinc dietary studies that increased zinc intake can protect against viral infections [20]. In fact, zinc supplements and lozenges are a popular remedy for the fight against colds and respiratory diseases.

Finally following any healthy diet at this time that provides adequate RNIs will support immune function. However, there is concern that people who are deficient in certain micronutrients, such as vitamin C, vitamin D or zinc, may warrant supplementation or change their dietary patterns in order to maintain nutritional status and promote healthy immune function [20]. This is especially important for our vulnerable populations, including the elderly. [38] Indeed, as many countries have established guidelines for people to "stay in place" and to limit travel, etc., it is important for everyone to consider vitamin D levels when exposure to sunlight is limited at this time. Similarly, although not always possible depending on the individual's personal circumstances, maintaining activity or moderate exercise at this time may be beneficial in the promotion of immune health [20]. Exercise in any form, even at home, will help to maintain one's health, as lack of exercise could significantly affect the cardiovascular risk of a person in the future. However, exercise may not be advisable in symptomatic patients with COVID-19, not only to reduce the transmission of the virus, but also because some infections with COVID-19 exhibit rare cardiac complications that could be aggravated by physical exercise.

13.3 Mental health and life quality

The feelings involved, such as fear and rage, must be understood and studied in order to recognize the psychological and clinical consequences of a pandemic. Fear is a system of animal safety adaptive which is crucial to survival in order to take account of future threats by many biological processes. But when it is chronic or disproportionate, it can be dangerous and can be a contributing factor in the development of various psychological disorders [16]. In a pandemic, uncertainty raises the anxiety and stress level of healthy individuals.

In the course of epidemics, the number of people with affected mental health appears to be greater than in the number of people affected by infection. Previous tragedies have demonstrated that the effects of mental health will last longer and have a higher incidence than the disease itself. [21].

Studies have found that patients who are or are suspected to be infected with COVID-19 can experience extreme reactions from emotions and behaviors, such as terror, boistery, lone-life, anxiety, insomnia, or angriness, as have been recorded in the past, in similar situations [22].

Psychological first aid is an integral component of the services given for populations that have been affected by crises and disasters, but the most effective psychosocial support practices cannot be directed by uniform procedures or guidelines. Although reports on local mental health interventions have been released, more detailed emergency guidelines are unknown.

Psychosocial tests, including the monitoring of questions concerning Covid-19-related stressors (for example, proximity to infected sources, infected relatives, lost loved ones and physical distance), secondary adverse factors (economic loss), psychosocial consequences (for example, depression, anxiety, psychosomatic problems, insomnia, increased use of substances), Covid-19-related stressors (such as preexisting physical or psychological conditions). Some patients may require referencing for formal assessment and treatment of mental wellbeing, and others may benefit from supporting programs to encourage well-being and improve treatment (such as psychoeducation or cognitive behavioral techniques). In the broader economic crisis, suicidal thoughts may arise and require urgent consultation with a specialist for mental wellbeing or referral to potential emergency psychiatrical hospitalization, as this pandemic is widespread.

The major stressors that inevitably lead towards widespread pollutants include unclear prognoses, extreme shortages of testing and treatment services and protecting health workers and providers against infection, the imposition of unfamiliar interventions in public health which are contrasting with personal freedoms, high, growing financial losses, and messages from authorities.

Public health emergencies can impact both individuals' health, safety and well-being (e.g., vulnerability, uncertainty, email isolation and stigma) (owing to economic loss, work and school closures, inadequate resources for medical response, and deficient distribution of necessities). Such effects can lead to various emotional responses (such as distress or mental illness), unsanitary behaviors in persons who are suffering from the disease and in the general public (such as inappropriate use of substances and noncompliance with public health guidelines such as in-house confines and vaccination).

Most people become resilient and do not succumb to psychopathology following disasters. Some people are actually seeking different strengths. Nevertheless, post-traumatic stress disorder (PTSD), [32] which is caused by exposure to trauma, is a priority issue in "conventional" natural disasters, technological events and deliberate acts of mass destruction [33]. The existing requirements for trauma needed for the diagnosis of PTSD cannot be met by natural circumstances such as life-threatening viral infections, but other psychopathology, such as depressive and anxiety disorders can also follow.

The WHO also shared concern about the emotional and psycho-social effects of the pandemic [34]. It speculates that new interventions, including self-isolation and quarantine, have changed people's usual lives and routines, which could lead to a rise in loneliness, anxiety, depression, insomnia, unhealthy alcohol and drug use, and self-harm or suicide [34].

Interestingly, in this social media era, the coronavirus pandemic is often characterized by rumors and disinformation not authentic and confirmed. With a sudden and almost continuous stream of news coverage about an outbreak, these rumors and unauthenticated details generate panic, anxiety and stress. Many people in India have been placed at risk by optimistic individuals or by those traveled abroad. However, because of social stigma and social isolation, they are not being evaluated [21]. They are frightened and worried that their family members will be blamed, isolated and taken away. It is crucial that psychologists and other practitioners warn these people not to do anything wrong, and this is a protocol to minimize and monitor the pandemic. In such a distressing situation, their fears must be explained and they should be helped, consulted, sympathized, and kind.

In order to alleviate these issues, the WHO has given a 31 point guidance. The recommendations provide guidance on the safety of the mental health of people of various ages impacted by COVID-19 who focus on children, women and service providers and recommend interventions to reduce anxiety, stigma and depression. Online tools are available to help control and deal with the stress resulting from the pandemic. It is critical that you, your family and friends take care of yourself. It improves the group to help others deal with their stress. Nevertheless, it is the shortage of mental health providers, practicers, psychologists, and healthcare services that poses the greatest obstacle in mitigating the mental health effects of the COVID 19 pandemic [21]. For a country like India this will be a challenge, where only 0,29 psychiatrists, 0,07 psychologists and 0,36 other paying psychiatrists for 100,000 people are available [39]. In such a scenario, a simple therapy kit that donors can deliver at home or in the hospital is valuable to create. The kit should include a variety of dimensions such as empathy and assistance for all affected individuals, with compassion and kindness you have to be understood. Development of online mental wellbeing and advice programs in hospitals, community health centers and psychiatric departments may be an opportunity to tackle this problem.

13.3.1 Long-term effects of lung rehabilitation

Rehabilitation after critical illness is a key component of the continuum of care [23]. The recovery is a dynamic procedure that refers to a longitudinal mechanism that seeks to mitigate the debilitating impact of a person's impairments, to promote and optimize the functionality of everyday activities, and to enhance prospects for significant involvement in community on the basis of some new functional basis.

13.3.2 Lung rehabilitation of COVID-19 patients

Rehabilitation specialists in China developed realistic and feasible recommendations for respiratory rehabilitation for COVID-19 patients based on the consensus and references given by front-line experts. The short-term aim of lung rehabilitation is to alleviate dyspnea and to relieve anxiety and depression, while the long-term objective is to optimize the patient's function, to enhance their quality of life and to allow them to return to society [35].

Thorough examinations should be carried out before the recovery program is started [19]. Medical and practice risk evaluations should also, for example, be carried out on the basis of clinical signs, vital signs, auxiliary tests, imaging, co-morbidities, contraindications, etc. for the patient, and psychological and nutritional assessments in the form of questionnaires should include the use of a quality of life, physical activity endurance, and other factors [21]. The outcomes of these measures can then be combined with the individualized and progressive recovery criteria for patients' aerobic endurance, muscle power, balance and flexibility.

The substance of the prescription is mainly:

A. Aerobic exercises: walk, quick walk, jogging, swimming from low intensity, incremental intensity and length increase. Every time 3–5 days a week, 20–30 minutes.
B. Power training: incremental strength training is advisable. Every target muscle group has a training load of 8–12 RM, 1–3 group/time. Each community has an exercise period of 2 minutes, 2–3 days a week, and the exercise charge is increased every week by 5–10%.
C. Balance preparation: balance training should provide equilibrium training in patients with balance problems like unarmed equilibrium and equilibrium training.
D. Respiratory training: if the patient has symptoms including breath loss, wheezing, and waiting problems after release, respiratory mode exercise should be planned in tandem with the assessment results, for example body posture handling, breathing rhythm modification, breathing muscle breathing exercise, and the expectoration exercise.

Under the principle of protection all recovery should be done. If the peripheral capillary oxygen saturation (SpO_2) displayed in <88%, the patient should quit the rehabilitation scheme immediately, or if signs are present such as palpitations, sweat, tightness of the head and shortness of the breath that the clinician finds unsuitable for rehabilitation.

Therapy should be initiated as soon as possible in mild and moderate situations. In comparison, life-saving measures should be given priority in serious and vital situations where the state of the patient is unstable or the disease persists. In these cases pulmonary therapy can only be done if the condition of the patient is stabilized. Furthermore, the movement of chronically or critically ill patients should be reduced to their bed or bed in view of their protection and human resources. When discharged, individualized care can proceed under the assumption of enhanced safety and prevention against other infectious diseases, such as cold.

The disease's infectiveness is the most defining feature for rehabilitation of COVID-19 patients as compared with general rehabilitation of patients with chronic disease. Operations, such as instructed cough, expiration training and tracheal compression, that may increase the risk of infection should also be reduced. In order to protect the mouth from infection, a sealed plastic bag should be used during sputum. Furthermore, Lung rehabilitation of COVID-19 patients is primarily conducted through instructional recordings, brochures, remote visits or online training in order to safeguard protective equipment and prevent cross-infection. The entire Lung rehabilitation program shall be assessed and tracked.

References

[1] V.M. Corman, O. Landt, M. Kaiser, R. Molenkamp, Meijer (2020).

[2] Y. Huang, C. Tan, J. Wu et al. Impact of coronavirus disease 2019 on pulmonary function in early convalescence phase. Respir. Res. 21 (1). doi:10.1186/s12931-020-01429-6.

[3] D.C. McKenzie. Respiratory physiology: Adaptations to high-level exercise. Br. J. Sports Med. 46 (6) 381–384. doi:10.1136/bjsports-2011-090824.

[4] N. Chen, M. Zhou, X. Dong, J. Qu, F. Gong, Y. Han, Y. Qiu, J. Wang, Y. Liu, Y. Wei, J. Xia, T. Yu, X. Zhang, L. Zhang, Epidemiological and clinical characteristics of 99 cases of 2019 novel coronavirus pneumonia in Wuhan, China: a descriptive study, The Lancet 395 (10223) (2020) 507–513 ISSN 0140-6736. https://doi.org/10.1016/S0140-6736(20)30211-7.

[5] P. Thomas, C. Baldwin, B. Bissett et al. Physiotherapy management for COVID-19 in the acute hospital setting: Clinical practice recommendations. Journal of Physiotherapy. 66 (2) 73–82. doi:10.1016/j.jphys.2020.03.011.

[6] L. Munshi, L. Del Sorbo, N.K.J. Adhikari et al. Prone position for acute respiratory distress syndrome: A systematic review and meta-analysis. Annals of the American Thoracic Society. 14 S280–S288. doi:10.1513/AnnalsATS.201704-343OT.

[7] A. Gugnani, C. Mehandiratta, Effect of Diaphragmatic Breathing and Pursed Lip Breathing In Improving Dyspnea-A Review Study, Eur. J. Mol. Clin. Med. 7 (6) (2020) 2643–2646.

[8] Stress Inoculation Training Supported by Physiology-Driven. (2022).

[9] M.A. Cottrell, O.A. Galea, S.P. O'Leary, A.J. Hill, T.G. Russell. Real-time telerehabilitation for the treatment of musculoskeletal conditions is effective and comparable to standard practice: A systematic review and meta-analysis. Clin. Rehabil. 31 (5) 625–638. doi:10.1177/0269215516645148.

[10] A.A.M. Castro, S.R. Calil, S.A. Freitas, A.B. Oliveira, E.F. Porto. Chest physiotherapy effectiveness to reduce hospitalization and mechanical ventilation length of stay, pulmonary infection rate and mortality in ICU patients. Respir. Med. 107 (1) 68–74. doi:10.1016/j.rmed.2012.09.016.

[11] A. Gugnani. Effects of breathing exercises and incentive spirometry in improving lung capacity on individuals with lung fibrosis. European Journal of Molecular and Clinical Medicine. 7 (2) 1407–1435. https://ejmcm.com/article_2349.html

[12] C.S. Castro, O. Freitas, E.F. Porto. Chest physiotherapy effectiveness to reduce hospitalization and mechanical ventilation length of stay, pulmonary infection rate and mortality in ICU patients. Respir. Med. doi:10.1016/j.ctcp.2020.10116631.
[13] J.-L. Romette, J. Ellis, M. Zambon, C. Drosten. Detection of 2019 novel coronavirus (2019-nCoV) by realtime RT-PCR. Eurosurveillance. 25.
[14] Australian Physiotherapy Association, Telehealth guidelines response to COVID-19 (2020).
[15] S.P. Adhikari, S. Meng, Y.J. Wu et al. Epidemiology, causes, clinical manifestation and diagnosis, prevention and control of coronavirus disease (COVID-19) during the early outbreak period: A scoping review. Infectious Diseases of Poverty. 9 (1). doi:10.1186/s40249-020-00646-x.
[16] M. Purgato, C. Gastaldon, D. Papola, M. van Ommeren, C. Barbui, W.A. Tol. Psychological therapies for the treatment of mental disorders in low- and middle-income countries affected by humanitarian crises. Cochrane Database of Systematic Reviews. 2018 (7). doi:10.1002/14651858.CD011849.pub2.
[17] T. Li, Y. Zhang, C. Gong et al. Prevalence of malnutrition and analysis of related factors in elderly patients with COVID-19 in Wuhan, China. Eur. J. Clin. Nutr. 74 (6) 871–875. doi:10.1038/s41430-020-0642-3.
[18] A. Boretti, B.K. Banik. Intravenous vitamin C for reduction of cytokines storm in acute. In: Could Reduce Risk of Influenza and COVID-19 Infections and Deaths. Vol. 12. doi:10.3390/nu12040988.91.
[19] M. O'Sullivan. Fast recovery or great depression? Three Scenarios for the Coronavirus Economic Crisis. Vol. 4, 210–212.
[20] A.J.W. te Velthuis, S.H.E. van den Worm, A.C. Sims, R.S. Baric, E.J. Snijder, M.J. van Hemert, Zn^{2+} Inhibits Coronavirus and Arterivirus RNA Polymerase Activity In Vitro and Zinc Ionophores Block the Replication of These Viruses in Cell Culture (11) (2010) e1001176. https://doi.org/10.1371/journal.ppat.1001176.
[21] J. Shigemura, R.J. Ursano, J.C. Morganstein, M. Kurosawa, D.M. Benedek. Public responses to the novel 2019 coronavirus (2019-nCoV) in Japan: Mental health consequences and target populations. Psychiatry Clin. Neurosci. 74 (4) 281–282. doi:10.1111/pcn.12988.
[22] L. Kang, Y. Li, S. Hu et al. The mental health of medical workers in Wuhan, China dealing with the 2019 novel coronavirus. The Lancet Psychiatry. 7 (3) e14. doi:10.1016/S2215-0366(20)30047-X.
[23] K.L. Piercy, R.P. Troiano, R.M. Ballard et al. The physical activity guidelines for Americans. JAMA - Journal of the American Medical Association. 320 (19) 2020–2028. doi:10.1001/jama.2018.14854.
[24] X. He, E.H. Lau, P. Wu, X. Deng, J. Wang, X. Hao, …, G.M. Leung, Temporal dynamics in viral shedding and transmissibility of COVID-19, Nat. Med. 26 (5) (2020) 672–675.
[25] Liu, Y., Yan, L.M., Wan, L., Xiang, T.X., Le, A., Liu, J.M., Zhang, W. (2020). Viral dynamics in mild and severe cases of COVID-19. The Lancet Infectious Diseases. doi:10.1016/S1473-3099(20)30232-2.
[26] C. Qin, L. Zhou, Z. Hu, S. Zhang, S. Yang, Y. Tao, D.S. Tian, Dysregulation of immune response in patients with COVID-19 in Wuhan, China, Clin. Infect. Dis. 71 (15) (2020) 762–768, doi:10.1093/cid/ciaa248.
[27] M. Kalluri, J. Richman-Eisenstat, From consulting to caring: Care redesign in idiopathic pulmonary fibrosis, NEJM Catalyst 5 (2) (2019), doi:10.1056/CAT.19.0682.
[28] A.K. Simonds, A. Hanak, M. Chatwin, M.J. Morrell, A. Hall, K.H. Parker, J.H. Siggers, R.J. Dickinson, Evaluation of droplet dispersion during non-invasive ventilation, oxygen therapy, nebuliser treatment and chest physiotherapy in clinical practice: implications for management of pandemic influenza and other airborne

infections, Health Technol. Assess. (Rockv) 14 (46) (2010) Article 2, Published in October. https://doi.org/10.3310/hta14460-02.

[29] G. Brankston, L. Gitterman, Z. Hirji, C. Lemieux, M. Gardam, Transmission of influenza A in human beings, Lancet Infect. Dis. 7 (4) (2007) 257–265 ISSN 1473-3099. https://doi.org/10.1016/S1473-3099(07)70029-4.

[30] A. Laviano, A. Koverech, M. Zanetti, Nutrition support in the time of SARS-CoV-2 (COVID-19), Nutrition 74 (2020 Jun) 110834 Epub 2020 Apr 2. PMID: 32276799; PMCID: PMC7132492, doi:10.1016/j.nut.2020.110834.

[31] W.B. Grant, H. Lahore, S.L. McDonnell, C.A. Baggerly, C.B. French, J.L. Aliano, H.P. Bhattoa, Evidence that Vitamin D Supplementation Could Reduce Risk of Influenza and COVID-19 Infections and Deaths, Nutrients 12 (2020) 988. https://doi.org/10.3390/nu12040988.

[32] R.A. Schmidt, T.D. Lee, C. Winstein, G. Wulf, H.N. Zelaznik, Motor control and learning: A behavioral emphasis, Human kinetics (2018).

[33] H. Mowbray, In Beijing, coronavirus 2019-nCoV has created a siege mentality, BMJ (2020) 368.

[34] WHO, C.O. (2020). World health organization. Responding to Community Spread of COVID-19. Reference WHO/COVID-19/Community_Transmission/2020.1.

[35] World Health Organization. Mental health and psychosocial considerations during the COVID-19 outbreak, 18 March 2020 (No. WHO/2019-nCoV/MentalHealth/2020.1), World Health Organization, 2020.

[36] Z. Xu, L. Shi, Y. Wang, J. Zhang, L. Huang, C. Zhang, …, F.S. Wang, Pathological findings of COVID-19 associated with acute respiratory distress syndrome, Lancet Respir. Med. 8 (4) (2020) 420–422.

[37] J. Yang, Y.A. Zheng, X. Gou, K. Pu, Z. Chen, Q. Guo, …, Y. Zhou, Prevalence of comorbidities and its effects in patients infected with SARS-CoV-2: a systematic review and meta-analysis, Int. J. Infect. Dis. 94 (2020) 91–95.

[38] E.D. Lewis, S.N. Meydani, D. Wu, Regulatory role of vitamin E in the immune system and inflammation, IUBMB Life 71 (4) (2019) 487–494.

[39] World Health Organization, Managing epidemics: key facts about major deadly diseases, World Health Organization, 2018.

CHAPTER

14

Aptasensor: Surface protein detection in case of coronavirus diagnosis

Nagappa L. Teradal[a,#], Ranjita D. Tandel[b,#], and Vishalkumar I. Naik[b]

[a]Department of Chemistry J. S. S. Arts, Science and Commerce College, Gokak, India, [b]Department of Chemistry, Gokhale Centenary College, Ankola, India

14.1 Introduction

The world has faced an infectious disease outbreak caused by a new virus, severe acute respiratory syndrome (SARS) is a respiratory viral disease caused by a coronavirus. SARS-CoV-2 is a member of Coronaviridae, which is the responsible agent for COVID-19 [1]. Coronaviridae, together with the Arteriviridae and Roniviridae families, belong to the order of Nidovirales. Coronaviridae family has two virus genus, Coronavirus (CoV) and Torovirus. Four distinct genera are part of the CoV genus, known as α, β, γ, and δ; however, only α-CoVs and β-CoVs infect mammalian hosts [2]. SARS-CoV-2 is an envelope and positively single-stranded RNA (ssRNA) virus with a typical genome size 29.7 kb, encoding directed RNA polymerase and structural proteins including the spike (S), envelope (E), membrane (M) and nucleocapsid (N) proteins. The N protein is a 422-amino-acid alkaline protein with a short lysine rich region suggested to be the nuclear localization signal. It plays an important role in the process of virus particle assembly by enveloping the entire genomic RNA. The S protein gives the spike-like projections on the surface of CoVs, conferring their distinctive aspect of a crown [4].

# Authors have contributed equally.

Sensing Tools and Techniques for COVID-19: Developments and Challenges in Analysis and Detection of Coronavirus
DOI: https://doi.org/10.1016/B978-0-323-90280-9.00010-9

Studies have shown that N and S structural proteins are closely related to CoVs pathogenicity. N protein plays an essential role both in viral genome packaging and viral transcription. In addition, N protein may also be associated with the host's immune response suppression, facilitating viral replication. S proteins have two subunits (S1 and S2), with distinct functions. S1 binds to the cell surface receptor host through the receptor binding domain (RBD) region present in this subunit. On the other hand, S2 mediates virus fusion with the host cell membrane [5,6].

Viruses with RNA genomes capable of having high mutation rates due to strange features of RNA genomes not found in DNA genomes, such as high rates of recombination, replication and segmentation in a stipulated time. Thus, these features enable a rapid viral genome evolution, favoring variants emerging that can present new characteristics as virulence increases. Thus, CoVs, RNA viruses, have innumerable mutations, mainly in the S protein, which could not be different. The mutations could favor the adaptation of this protein to various host receptor types, causing changes in the tropism and in the pathogenicity of CoVs. The viral infection starts with the entrance of the virus into host cells. In the case of CoVs, the RBD domain present in the S1 subunit of S protein recognizes the receptor located on the host's cell surface, initiating virus entrance into the cytosol. The entry of the SARS-CoV-2 into the host's cell occurs through the attachment between angiotensin-converting enzyme-2 (ACE2) receptor and S protein, new researches have also shown that S protein can bind to other receptors, such as NRP1 (neuropilin-1) receptor; however, its mechanism is not fully understood [7–10].

Native S protein is present on the viral envelope surface in its closed form (inactive S protein), in which RBDs domains are inaccessible to the cell receptor. After protein cleavage, the S1 subunit opens (it becomes active), exposing the RBD domain. Depending on the type of protease present in the host's tissue, SARS-CoV-2 can enter cells by two distinct pathways, the nonendocytic and the endocytic pathways. Proteases expressed in the cellular membrane, such as trypsin, furin, and transmembrane protease (TMPRSS2), lead SARS-CoV-2 to the nonendocytic pathway. These host's enzymes cleave the S protein, causes conformational changes that expose the S2 subunit and, consequently, virus fusion immediately with the host's plasma membrane. In the absence of those enzymes, the virus enters by the endocytic pathway, within vesicles (endosomes) after S protein cleavage by endosomal proteases (clathrin and nonclathrin types). ACE2 and NRP1 receptors are expressed in multiple tissues from the human body, as lungs, oral mucosa, kidneys, liver, intestines, heart, brain, blood vessels, and others. That is why SARS-CoV-2 can invade these tissues and induce systemic inflammation, leading to multiple organ dysfunction syndromes [7–10].

Therefore, SARS-CoV N/S protein has long been thought an ideal target for its detection. The monitoring of SARS-CoV N/S protein has been of considerable interest in the development of simple and reliable analytical tools for the detection of SARS [7–10]. Currently, the most widely used technology for diagnosing SARS is enzyme-linked immunosorbent assay (ELISA), based on recombinant proteins from the genome of SARS. Although the existing techniques are highly specific but suffering from some serious limitations. For example, these methods cannot be used to detect viruses in its premature stage, at a time when antibodies against SARS antigens are not produced. An early diagnosis of the disease is need of the hour and requirement for effective treatment and, management of diseases. In most cases, the diagnosis of viral diseases is done based on symptoms and microbial, basic biochemical, serological and parasitological tests. Additional confirmatory tests include ELISA, PCR, DNA chips and chromatographic techniques, which require sophisticated instruments, professional operators, and need of costly solvents. Aptamer-based biosensors are generally indicated as aptasensors, rapidly becoming a cutting-edge novel technology in disease diagnosis. Aptasensors are bioanalytical devices that use biological molecules as recognition elements in their fabrication to generate signals that can be used to detect the status of the disease. Recently, researchers achieved the specificity and sensitivity of the bioanalytical tools through the use of aptasensors for the detection of COVID-19. With the accumulating knowledge on the development of aptamers and aptasensors, they have been increasingly explored and tested for the detection of various infectious agents, including bacteria and viruses, as well as their toxins.

14.2 Aptamers

The aptamers are single-stranded nucleic acids (DNA, RNA) having molecular recognition properties similar to antibodies, 20–100 nucleobases long and consists of variable region edged by constant regions, which allow for amplification and identification of sequences [11,12]. Aptamers are excellent receptors because they bind with high affinity and specificity to various targets such as viruses, proteins, peptides and small organic molecules. Aptamers bind their target via an induced fit mechanism involving van der Waals forces, hydrogen bonding and electrostatic interactions. In addition, aptamers offer more attractive features compared to antibodies. They have a less expensive production process since they use in vitro synthesis, are easily modifiable with functional groups (facilitating immobilization process), and they provide long-term storage. Aptamers have become increasingly important molecular tools for diagnostics and therapeutics. In particular, Aptamers exhibit advantages as recognition elements in biosensing when compared with antibodies and enzymes, thus making it possible to develop a wide range of aptamer-based biosensors. Chemically stable and cost-effective aptamers offer great flexibility in the design of novel biosensors with high detection sensitivity and specificity [13–15]. Aptamer-based biosensors offer promising platforms to substitute conventional technologies. Nucleic acid functionalized sensors play an important role in the development of the biosensing field. Aptamer-functionalized devices are one unique type for the outstanding advantages of aptamers.

Electrochemical aptasensors: Electrochemical aptasensors detection is based on the electrical signal changes caused by the specific interaction and combination between biorecognition element (aptamer) and the target analyte. The relationships among targets, aptamers, redox probes, electrodes, and instruments are the keys to determine the specificity and sensitivity of aptasensors [16]. Based on label free strategies these electrochemical aptasensors are categorized as, enzyme amplification methods, and field-effect transistor (FET) techniques. The label free electrochemical transducer is performed by the change of the resistance and the capacitance of the solution electrode interface based on the aptamer-target complex on conductive or semi conductive surfaces. However, field effect transistors (FETs) can transform the interaction with target molecules on FET surface into a readable signal [17].

Optical aptasensors: The fluorescence and colorimetric platforms are most commonly applied in conjunction with aptamers to develop aptasensors. Commercialized antibody based test strips and reagents will direct aptamer products [18]. Surface-enhanced Raman scattering spectroscopy (SERS) is a powerful vibrational spectroscopy technique that can overcome the limitations of Raman spectroscopy's low sensitivity. The two main mechanisms of SERS are considered to be electromagnetic (EM) and chemical effects. The EM was considered to be the prior effect to the enhancement. SERS could be used for sensitive structural detection of low concentration, including single molecules, biomolecules, biological cells, and even viruses [19].

At present, coronavirus diagnostics are based on direct detection of viral components or indirect detection by measuring antibodies generated in response to viral infection. Compared to traditional antibody based detection, aptamers could provide faster adaptation to continuously evolving virus strains and higher discriminating capacity between specific virus stereotypes. Aptamers are very stable and easily modifiable, so are ideal molecules for detection and chemical sensing applications. Herein, we summarize the use of aptasensors for detection of coronavirus and consider the feasibility of aptasensors to become portable devices for point of care diagnostics of COVID 19.

Tian et al. fabricated an electrochemical dual-aptamer biosensor based on the metal-organic frameworks MIL-53(Al) decorated with silver/platinum (Au/Pt) nanoparticles and enzymes to determine SARS-CoV-2 nucleocapsid protein (2019-nCoV-NP) via cocatalysis of the nanomaterials, horseradish peroxidase (HRP) and G-quadruplex DNAzyme [20]. In the present work, the recognition elements, thiolated DNA aptamers (N48 and N61) immobilized on the surface of gold substrate through Au-S bonds for the detection of COVID-19 biomarker, 2019-nCoV-NP. The nanocomposites, Au@Pt/MIL-53 (Al) are designed with HRP and hemin/G-quadruplex DNAzyme as signal nanoprobe. Nanoprobe was then applied to amplify the aptasensor signal *via* cocatalyzing the oxidation of hydroquinone in the presence of hydrogen peroxide. Further, the aptamer-protein-nanoprobe sandwich electrochemical biodetection system is assembled on the surface of gold substrate. Under optimal conditions and parameters the demonstrated aptasensor showed wide linear range from 0.025 to 50 ng mL^{-1} with the detection limit of 8.33 pg mL^{-1} for 2019-nCoV-NP. The presented aptasensor has great potential in the early diagnosis of COVID-19 with high sensitivity and reliability [20].

Yet in another work, aptamer based electrochemical impedimetric biosensor is reported for the detection of SARS-CoV-2 S protein [21]. In the present work, gold nanoparticles (AuNPs)-modified screen-printed carbon electrode (SPCE) and an aptamer targeting the RBD of the SARS-CoV-2 is utilized for the development of biosensor. The fabrication procedure involves the immobilization of the ssDNA aptamer on gold nanoparticles (AuNPs) and the binding of S-protein with the aptamer are investigated through photo-induced force microscopy (PiFM) and electrochemical techniques. The disulfide-modified aptamer was homogeneously immobilized on the surface of the AuNPs, which allowed the probe to capture S-protein atop practically all the AuNPs-modified surface, as revealed by the PiFM mapping imaging in the 1400 cm^{-1} region. The detection of SARS-CoV-2 S-protein is achieved by electrochemical impedance spectroscopy after 40 min incubation with several analyte concentrations, yielding a limit of detection of 1.30 pM (66 pg/mL). Moreover, the aptasensor is successfully applied for the detection of a SARS-CoV-2 pseudovirus, thus suggesting it a promising tool for the diagnosis of COVID-19. The selectivity studies showed that the aptasensor is active to both SARS-CoV and SARS-CoV-2 S proteins [21].

In order to facilitate point-of-care diagnosis of SARS-CoV-2, Rehmati et al., have developed a label-free ultrasensitive electrochemical nanobiodevice for SARS-CoV-2 spike protein detection using Protein A/Cu_2O modified SPCE as a substrate for the ordered orientation of IgG antibodies as a specific receptor [22]. By utilizing the IgG anti-SARS-CoV-2 spike antibody onto the electrode surface, researchers have constructed the screen-printed carbon electrode modified with Cu_2O nanocubes

(Cu_2O NCs), which provide a large surface area to increase the protein A (ProtA) loading on the electrode surface. The modified aptasensor is used in clinical samples to detect the SARS-CoV-2 virus in less than 20 min. The electrochemical evaluations proved the very good linear response between the charge transfer resistance (Rct) and spike protein contents *via* a specific binding reaction in the range 0.25 fg mL^{-1} to 1 μg mL^{-1} [22]. In another work, Jiang et al., have developed highly sensitive and stable two-dimensional metal-organic framework (2D-MOF) based photoelectrochemical (PEC) aptasensor for the detection of SARS-CoV2 spike glycoprotein [23]. The PEC aptasensor is developed by plasmon-enhanced photoactive material, AuNPs/Yb-TCPP with a specific DNA aptamer against S protein. The Au Nps/Yb-TCCP fabrication is done by in situ growth of Au NPs on the surface of 2D Yb-TCPP nanosheets. This showed excellent photoelectric performance. The modified DNA aptamer on the surface of Au NPs/Yb-TCCP binds with S protein with high selectivity, photoelectric current is found to decrease this might be due to the high steric hindrance and low conductivity of S protein. The developed PEC aptamer showed high S protein detection with linear range of 0.5–8.0 μg mL^{-1} with limit of detection of 72 $ngmL^{-1}$ [23].

A highly sensitive and specific optical aptasensor for the detection of receptor-binding domain (RBD) of the SARS-CoV-2 spike glycoprotein is developed [24]. Immobilization of specific aptameric sequence on short polyethyleneglycol (PEG) interface on gold nanofilm deposited on a D-shaped plastic optical fiber (POFs) probe, and the protein binding was monitored by exploiting the very sensitive surface plasmon resonance (SPR) phenomenon. In the present work receptor-binding domain (RBD) of the SARS-CoV-2 spike glycoprotein is used to develop an aptasensors [24]. The prepared bointerface is characterized by surface analysis techniques coupled to fluorescence microscopy and plasmonic analysis. Spanning a wide protein range (25÷1000 nM), the SARS-Cov-2 spike protein is detected with a limit of detection of about 37 nM. The specificity of aptasensor is also confirmed by testing different interfering agents (BSA, AH1N1 hemagglutinin protein and MERS spike protein) [24].

Tabrizi and coworkers for the first time developed a photoelectrochemical aptasensor for the quantitative determination of the SARS-Cov-2 RBD (receptor-binding domain) [25]. For the fabrication of aptasensor researchers have used CdS quantum dots and gC_3N_4 as photo-active materials, ascorbic acid as an electron donor molecule, chitosan as a binder, and an amine terminal aptamer as a biorecognition element. The development process involves the following steps: firstly, graphitic carbon nitride and (gC_3N_4) and cadmium sulfide (CdS) quantum dots are fabricated. Secondly, gC_3N_4 and CdS quantum dots mixed well to produce CdS QDs-gC3N4 nanocomposite and then added to the solution containing chitosan as an amine-rich polymer to generate a Chitosan/CdS-gC_3N_4

nanocomposite. Subsequently, the surface of the ITO electrode is modified with Chitosan/CdS-gC_3N_4. After that, the amine-terminal aptamer probes are immobilized on the surface of the Chitosan/CdS QDs-gC_3N_4/ITO electrode by using glutaraldehyde as an amine-amine crosslinker [25]. The mechanism of the aptasensor is based on the change in the photoactive response of the surface of the ITO electrode modified with the Aptamer/Chitosan/CdS QDs-gC_3N_4. Upon the interaction of SARS-Cov-2 RBD with the immobilized aptamer probe on the surface of the electrode, the mass transfer limitation for AA as an electron donor molecule to the photoactive nanocomposite (CdS QDs-gC_3N_4) increased. Consequently, the intensity of the photo-current decreased. Therefore, the demonstrated aptasensor is considered as a signal-off-based aptasensor. The reported aptasensor is used for the determination of SARS-Cov-2 RBD in the dynamic range of 0.5–32 nM with a detection limit of 0.12 nM [25].

A novel label-free surface plasmon resonance aptasensor is fabricated by Chen et al., for the detection of N-gene of SARS-CoV-2 by using thiol-modified niobium carbide MXene quantum dots (Nb_2C-SH QDs)/N-gene targeted N58 aptamer [26]. In the present study, the gold chip for surface plasmon resonance measurements is modified with Nb_2C-SH QDs via Au-S covalent binding and self-assembling. Nb_2C-SH QDs not only resulted in high bioaffinity toward aptamer but also enhanced the surface plasmon resonance response. The reported aptamer showed low detection limit (4.9 pg mL^{-1}) with a linearity range of 0.05–100 ng mL^{-1} N-gene concentration. High selectivity, excellent repeatability results make the developed aptamer good applicability in practical usage. Moreover, the presented Nb_2C-SH QD-based surface plasmon resonance response aptasensor displayed a huge practical application in terms of qualitative analysis of N-gene from different samples, including seafood, seawater, and human serum [26]. Yet in another study, Chen and co-workers used surface-enhanced Raman scattering (SERS) technique to fabricate aptasensor platform capable of quantifying (SARS-CoV-2) lysates with a high sensitivity [27]. In the report, a spike protein DNA aptamer is used as a receptor, and a self-grown Au nanopopcorn surface is used as a SERS detection substrate for the sensitive detection of SARS-CoV-2. A quantitative analysis of the SARS-CoV-2 lysate is performed by monitoring the change in the SERS peak intensity caused by the new binding between the aptamer DNA released from the Au nanopopcorn surface and the spike protein in the SARS-CoV-2 virion. From this technique, the detection limit is noticed to be less than 10 PFU mL^{-1} within 15 min [27].

Cho et al., demonstrated an ssDNA aptamer that specifically binds to SARS CoV nucleocapsid protein, it is isolated from a DNA library containing 45-nuceotide random sequences in the middle of an 88mer single-stranded DNA [28]. They have used an Enzyme-linked immunosorbent assay (ELISA) to identify the aptamer with the highest binding affinity to the SARS CoV nucleocapsid protein. Their results concluded that ssDNA aptamer binds to the nucleocapsid protein with a Kd of 4.93 ± 0.30 nM. The Western blot analysis using the selected ssDNA aptamer demonstrated that the aptamer is a good alternative probe to monoclonal antibodies. When considering the advantages of aptamers relative to antibodies, including their stability and facile synthesis, the selected ssDNA aptamer showed good detection of SARS CoV N protein [28].

Researchers have used three different optical techniques, viz., biolayer interferometry, surface plasmon resonance and surface enhanced Raman spectroscopy, for quantifying the binding of recombinant SARS-CoV-2 spike protein to surface immobilized oligonucleotide aptamers [29]. The demonstrated biosensing platform using different techniques, functionalizing the surface with DNA aptamers that bind specifically to the SARS-CoV-2 spike protein receptor. The change in surface property of biosensing platform is identified by refractive index and the vibrational spectrum of probe aptamer. The results of the study indicated that, biolayer interferometry showed detection limit of 250 nM. Surface plasmon resonance is a more sensitive technique the detection limit for this was found to be 5 nM. However, surface-enhanced Raman spectroscopy enabled detection of spike protein to sub-picomolar concentrations [29]. The control experiments are performed using scrambled aptamers and bovine serum albumin as an analyte, results of this study showed the demonstrated aptasensing approach is both sensitive and selective, with no appreciable response observed for any of the controls [29].

The SARS-CoV (N) protein is one of the most abundant structural proteins and serves as a diagnostic marker for accurate and sensitive detection of the virus. Using a SELEX (systematic evolution of ligand by exponential enrichment procedure) and recombinant N protein, Ahn and coworkers selected a high-affinity RNA aptamer capable of binding to N protein with a dissociation constant of 1.65 nM [30]. The results of the studies of RNA competition experiments and electrophoretic mobility shift assays showed that the demonstrated aptamer recognized selectively the C-terminal region of N protein with high specificity. Further, by using a chemiluminescence immunosorbent assay and a nanoarray aptamer chip with the selected aptamer as an antigen-capturing agent, they have detected N protein at a concentration of 2 pg/mL. The prepared aptamers acted as antigen capturing molecule in chemiluminescence immunosorbent assay. Also reported that the aptamer-antibody hybrid immunoassays were useful for rapid, sensitive detection of SARS-CoV N protein [30].

In response to the current need for highly sensitive, quick and on-site detection of the SARS-CoV-2 virions in different aqueous solutions, two different nanolayer biorecognition systems separately combined with an adaptable optical fiber sensor have is reported by Cennamo et al [31].

The demonstrated sensors is tested by exploiting a plasmonic plastic optical fiber (POF) sensor coupled with two different receptors, both designed for the specific recognition of the SARS-CoV-2 Spike protein; one is aptamer-based and the other is molecular imprinted polymer-based. The preliminary experimental results showed that the sensitivity of the proposed SARS-CoV-2 optical fiber sensors proved to be high enough to detect virions. The results of the work showed that molecularly imprinted polymer receptor (MIP) combined with the SPR-POF platform, exhibited good performances in the SARS-CoV-2 detection. Both sensors resulted in a nanomolar range of detection of spike protein [31].

Lasserrea and coworkers have reported an impedimetric SARS-CoV-2 biosensor using SARS-CoV-2 Optimers [32]. In the report, thin film gold electrodes are fabricated, by sputter coating gold onto a polyester substrate, and used for the development of biosensor. Further, the functionalization of thin film gold electrodes with aptamer is done by temperature treatment. Further, reduced oligonucleotide mixtures are filtered using NAP-10 columns to 30 nM for aptamers and 32.4 nM for optimers, prior to an hour long immobilization onto electrodes. The results demonstrated that the interaction between the receptor and the spike protein is much stronger following binding with SARS-CoV-2 S1 than following incubation with IL-6 protein. Hence, it is concluded that following the successful development of an antispike protein Optimer sequence, sensitive, and selective detection of recombinant SARS-CoV-2 spike protein could be achieved with modified aptamer [32].

The applicability optical property of quantum dots (QD) in the detection of COVID-19 is established through the QDs surface conjugation with aptamers [33]. In the report researchers showed that QDs-conjugated RNA oligonucleotide could specifically recognize the SARS-CoV N protein. The fabricated biosensing platform, QDs-conjugated RNA aptamer, could be used as biosensor prototype for SARS-CoV N protein diagnosis. Advantageous binding of QDs-conjugated RNA aptamer for the detection of SARS-CoV N protein is determined by analyzing the fluorescence intensity through the use of confocal microscopy. Form the results of the analysis, the QDs-conjugated RNA aptamer could recognize the SARS-CoV N protein with a detection limit of 0.1 pg mL^{-1} on a designed chip [33]. A study performed by Vlamir et al. fabricated an aptamer-based nanosensor containing silver nanoparticles as SERS substrate to detect SARS-CoV-2 [34]. It is a one-step protocol that demonstrated the aggregation of silver nanoparticles after the binding of labeled aptamer with incubated virus and showed an intense SERS signal at 587 cm^{-1}. The reported detection system showed high selectivity and sensitivity of 5.5×10^4 TCID50/mL for the determination of SARS-CoV-2 [34]. In another work, Rhodamine 6 G-linked DNA aptamer functionalized gold nanostars (GNS) are employed to detect SARS-CoV-2 using the distance-dependent nanoparticle surface energy transfer (NSET) spectroscopy [35], the study is explored by Pramanik et al. [35]. In the present work, GNS possess high extinction co-efficient which is quite similar to organic (106) molecules. The binding between rhodamine 6G-ssDNA aptamer displayed 99% quenching as a result of NSET from dye to GNS. These functionalized nanoparticles possess high sensitivity to SARS-CoV-2 spike protein with limit of detection of 130 fg mL^{-1}. Also, these aptamer-GNS particles inhibited the penetration of SARS-CoV-2.

14.3 Conclusions and future perspectives

The recent developments in the aptamer-based biosensing platforms for the detection of SARS-CoV-2 are briefly discussed in this chapter. Affinity-based aptameric nanobiosensors have the ability to detect the virus or their entities, as low as pico molar and even up to femto molar concentrations, in few cases. Proper selection of aptamers is a critical step in the development of aptameric nanobiosensors. Large-scale clinical validation studies and the processing of complex samples like blood, urine, sweat, feces, exhaled breath, and few others are necessitated to understand the sensor robustness and commercial viability. Development of smart watches-based diagnostic platforms may be advantageous for continuous monitoring of COVID-19.

The aptasensors can monitor viral spread in real-time, promoting more accurate and faster diagnostic solutions. The importance of investment in science and technology for developing innovative ways to deal with tragedies as COVID-19 pandemics is urgent, mainly in developing countries. Public authorities, thinking in general health and the common good, should base their decisions on science and educate their population, leading to good practices, health security and the pandemic's end. Indeed, with the advanced platform of 3D printing, a wide range of nanostructures can be prepared for the smart manufacturing of aptameric nanobiosensors for the diagnosis of COVID-19. However, this virus is more complex and hence it is necessary to invent rapid, sensitive and low cost technique for the early diagnosis of this infection. The sensors have shown their potential towards the diagnosis and sensing of viral infections and hence may perhaps fulfill the current demand for early diagnosis of SARS-CoV-2 cases. A number of biosensing methods including impedimetric, amperometric, and colorimetric-mediated methods have been established. The reported method could be used to detect the SARS-CoV-2 within short period of time. Though, fewer studies on the development of sensors for SARS-CoV-2 detection have been reported, the biosensing techniques offer alternative approach to PCR based testing for SARS-CoV-2 infection.

In recent developed medicine technology, there is still lot to be done for the developments in the diagnostic techniques to meet the demand of dairy farmers. We expect that this new field of aptamer-based sensors and biosensors in the diagnosis and treatment of viral diseases will eventually become a real-world tool. This would help in encountering challenges that could be difficult with currently available conventional technologies. The success of aptamers in clinics depends on their stability in blood and other biological fluids upon injection into the body. The general use of aptamers is also hindered by their incapability of intracellular accumulation at the site of action. While several studies have shown success with the use of aptamers under in vitro conditions, their use in human system requires special considerations.

References

[1] Y. Song, X.J. Song, X. Wei, et al., Discovery of aptamers targeting the receptor-binding domain of the SARS-CoV-2 spike glycoprotein, Anal. Chem. 92 (2020) 9895–9900.

[2] B. Balasubramaniam, S. Prateek, M. Ranjan, et al., Antibacterial and antiviral functional materials: Chemistry and biological activity toward tackling COVID-19-like pandemics, ACS Pharmacol. Transl. Sci. 4 (2021) 8–54.

[3] A.R. Fehr, S. Perlman, Coronaviruses: An overview of their replication and pathogenesis, Methods Mol. Biol. 1282 (2015) 1–23. https://doi.org/10.1007/978-1-4939-2438-7_1.

[4] D. Forni, R. Cagliani, M. Clerici, et al., Molecular evolution of human coronavirus genomes, Trends Microbiol. 25 (2017) 35–48. https://doi.org/10.1016/j.tim.2016.09.001.

[5] C.-Y. Chen, C. Chang, Y.-W. Chang, et al., Structure of the SARS coronavirus nucleocapsid protein RNA-binding dimerization domain suggests a mechanism for helical packaging of viral RNA, Journal. Mol. Biol. 368 (2007) 1075–1086. https://doi.org/10.1016/j.jmb.2007.02.069.

[6] J. Woo, E.Y. Lee, M. Lee, et al., An in vivo cell-based assay for investigating the specific interaction between the SARS-CoV N-protein and its viral RNA packaging sequence, Biochem. Biophys. Res. Commun. 520 (2019) 499–506. https://doi.org/10.1016/j.bbrc.2019.09.115.

[7] L. Duan, Q. Zheng, H. Zhang, et al., The SARS-CoV-2 spike glycoprotein biosynthesis, structure, function, and antigenicity: Implications for the design of spike-based vaccine immunogens, Front. Immunol. 11 (2020) 576622.

[8] J. Shang, G. Ye, K. Shi, et al., Structural basis of receptor recognition by SARS-CoV-2, Nature 581 (2020) 221–224.

[9] M. Sun, S. Liu, X. Wei, S. Wan, M. Huang, T. Song, et al., Aptamer blocking strategy inhibits SARS-CoV-2 virus infection, Angew. Chem. 60 (2021) 10266–10272.

[10] D. Wrapp, N. Wang, K.S. Corbett, et al., Cryo-EM structure of the 2019-nCoV spike in the prefusion conformation, Science 367 (2020) 1260–1263.

[11] A.D. Ellington, J.W. Szostak, In vitro selection of RNA molecules that bind specific ligands, Nature 346 (1990) 818–822.

[12] C. Tuerk, L. Gold, Systematic evolution of ligands by exponential enrichment: RNA ligands to bacteriophage T4 DNA polymerase, Science 249 (1990) 505–510.

[13] S. Afrasiabi, M. Pourhajibagher, R. Raoofian, et al., Therapeutic applications of nucleic acid aptamers in microbial infections, J. Biomed. Sci. 27 (2020) 6. https://doi.org/10.1186/s12929-019-0611-0.

[14] A. Díaz-Fern_andez, R. Lorenzo-G_omez, R. Miranda-Castro, et al., Electrochemical aptasensors for cancer diagnosis in biological fluids e a review, Anal. Chim. Acta. 1124 (2020) 1–19. https://doi.org/10.1016/j.aca.2020.04.022.

[15] X. Zou, J. Wu, J. Gu, L. Shen, L. Mao, Application of aptamers in virus detection and antiviral therapy, Front. Microbiol. 10 (2019) 1462. https://doi.org/10.3389/fmicb.2019.01462.

[16] J.A. Park, J. Kim, S.M. Kim, et al., Fabrication of electrochemical influenza virus (H1N1) biosensor composed of multifunctional DNA four-way junction and molybdenum disulfide hybrid material, Materials 14 (2021) 343.

[17] Z. Rahmati, M. Roushani, H. Hosseini, Electrochemical immunosensor with Cu2O nanocube coating for detection of SARS-CoV-2 spike protein, Surf. Interfaces. 22 (2021) 100813.

[18] P. Weerathunge, R. Ramanathan, V.A. Torok, et al., Ultrasensitive colorimetric detection of murine norovirus using nanozyme aptasensor, Anal. Chem. 91 (2019) 3270–3276.

[19] Y.Y. Wu, B.W. Liu, P.C. Huang, A novel colorimetric aptasensor for detection of chloramphenicol based on lanthanum ion-assisted gold nanoparticle aggregation and smartphone imaging, Anal. Bioanal. Chem. 411 (2019) 7511–7518.

[20] T. Jinju, L. Zhixian, H. Ou, et al., An electrochemical dual-aptamer biosensor based on metal-organic frameworks MIL-53 decorated with Au@Pt nanoparticles and enzymes for detection of COVID-19 nucleocapsid protein, Electrochim. Acta 387 (2021) 138553.

[21] J.C. Abrego-Martinez, J. Maziar, C. Siham, et al., Aptamer-based electrochemical biosensor for rapid detection of SARS-CoV-2: Nanoscale electrode-aptamer-SARS-CoV-2 imaging by photo-induced force microscopy, Biosens. Bioelectron. 195 (2022) 113595. doi.org/10.1016/j.bios.2021.113595.

[22] Z. Rahmati, R. Mahmoud, H. Hadi, et al., Electrochemical immunosensor with Cu2O nanocube coating for detection of SARS-CoV-2 spike protein, Microchim. Acta 188 (2021) 105.

[23] W.J. Zhong, Z. Ting, M.L. Chun, et al., 2D MOF-based photoelectrochemical aptasensor for SARS-CoV-2 spike glycoprotein detection, ACS Appl. Mater. Interfaces. 42 (2021) 49754–49761. https://doi.org/10.1021/acsami.1c17574.

[24] C. Nunzio, P. Laura, A. Francesco, et al., SARS-CoV-2 spike protein detection through a plasmonic D-shaped plastic optical fiber aptasensor, Talanta 233 (2021) 122532, doi:10.1016/j.talanta.2021.122532.Talanta.

[25] A.T. Mahmoud, N. Leila, A. Pablo, A photo-electrochemical aptasensor for the determination of severe acute respiratory syndrome coronavirus 2 receptor-binding domain by using graphitic carbon nitride-cadmium sulfide quantum dots nanocomposite, Sensors & Actuators: B. Chemical. 345 (2021) 130377.

[26] Surface plasmon resonance aptasensor based on niobium carbide MXene quantum dots for nucleocapsid of SARS-CoV-2 detection, Microchim. Acta 188 (2021) 316.

[27] C. Rongyuan, K. Lun, D. Fenghe, et al., Sensitive Detection of SARS-CoV-2 Using a SERS-Based Aptasensor, ACS Sens 6 (2021) 2378–2385.

[28] C. Seong-Je, W. Hye-Min, K. Ki-Sun, et al., Novel system for detecting SARS coronavirus nucleocapsid protein using an ssDNA aptamer, J. Biosci. Bioeng. 112 (2011) 535–540.

[29] S. Tamsyn, M.G. Fiona, K. Barbara, et al., optical detection of CoV-SARS-2 viral proteins to sub-picomolar concentrations, ACS Omega 6 (2021) 6404–6413.

[30] A. Dae-Gyun, J. Il-Ji, D. Jung et al. RNA aptamer-based sensitive detection of SARS coronavirus nucleocapsid protein. 10.1039/b906788d.

[31] C. Nunzio, D. Girolamo, P. Laura, et al., Quantitative detection of SARS-CoV-2 virions in aqueous mediums by IoT optical fiber sensors, Results in Optics 5 (2021) 100177.

[32] L. Perrine, B. Banushan, J.V. Vincent et al. A SARS-CoV-2 aptasensor based on electrochemical impedance spectroscopy and low-cost gold electrode substrates.

[33] R. Changhyun, K.J. Sung, Quantitative and sensitive detection of SARS coronavirus nucleocapsid protein using quantum dots-conjugated RNA aptamer on chip, Technol. Biotechnol. 86 (2011) 1475–1479.

[34] E. Zavyalova, O. Ambartsumyan, G. Zhdanov, et al., SERS-based aptasensor for rapid quantitative detection of SARS-CoV-2, Nanomaterials 11 (2021) 1394.

[35] A. Pramanik, Y. Gao, S. Patibandla, et al., Aptamer conjugated gold nanostar-based distance- dependent nanoparticle surface energy transfer spectroscopy for ultrasensitive detection and inactivation of corona virus, J. Phys. Chem. Lett. 12 (2021) 2166–2171.

Index

Page numbers followed by "*f*" and "*t*" indicate, figures and tables respectively.

N

O

P

Q

R

S

FSC
www.fsc.org
MIX
Papier | Fördert
gute Waldnutzung
FSC® C083411

Kolibri 360 GmbH
Ferdinand-Jühlke-Straße 7
99095 Erfurt, Deutschland
produktsicherheit@kolibri360.de